MATHEMATICS

FUNCTIONAL CALCULUS

(12th , Engineering)

Content: —

- ➤ Function
- ➤ Limit
- ➤ Continuity and Differentiability
- ➤ Maxima and Minima
- ➤ **Tangent** & *Normal*
- ➤ Differentiation

Function

Important Definitions: − (1) Defined functions, Dependent and independent variables. A function f(x) of x is defined for a certain value of x in its domain provided it attains a unique and definite value for that value of x.

e.g. $y = \dfrac{3x}{2x-3}$ is not defined for $x = \dfrac{3}{2}$ (not definite)

$y = \dfrac{\sin x}{x}$ is not defined for $x = 0$ (Indeterminate)

$y = \sqrt{5x^2}$ is not defined for any $x \in R$ (not unique)

If $y = f(x)$, then x is called independent variable and y the dependent variable. Just as a area of a circle is a function of radius r.

i.e. $A = \pi r^2 = f(r)$, Area is dependent on r, r is independent variable and A is dependent.

(2) Range and Domain of a function: − Let $y = f(x)$ and if this function is defined for all values of x which lie between a and b

if $a < x < b$ *then the open interval* (a, b) or $]a, b[$ will constitute the domain. If, however, $f(x)$ is also defined both for $x = a$ and b, then the domain will consist of the closed interval $[a, b]$.

Range of $f(x) = \{y: y = f(x), x \in domain\}$.

e.g. $y = \dfrac{3x-2}{x-1}$, let $y = f(x) = \dfrac{3x-2}{x-1}$, $f(x)$ is not defined $x - 1 = 0$ or $x = 1$

Hence, domain consist of all real value of x except $x = 1$. or domain $= R - \{1\}, D = x \in R - \{1\}$, whereas domain denoted by D

Range, $y = f(x) = \dfrac{3x-2}{x-1}$ or $(x-1)y = 3x - 2$ or $xy - y = 3x - 2$ or $xy - 3x = y - 2$ or $x(y-3) = y - 2$ or $x = \dfrac{y-2}{y-3}$ or $x = f(y)$

or x is not defined for value $y = 3$, x is defined all real values of y except $y = 3$ Finally, Range $= R - \{3\}$

The following formula will be useful for finding the domain: −

Let $f_1(x)$ and $f_2(x)$ be two functions whose domain are D_1 and D_2 respectively, then   (i) $Dom[f_1(x) + f_2(x)] = D_1 \cap D_2$

(ii) $Dom[f_1(x) . f_2(x)] = D_1 \cap D_2$ (iii) $Dom\left[\dfrac{f_1(x)}{f_2(x)}\right] = D_1 \cap D_2 \cap \{x: f_2(x) \neq 0\}$ (iv) $Dom\left[\sqrt{f_1(x)}\right] = D_1 \cap \{x: f_1(x) > 0\}$

(3) Periodic function: − If $f(x + T) = f(x) \, \forall \, x, T$
> 0 *then* $f(x)$ is called a periodic function and T(Least) is called its fundamental period.

Since T is period of f(x), we have $f(x) = f(x + T) = f(x + 2T) = f(x + 3T) = \cdots \ldots \ldots \ldots \ldots \ldots$

We can say $T, 2T, 3T, \ldots \ldots \ldots \ldots$ are all periods of f(x) but only smallest of these number i.e T will be called the fundamental period.

we know that $\sin x = \sin(x + 2\pi) = \sin(x + 4\pi) = \cdots \ldots \ldots \ldots \ldots \ldots \ldots \ldots \ldots$

$\cos x = \cos(x + 2\pi) = \cos(x + 4\pi) = \cdots \ldots \ldots \ldots \ldots \ldots \ldots \ldots \ldots$
$\tan x = \tan(x + \pi) = \tan(x + 2\pi) = \cdots \ldots \ldots \ldots \ldots \ldots \ldots \ldots \ldots \ldots$

Hence, sinx, cosx and tanx are periodic functions and their periods are $2\pi, 2\pi$ and π respectively.   i.e. least value of T.

Period of (ax + b): − If period of f(x) is T, then period of f(ax + b) is $\dfrac{T}{|a|}$.

Hence, period of **sin2x, cos$\dfrac{3x}{2}$** and **tan$\dfrac{x}{2}$** will be $\pi, \dfrac{4\pi}{3}$ and $\dfrac{\pi}{2}$ respcetively.

Period of $f(ax + b)$ is $\frac{T}{|a|}$. $\therefore$ Period of $\sin 3x = \frac{\text{period of } \sin x}{|a|}$, $\therefore a = 3$ or Period of $\sin 3x = \frac{2\pi}{|3|} = \frac{2\pi}{3}$.

Period of $f(x) + g(x)$: — If period of $f(x)$ and $g(x)$ be T and T' respectively then the L. C. M of T and T' will be period of $f(x) + g(x)$.

Note: —If there exists a number $< L.C.M.$ (T, T') such that $f(x + r) = g(x)$ and $g(x + r) = f(x)$

then r itself will be the period instead of L. C. M of T, T'.

Consider $|\sin x| = \sqrt{\sin^2 x} = \sqrt{\frac{1 - \cos 2x}{2}}$, $\therefore$ Period is $\frac{2\pi}{2} = \pi$ Similarly, Period of $|\cos x| = \pi$

Now if $F(x) = |\cos x| + |\sin x|$ then its period is not π because there exists $\frac{\pi}{2} < \pi \; such \; that \; \left|\cos\left(x + \frac{\pi}{2}\right)\right| = |\sin x|$

and $\left|\sin\left(x + \frac{\pi}{2}\right)\right| = |\cos x|$ Hence, for $F(x)$ the period will be $\frac{\pi}{2}$ and not π

(4) Odd and Even functions: — A function $f(x)$ is said to be odd if it changes sign when the sign of independent variable x is changed

i.e. if $f(-x) = -f(x)$.

For Example: — $f(x) = \sin x . \cos x$, $f(-x) = \sin(-x) . \cos(-x)$ or $f(-x) = -\sin x . \cos x$ or $f(-x) = -f(x)$

so, $f(x) = \sin x . \cos x$ is odd function.

A function $f(x)$ is said to be even if its sign does not change when the sign of the independent variable x is changed.

i.e. $f(-x) = f(x)$ For example, $f(x) = ax^4 + bx^2 + c$ and $f(-x) = a(-x)^4 + b(-x)^2 + c = ax^4 + bx^2 + c$

or $f(-x) = f(x)$ so, $f(x) = ax^4 + bx^2 + c$ is an even function.

(5) Increasing and Decreasing function: — Let $y = f(x)$ is an increasing or decreasing function a certain interval. if $\frac{dy}{dx} = +ve$ is an

increasing and $\frac{dy}{dx} = -ve$ is an decreasing in that interval. Because the tangent will make an acute angle for increasing function and will

make an obtuse angle for decreasing function.

Ex. (1) Determine the intervals of monotonocity of the function $f(x) = y = 3x^4 - 24x^3 + 66x^2 - 72x + 36 = 0$

or $\frac{dy}{dx} = 12x^3 - 72x^2 + 132x - 72 = 12(x^3 - 6x^2 + 11x - 6) = 12(x - 1)(x^2 - 5x + 6)$ or $\frac{dy}{dx} = 12(x - 1)(x - 2)(x - 3) = 0$

or $x = 1,2,3$ Ans.

Very important rule: — I stands for increasing and D stands for decreasing.

+ve		−ve		+ve		−ve		+ve
I	2	D	3	I	4	D	5	I

Let function $f(x) = (x - 2)(x - 3)(x - 4)(x - 5)$. Hence , for $x > 5$, $f(x)$ is increasing, $4 < x < 5, f(x)$ is decreasing,

$3 < x < 4, f(x)$ is increasing, $2 < x < 3, f(x)$ is decreasing, $x < 2, f(x)$ is increasing

or **$f(x)$ is increasing in $(5, \infty) \cup (3, 4) \cup (2, -\infty)$ and $f(x)$ is decreasing in $(4, 5) \cup (2, 3)$**

(6) Composite functions: — $(fog)x = f(g(x)), (gof)x = g(f(x))$.

Properties of functions: — **(a.)** Linear property for algebraic function: — If $y = f(x)$ is a linear algebraic function of the form

$y = mx + c = f(x)$ then $f(ax + b) = af(x) + f(b)$, $f(x + y) = f(x) + f(y)$, $f(x - z) = f(x) - f(z)$ $\therefore f(2x + 3) = 2 f(x) + f(3)$

(b) Logarithmic and Exponential functions: – We know that, $\log(pq) = \log p + \log q$ and $a^{p+q} = a^p . a^q$

Hence for logarithmic functions, $f(xy) = f(x) + f(y)$ For exponential functions, $f(x + y) = f(x).f(y)$

Closed and open intervals: – If $a \le x \le b$ then we say $x \in [a, b]$ is closed interval if $a < x < b$ *then we say* $x \in (a, b)$ *is open interval*

Similarly, $x \in [a, b)$ or $[a, b[$ *or* $a \le x < b$ *is semi* $-$ *closed or open and* $x \in (a, b]$ *or* $]a, b]$ *or* $a < x \le b$ *is semi* $-$ *open or closed.*

if $x > a$, *or* $x \in (a, \infty)$, $x \ge a$, *or* $x \in [a, \infty)$ *or* $x \in [a, \infty[$, $x < b$, *or* $x \in (-\infty, b)$ *or* $x \le b$, *or* $x \in (-\infty, b]$ *or* $x \in]-\infty, b]$

Example: – $x > 2$, —————————————— or $x \in (2, \infty)$
$\quad\quad\quad\quad\quad -\infty \quad\quad\quad\quad 2 \quad\quad\quad\quad \infty$

$x \ge 5$, —————————————— or $x \in [5, \infty)$
$\quad\quad -\infty \quad\quad\quad\quad 5 \quad\quad\quad\quad \infty$

$[\therefore \bullet \to$ closed, $\quad$ o $\to$ open$]$

$x < 3$, —————————————— or $x \in (-\infty, 3)$
$\quad\quad -\infty \quad\quad\quad\quad 3 \quad\quad\quad\quad \infty$

$x \le 4$, —————————————— or $x \in (-\infty, 4]$
$\quad\quad -\infty \quad\quad\quad\quad 4 \quad\quad\quad\quad \infty$

$2 < x < 3$, —————————————— or $x \in (2, 3)$
$\quad\quad\quad -\infty \quad\quad 2 \quad\quad 3 \quad\quad\quad \infty$

(8) Modulus: – Modulus of x i.e $|x|$ $\quad\quad \{|a| =$ modulus of a. $\}$

$|x| = x$, if x is $+$ ve i.e. $|5| = 5$, $|x| = -x$, if x is $-$ ve i.e $|-3| = -(-3) = 3$

$|x - a| = x - a$, if $x - a$ is $+$ ve , $x > a$, $|x - a| = -(x - a)$, if $x - a$ is $-$ ve , $x < a$

or $|X| = 3$ or $x = \pm 3$

Properties of Modulus function: – (a) $|x^n| = |x|^n$, $|2^3| = |2|^3$, $|8| = (|2|)^3$ $\quad$ (b) $|xy| = |x||y|$ $\quad$ (c) $\left|\dfrac{x}{y}\right| = \dfrac{|x|}{|y|}$

(d) modulus of $(|x| - |y|) \le |x + y| \le |x| + |y|$

Example: – $f(x) \doteq |x - 3| = x - 3$, if $x - 3$ is $+$ ve or $x \ge 3$, $x - 3 = +$ve and $f(x) = -(x - 3)$, if $x - 3 = -$ve or $x < 3$

Consider, $f(x) = |x - 3| + |x + 1|$

The critical points are $-1, 3$

	$-$ve		$+$ ve		$+$ ve	
$-\infty$	$-$ ve	-1	$-$ ve	3	$+$ ve	∞

case I: – $x < -1$, *so that* $x - 3 < 0$, $x + 1 < 0$ $\therefore |x - 3| = -(x - 3)$, $|x + 1| = -(x + 1)$

$f(x) = -(x - 3) - (x + 1) = -x + 3 - x - 1 = -2x + 2 = -2(x - 1)$

case II: – $-1 \le x < 3$, *so that* $x - 3 < 0$, $x + 1 \ge 0$ *or* $f(x) = -(x - 3) + (x + 1) = -x + 3 + x + 1 = 4$

case III: – $x \ge 3$, then $x - 3 > 0$, $x + 1 > 0$ *or* $f(x) = x - 3 + x + 1 = 2x - 2$

Hence, Combined function $f(x) = \begin{cases} -(x+1), & \text{if } x < -1 \\ 4, & \text{if } -1 \leq x < 3 \\ 2x - 2 & \text{if } x \geq 3 \end{cases}$

The graph of modulus function $y = |x|$ The critical point $x = 0$

$\begin{array}{ccc} & -ve & +ve \\ \hline -\infty & 0 & \infty \end{array}$

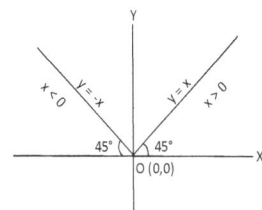

If $y = x$, $x > 0$ and if $y = -x$, $x < 0$ or $y = \begin{cases} x, & \text{if } x > 0 \\ -x, & \text{if } x < 0 \end{cases}$

(9) **Greatest Integer Function** $[x]$: $-$ $[x]$ means the greatest integer not exceeding x,

if, $x = 3\dfrac{1}{2}, 2\dfrac{1}{4}, \dfrac{1}{3}, 5$ then $[x] = 3, 2, 0, 5$ respectively.

If $x = -3\dfrac{1}{2}, -2\dfrac{1}{4}, -\dfrac{1}{3}, -5$ then $[x] = -4, -3, -1, -5$ respectivley. Domain and Range of $[x]$.

The domain of $[x]$ is set of all real numbers its Range $y = [x]$ is set of all integers.

Properties of $[x]$

(a) $[x] = a$, $\Leftrightarrow a \leq x < a + 1$ i.e. $[x] = 3, \Rightarrow 3 \leq x < 3 + 1 \Rightarrow 3 \leq x < 4$

(b) $[x] > a$, $\Rightarrow [x] = a + 1 \Rightarrow x \geq a + 1$ (c) $[x] < a$, $\Rightarrow [x] = a - 1 \Rightarrow x < a$ (d) $a \leq [x] \leq b$, then $x \in [a, b + 1]$

e.g. $-2 \leq [x] \leq 5$, $\Rightarrow x \in [-2, 5 + 1]$, $\Rightarrow x \in [-2, 6]$ Ans.

Example: $-$ (1) If $[x]^2 - 5[x] + 6 \leq 0$ then $[x]^2 - 3[x] - 2[x] + 6 \leq 0$ or $([x] - 2)([x] - 3) \leq 0$, $\Rightarrow x \in [2, 3)$ Ans.

Example: $-$ (2) $[x]^2 - 8[x] + 15 > 0$ then $[x]^2 - 5[x] - 3[x] + 15 > 0$

or $[x]\{[x] - 5\} - 3\{[x] - 5\} > 0$ or $([x] - 3)([x] - 5) > 0$, $\Rightarrow [x] < 3$ or $[x] > 5$

$\Rightarrow [x] < 3$ or $x \geq 5 + 1$, $\Rightarrow x < 3$ or $x \geq 6$ $\therefore x \in (-\infty, 3) \cup [6, \infty)$ Ans.

(10) **Sign of** $(x - a)(x - b), x < b$ (*Important*)

$(x - a)(x - b)$ is $+$ ve, if $x < a$ or $x > b$ *i.e. x does not lie between a and b*

$(x - a)(x - b)$ is $-$ ve, if $a < x < b$ *i.e. x lies between a and b*

(a) $|x| < a \Rightarrow x^2 < a^2 \Rightarrow x^2 - a^2 < 0$ is $-$ ve   or $x^2 < a^2$, or $x < \pm a$, or $-a < x < a$

(b) $|x| \geq a$, or $x^2 \geq a^2$, or $x \geq \pm a$ $\therefore x \leq -a$ or $x \geq a$

(11) **Inequalities**: $-$ Triangular inequalities: $-$ (a) $|x + y| \leq |x| + |y|$ (b) $|x - y| \geq |x| - |y|$

Example: $- |x - 3| > 5$, Let $y = x - 3$, $|y| > 5$, $y < -5$ or $y > 5$ or $x - 3 < -5$ or $x - 3 > 5$

or $x < -5 + 3$ or $x > 8$, $\therefore x < -2$ or $x > 8$ $\begin{array}{cccc} \hline -\infty & -2 & 8 & \infty \end{array}$

$$x \in (-\infty, 2) \cup (8, \infty)$$

Remember: $-$ (i) $|x| = a$ or $x = \pm a$ (ii) $|x| > a$, $\therefore x < -a$ or $x > a$ (iii) $|x| \leq a$, $\therefore -a \leq x \leq a$

(iv) $a < |x| \leq b$, $\therefore |x| > a$ or $|x| \leq b$ (v) $a \geq |x| > b$, $\therefore |x| \leq a$ or $|x| > b$

(vi) $a \leq |x| \leq b$, $\therefore |x| \geq a$ or $|x| \leq b$ (vii) $a \geq |x| \geq b$, $\therefore |x| \leq a$ or $|x| \geq b$

(12) $\log_a x$ is defined if both x and a are + ve and $x \neq 0$ and $a \neq 1$, Also $\log_a x = y$ then $x = a^y$ exponential form

∗ Exponential function is always positive

∗ $\log_a x > \log_a y$ or $x > y$, $if\ a > 1$ or $x < y$, $if\ a < 1$

Even and odd Extension

Let $f(x)$ be a function defined on $A = [0, a]$ and $B = [-a, a]$ is a super set of A·then an extension of $f(x)$ on $B = [-a, a]$ will be even or odd extension if $f(x)$ becomes an even or odd function on B.

Solved Example

(1) If $f(x) = ax^2 + bx + c$ then find $f(0), f(a), f(b)$ and $f(1)$.

Solution: $-\ f(x) = ax^2 + bx + c$ At $x = 0$, then $f(0) = a \times 0 + b \times 0 + c$ $\therefore f(0) = c$ Ans.

At $x = a$, then $f(a) = a \times a^2 + b \times a + c$ $\therefore f(a) = a^3 + ab + c$ Ans.

At $x = b$, then $f(b) = a \times b^2 + b \times b + c$ $\therefore f(b) = ab^2 + b^2 + c$ Ans.

At $x = 1$, then $f(1) = a \times 1 + b \times 1 + c$ $\therefore f(1) = a + b + c$ Ans.

(2) If $f(x) = x - \dfrac{1}{x}$, prove that $f(x) = -f\left(\dfrac{1}{x}\right)$.

Solution: $-\ f(x) = x - \dfrac{1}{x}$, $f\left(\dfrac{1}{x}\right) = \dfrac{1}{x} - \dfrac{1}{\frac{1}{x}} = \dfrac{1}{x} - x = -\left(x - \dfrac{1}{x}\right) = -f(x)$ or $f\left(\dfrac{1}{x}\right) = -f(x)$, $\therefore f(x) = -f\left(\dfrac{1}{x}\right)$ Proved.

Type of functions

(i) Rational function: $-$ This function is defined as the ratio of two polynomials

$$y = \frac{a_0 x^n + a_1 x^{n-1} + \cdots \ldots \ldots \ldots \ldots \ldots \ldots \ldots \ldots \ldots . + a_n}{b_0 x^m + b_1 x^{m-1} + \cdots \ldots \ldots \ldots \ldots \ldots \ldots \ldots \ldots . + b_m}$$

For Example, $y = \dfrac{x^2 + 1}{x^3 + 3}$ is rational function

(ii) Irrational function: $-$ If in the $y = f(x)$, the operations of addition, subtraction, multiplication, division and raising to a power with rational non $-$ integral exponents are performed on the right $-$ hand side the function $y = f(x)$ is said to be irrational.

Example: $-\ y = \dfrac{3x^2 + \sqrt{x}}{\sqrt{1 + 2x}}$ and $y = \sqrt{x}$.

(iii) Explicit function: $-$ If the dependent variable, say y, is expressible explicitly in terms of the independent variable, the function

$y = f(x)$ is called an explicit function. otherwise it is said to be an implicit function.

Example: $-$ $y = \cos^3 x + x^3$ is an explicit function whereas $y = x^3 + y^3 - 3yx = 0$ is an explicit function

(iv) Onto function (or Surjective function): $-$ If a fucntion f: A $\rightarrow$ B is such that each element in B is the f $-$ image of at least one

element in A , then we say that f is a function of A "onto" B equivalently a function f is an onto function if co $-$ domain of f = Range of f.

Example: $-$ f: R $\rightarrow$ [$-1,1$] defined by $f(x) = \sin x$ is an onto function but f: R $\rightarrow$ R defined by

$f(x) = \sin x$ is not onto since Range of f = [$-1,1$]and co $-$ domain of f = R.

(v) One $-$ to $-$ one function (or injective function): $-$ A function f is said to be one $-$ to $-$ one if it does not take the same

values at two distinct points in its domain.

Example: $-$ $f(x) = x^3$ is one $-$ to $-$ one. whereas $f(x) = x^2$ is not, as $f(1) = 1$ and $f(-1) = 1, f(-1) = f(1)$

so, $f(x) = x^2$ is not a one $-$ to $-$ one function.

(vi) Bijective function (or one $-$ to $-$ one and onto): $-$ If a function f is both one $-$ to $-$ one and onto, then f is said to be a

bijective function.

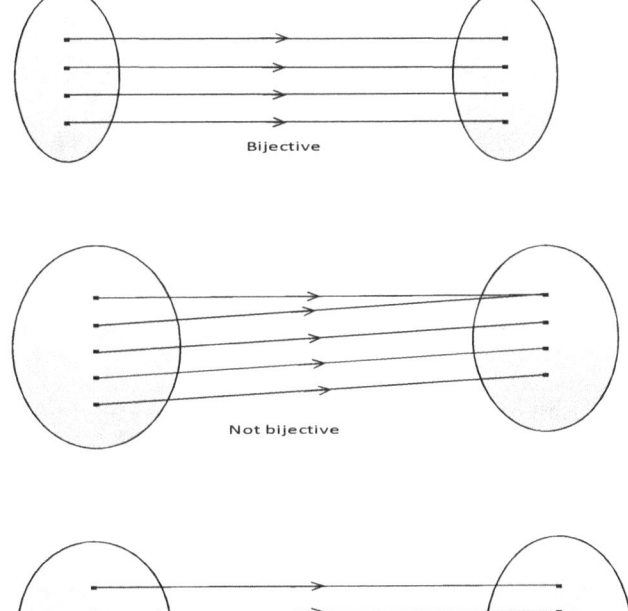

Important Function

\# . The domain and Range of trigonometric function are as follows: $-$

(i) The domain of $f(x) = \sin x$ is R and its Range is $[-1,1] = \{x \in R; -1 \le x \le 1\}$

(ii) The domain of $f(x) = \cos x$ is R and its Range is $[-1,1]$.

(iii) The domain of $f(x) = \tan x$ is $R - \{(2m+1)\frac{\pi}{2}, m \in I\}$ or $f(x) = \{x \in R: x \ne (2m+1)\frac{\pi}{2}, \ m \text{ is an integer}\}$ and its Range is R.

(iv) The domain of $f(x) = \cot x$ is $R - \{m\pi: m \in I\}$ or $f(x) = \{x \in R, x \neq m\pi, m \text{ is an integer}\}$ and its Range is R.

(v) The domain of $f(x) = \sec x$ is $R - \left\{(2m+1)\dfrac{\pi}{2}, m \in I\right\}$ and its Range is $(\infty, -1] \cup [1, \infty) = \{x \in R, x \notin (-1,1)\}$ so, $|\sec x| \geq 1$

(vi) The domain of $f(x) = \csc x$ is $R - \{m\pi, m \in I\}$ and its Range is the same as that of $\sec x$.

♯ . The domain and Ranges of the inverse trigonometric functions are as follows: −

(i) The domain of $f(x) = \sin^{-1} x$ is $[-1,1]$ and its Range is $\left[-\dfrac{\pi}{2}, \dfrac{\pi}{2}\right]$.

(ii) The domain of $f(x) = \cos^{-1} x$ is $[-1,1]$ and its Range is $[0, \pi]$.

(iii) The domain of $f(x) = \tan^{-1} x$ is R and its Range is $\left(-\dfrac{\pi}{2}, \dfrac{\pi}{2}\right) = \{x \in R, -\dfrac{\pi}{2} < x < \dfrac{\pi}{2}\}$.

(iv) The domain of $f(x) = \cot^{-1} x$ is R and its Range is $(0, \pi)$.

(v) The domain of $f(x) = \sec^{-1} x$ is $R - (-1,1)$ and its Range is $\left[0, \dfrac{\pi}{2}\right) \cup \left(\dfrac{\pi}{2}, \pi\right] = \{x \in R, 0 \leq x \leq \pi, x \neq \dfrac{\pi}{2}\}$.

(vi) The domain of $f(x) = \csc^{-1} x$ is $R - (-1,1)$ and its Range is $\left[-\dfrac{\pi}{2}, 0\right) \cup \left(0, \dfrac{\pi}{2}\right] = \{x \in R, -\dfrac{\pi}{2} \leq x \leq \dfrac{\pi}{2}, x \neq 0\}$.

Solved Example

(1) The domain of the function $f(x) = \sqrt{x-2} + \sqrt{5-x}$ is (a) $[2, \infty)$ (b) $[2,5]$ (c) $(-\infty, 5)$ (d) $(2,5)$

Solution: − $f(x) = \sqrt{x-2} + \sqrt{5-x}$, function $f(x)$ to be defined as $f(x) = \begin{cases} x - 2 \geq 0 \\ 5 - x \geq 0 \end{cases} = \begin{cases} x \geq 2 \\ -x \geq -5 \end{cases} = \begin{cases} x \geq 2 \\ x \leq 5 \end{cases}$

domain of $f(x)$ is $x \in [2,5]$, ∴ domain $= [2,5]$ Ans. (b) $[2,5]$

(2) The domain of the function $f(x) = \sqrt{\dfrac{1-x}{1+x}}$ is (a) $(-\infty, 1]$ (b) $(-\infty, -1) - \{1\}$ (c) $(-\infty, 1) - \{-1\}$ (d) $R - (-1,1)$

Solution: − $f(x) = \sqrt{\dfrac{1-x}{1+x}}$ to be defined as $f(x) = \begin{cases} \dfrac{1-x}{1+x} > 0 \\ 1+x \neq 0 \end{cases} = \begin{cases} 1 - x > 0 \\ x \neq -1 \end{cases} = \begin{cases} x < 1 \\ x \neq -1 \end{cases}$

domain $= (-\infty, 1) - \{-1\}$, Ans. (c)

(3) If D is the domain of the function $f(x) = \sqrt{3-4x} + 2\sin^{-1}\dfrac{3x-2}{5}$ then D contains (a) $\left[-1, \dfrac{3}{4}\right)$ (b) $\left[-1, \dfrac{3}{4}\right]$ (c) $(-1,0)$ (d) $[-1,0]$

Solution: − $f(x) = \sqrt{3-4x} + 2\sin^{-1}\dfrac{3x-2}{5}$

$f(x)$ to be defined as $f(x) = \begin{cases} 3 - 4x \geq 0 \\ -1 \leq \dfrac{3x-2}{5} \leq 1 \end{cases} = \begin{cases} -4x \geq -3 \\ -5 \leq 3x - 2 \leq 5 \end{cases} = \begin{cases} 4x \leq 3 \\ -3 \leq 3x \leq 7 \end{cases} = \begin{cases} x \leq \dfrac{3}{4} \\ -1 \leq x \leq \dfrac{7}{3} \end{cases}$

domain(D) $= \left(-\infty, \dfrac{3}{4}\right] \cap \left[-1, \dfrac{7}{3}\right]$

$$D = [-1,0] \cup \left[0,\frac{3}{4}\right] = [-1,0], \left[-1,\frac{3}{4}\right] \qquad \text{Ans. (b) } \left[-1,\frac{3}{4}\right], \text{(d) } [-1,0]$$

(4) The function $f(x) = \log\left(\dfrac{1-x}{1+x}\right)$ satisfies the equation (a) $f(x+2) - 2f(x+1) + f(x) = 0$

(b) $f(x) + f(x-1) = f\big(x(x+1)\big)$ (c) $f(x_1) \cdot f(x_2) = f(x_1 + x_2)$ (d) $f(x_1) + f(x_2) = f\left(\dfrac{x_1 + x_2}{1 + x_1 \cdot x_2}\right)$

Solution: — $f(x) = \log\dfrac{1-x}{1+x}$, $f(x_1) = \log\dfrac{1-x_1}{1+x_1}$, $f(x_2) = \log\dfrac{1-x_2}{1+x_2}$

$$f(x_1) + f(x_2) = \log\frac{1-x_1}{1+x_1} + \log\frac{1-x_2}{1+x_2} = \log\left(\frac{1-x_1}{1+x_1} \times \frac{1-x_2}{1+x_2}\right) = \log\left(\frac{1 - x_2 - x_1 + x_1 x_2}{1 + x_1 + x_2 + x_1 x_2}\right)$$

$$f(x_1) + f(x_2) = \log\left(\frac{1 - \dfrac{x_1 + x_2}{1 + x_1 x_2}}{1 + \dfrac{x_1 + x_2}{1 + x_1 x_2}}\right) \qquad \therefore \ f(x_1) + f(x_2) = f\left(\frac{x_1 + x_2}{1 + x_1 x_2}\right) \quad \text{Ans. (d)}$$

(5) Which of the following functions is an even function: — (a) $f(x) = x\log\left(\dfrac{1-x}{1+x}\right)$ (b) $f(x) = \dfrac{a^x - 1}{a^x + 1}$

(c) $f(x) = x\dfrac{a^x + 1}{a^x - 1}$ (d) $f(x) = \dfrac{a^{2x} + 1}{a^{2x} - 1}$

Solution: — (a) $f(x) = x\log\left(\dfrac{1-x}{1+x}\right)$, $f(-x) = -x\log\left(\dfrac{1+x}{1-x}\right) = x\log\left(\dfrac{1+x}{1-x}\right)^{-1} = x\log\left(\dfrac{1-x}{1+x}\right) = f(x)$

$$f(-x) = f(x) \text{ is an even function.}$$

(b) $f(x) = \dfrac{a^x - 1}{a^x + 1}$, $f(-x) = \dfrac{a^{-x} - 1}{a^{-x} + 1} = \dfrac{1 - a^x}{1 + a^x} = -\left(\dfrac{a^x - 1}{a^x + 1}\right) = -f(x)$ $\therefore$ $f(-x) = -f(x)$ is an odd function.

(c) $f(x) = x\dfrac{a^x + 1}{a^x - 1}$, $f(-x) = -x\dfrac{a^{-x} + 1}{a^{-x} - 1} = -x\dfrac{1 + a^x}{1 - a^x} = -x\dfrac{1 + a^x}{-(a^x - 1)} = x\dfrac{a^x + 1}{a^x - 1} = f(x)$

$$f(-x) = f(x) \text{ is an even funciton.}$$

(d) $f(x) = \dfrac{a^{2x} + 1}{a^{2x} - 1}$, $f(-x) = \dfrac{a^{-2x} + 1}{a^{-2x} - 1} = \dfrac{1 + a^{2x}}{1 - a^{2x}} = -\dfrac{a^{2x} + 1}{a^{2x} - 1} = -f(x)$ or $f(-x) = -f(x)$ is an odd funciton. Ans: — (a) and (c)

(6) The domain of definition of the function $y = \dfrac{1}{\log_2(3-x)} + \sqrt{1-x}$ is (a) $(-\infty, 1) - \{2\}$ (b) $(-\infty, 1)$ (c) $(-\infty, 2)$ (d) $(3, \infty)$

Solution: — $y = f(x) = \dfrac{1}{\log_2(3-x)} + \sqrt{1-x}$

$f(x)$ to be defined as $f(x) = \begin{cases} \log_2(3-x) \neq 0 \\ 3-x > 0 \\ 1-x \geq 0 \end{cases} = \begin{cases} 3-x \neq 1 \\ x < 3 \\ x \leq 1 \end{cases} = \begin{cases} x \neq 2 \\ x < 3 \\ x \leq 1 \end{cases}$

Remember
$f(x) = \log_a x$
$f(x)$ to be defined as
$f(x) = \begin{cases} a > 0, a \neq 1 \\ x > 0, x \neq 0 \end{cases}$

$$\begin{array}{c|c|c|c|c} \hline -\infty & 1 & 2 & 3 & \infty \end{array}$$

or $D = (-\infty, 1)$ (b) Ans.

(7) Find domain and Range of the function: — (a) $f(x) = \dfrac{4x}{2x - 3}$ (b) $f(x) = \dfrac{3x - 2}{2x - 3}$ (c) $f(x) = \dfrac{1}{\sqrt{1 - x^2}}$

Solution: — (a) $f(x) = \dfrac{4x}{2x - 3}$ function $f(x)$ to be defined $2x - 3 \neq 0$ $\therefore x \neq \dfrac{3}{2}$

or $\boxed{\text{Domain (D)} = R - \left\{\dfrac{3}{2}\right\}}$ Ans.

$$-\infty \qquad \dfrac{3}{2} \qquad \infty$$

Range:– Let $y = f(x) = \dfrac{4x}{2x - 3}$, $\therefore y(2x - 3) = 4x$

or $2xy - 3y = 4x$ or $2xy - 4x = 3y$ or $x(2y - 4) = 3y$ or $x = \dfrac{3y}{2y - 4}$, $x = f(y)$

function $f(y)$ to be defined $2y - 4 \neq 0, y \neq 2$

$$-\infty \qquad 2 \qquad \infty$$

$\boxed{\text{Range} = R - \{2\} \quad \text{Ans.}}$

(b) $f(x) = \dfrac{3x - 2}{2x - 3}$ function $f(x)$ to be not defined is $2x - 3 = 0$ or $x = \dfrac{3}{2}$ Domain (D) $= R - \left(\dfrac{3}{2}\right)$

Range:– $y = f(x) = \dfrac{3x - 2}{2x - 3}$ or $2xy - 3y = 3x - 2 = x(2y - 3) = 3y - 2$ $\therefore$ $x = \dfrac{3y - 2}{2y - 3}$

function $f(y)$ to be not defined as $2y - 3 = 0$ or $y = \dfrac{3}{2}$

$$-\infty \qquad \dfrac{3}{2} \qquad \infty$$

$\boxed{\text{Range} = R - \left\{\dfrac{3}{2}\right\} \quad \text{Ans.}}$

(c) $f(x) = \dfrac{1}{\sqrt{1 - x^2}}$ function $f(x)$ to be defined $f(x) = \begin{cases} \sqrt{1 - x^2} > 0 \\ 1 - x^2 \neq 0 \end{cases}$ or $\begin{cases} 1 - x^2 > 0 \\ x^2 \neq 1 \end{cases}$

$$-\infty \qquad -1 \qquad 1 \qquad \infty$$

Domain (D) $= -1 < x < 1$

Range:– $y = f(x) = \dfrac{1}{\sqrt{1 - x^2}}$ or $y^2 = \dfrac{1}{1 - x^2}$ or $1 - x^2 = \dfrac{1}{y^2}$ or $x^2 = 1 - \dfrac{1}{y^2}$ or $x^2 = \dfrac{y^2 - 1}{y^2}$ or $x = \sqrt{\dfrac{y^2 - 1}{y^2}}$

function $f(y)$ to be defind $\begin{cases} \dfrac{y^2 - 1}{y^2} \geq 0 \\ y^2 \neq 0 \end{cases} = \begin{cases} y^2 - 1 \geq 0 \\ y \neq 0 \end{cases} = \begin{cases} y^2 \geq 1 \\ y \neq 0 \end{cases} = \begin{cases} y \geq \pm 1 \\ y \neq 0 \end{cases}$

$$-\infty \qquad -1 \qquad 0 \qquad 1 \qquad \infty$$

$\boxed{\text{Range} = (-\infty, -1) \cup (1, \infty) \quad \text{Ans.}}$

(8) find the range of each of the following function:– (a) $y = \dfrac{x - 3}{2x - 5}$ (b) $y = 2 - \dfrac{1}{1 + x}$ (c) $y = \sin^{-1} x$ (d) $y = x - |x|$

Solution : –(a) $y = \dfrac{x - 3}{2x - 5}$ or $2xy - 5y = x - 3$ or $x(2y - 1) = 5y - 3$ $\therefore$ $x = \dfrac{5y - 3}{2y - 1} = f(x)$

Hence, function $f(x)$ to be defined $2y - 1 \neq 0$ or $y \neq \dfrac{1}{2}$

$$-\infty \qquad \dfrac{1}{2} \qquad \infty$$

$\boxed{\text{Range} = R - \left\{\dfrac{1}{2}\right\} \quad \text{Ans.}}$

(b) $y = 2 - \dfrac{1}{1 + x} = \dfrac{2 + 2x - 1}{1 + x} = \dfrac{2x + 1}{1 + x}$

Hence y to be defined $1 + x \neq 0$ or $x \neq -1$

$$-\infty \qquad -1 \qquad \infty$$

$$\boxed{\text{Domain (D)} = R - \{-1\}}$$

Now, $y = \dfrac{2x+1}{1+x}$ or $y + xy = 2x + 1$ or $x(y-2) = 1 - y$ $\therefore x = \dfrac{1-y}{y-2}$

Hence x to be defined $y - 2 \neq 0$ or $y \neq 2$

$$\underset{-\infty \qquad\qquad 2 \qquad\qquad \infty}{\rule{300pt}{0.5pt}} \qquad \text{or Range} = R - \{2\}$$

(c) $y = \sin^{-1} x$, Here y to be defined $-1 \leq x \leq 1$ $\qquad \boxed{\text{Domain (D)} = |-1,1|}$

Now, $y = \sin^{-1} x$ $\quad \therefore x = \sin y = f(x)$ $\quad$ Hence f(x) to be defined of all real value.

$$\boxed{\text{Range} = R}$$

(d) $y = x - |x|$ $\qquad$ Ans: $-$ Range $= R$

(9) If $f: R \rightarrow R$ is defined by $f(x) = x^3 + 2$, the value of $f^{-1}(29)$ and $f^{-1}(10)$ are respectively.

Solution: $-$ $f(x) = x^3 + 2$ $\therefore f(x) = y = x^3 + 2$ or $x^3 = y - 2$ or $x = \sqrt[3]{y-2}$ $\therefore y = f(x)$

or $x = f^{-1}(y)$ or $f^{-1}(y) = \sqrt[3]{y-2}$ $\therefore f^{-1}(29) = \sqrt[3]{29-2} = \sqrt[3]{27} = 3$ Ans. $\therefore f^{-1}(10) = \sqrt[3]{10-2} = \sqrt[3]{8} = 2$ Ans.

(10) find the period of the following function: $-$ (a) $y = \sin\left(\dfrac{2t-5}{4\pi}\right)$ $\quad$ (b) $y = \sin|x| + \cos|x|$ $\quad$ (c) $y = \dfrac{\sin x + \sin 2x}{\cos x + \cos 2x}$ $\quad$ (d) y $= \dfrac{\tan x - \cot x}{\tan x + \cot x}$

(e) $y = \cos x + \cot\dfrac{x}{2} + \cos\dfrac{x}{2^2} + \cot\dfrac{x}{2^3}$ $\quad$ (f) $y = \sin\left|2x + \dfrac{\pi}{2}\right|$ $\quad$ (g) $y = \tan\left(\dfrac{3x-2}{5}\right)$

Solution: $-$ (a) $y = \sin\left(\dfrac{2t-5}{4\pi}\right)$ $\qquad \therefore y = \sin(ax+b)$ then period $= \dfrac{|T|}{|a|}$ $\qquad \boxed{\text{Time Period of } y = \dfrac{2\pi}{\dfrac{1}{2\pi}} = 4\pi^2 \text{ Ans.}}$

(b) $y = \sin|x| + \cos|x|$, Period of $y = \pi$

(c) $y = \dfrac{\sin x + \sin 2x}{\cos x + \cos 2x}$ Formula, $\sin C + \sin D = 2\sin\dfrac{C+D}{2} \cdot \cos\dfrac{C-D}{2}$ and $\cos C + \cos D = 2\cos\dfrac{C+D}{2} \cdot \cos\dfrac{C-D}{2}$

$$y = \tan\dfrac{3x}{2}, \text{ Period} = \dfrac{T}{|a|} = \dfrac{\pi}{\dfrac{3}{2}} = \dfrac{2\pi}{3} \quad \text{Ans.}$$

(d) $y = \dfrac{\tan x - \cot x}{\tan x + \cot x} = \sin^2 x - \cos^2 x = -\cos 2x = \cos(\pi - 2x)$

$$y = \cos[\pi + (-2x)] \quad \therefore a = -2, \text{ Period (T)} = \dfrac{T}{|a|} \quad \boxed{\therefore \text{Period} = \dfrac{2\pi}{|-2|} = \pi \quad \text{Ans.}}$$

(e) $y = \cos x + \cot\dfrac{x}{2} + \cos\dfrac{x}{2^2} + \cot\dfrac{x}{2^3}$

$$\boxed{\text{Period of } f(x) + g(x) \text{ are T and } T' \text{ respectively. then, period of } f(x) + g(x) = \text{L. C. M of T and } T'}$$

$$\text{Period of } y = 2\pi + \dfrac{\pi}{\dfrac{1}{2}} + \dfrac{2\pi}{\dfrac{1}{4}} + \dfrac{\pi}{\dfrac{1}{8}} = 2\pi + 2\pi + 8\pi + 8\pi = \text{L. C. M of all} = 4\pi \quad \text{Ans.}$$

(f) $y = \sin\left|2x + \dfrac{\pi}{2}\right|$ period of $\sin|x| = \pi$, $a = 2$ Period of $y = \dfrac{T}{|a|} = \dfrac{\pi}{2}$ Ans.

(g) $y = \tan\left(\dfrac{3x-2}{5}\right) = \tan\left(\dfrac{3x}{5} + \dfrac{-2}{5}\right), a = \dfrac{3}{5}$ $\quad$ Period of $y = \dfrac{T}{|a|} = \dfrac{\pi}{\dfrac{3}{5}} = \dfrac{5\pi}{3}$ Ans.

(11) If f is an even function defined on the interval $(-3,3)$, then the one value of x satisfying the equation $f(x) = f\left(\dfrac{x+3}{x-1}\right)$ are ... -1 ... ...

Solution: $-$ $f(x) = f\left(\dfrac{x+3}{x-1}\right)$ or $x = \dfrac{x+3}{x-1}$ or $x^2 - x = x+3$ or $x^2 - 2x - 3 = 0$ or $x = \dfrac{2 \pm \sqrt{4+12}}{2} = \dfrac{2 \pm 4}{2}$ $\therefore x = 3, -1$

Again $f(-x) = f(x)$ as f is an even function. $f(-x) = f(x) = f\left(\dfrac{x+3}{x-1}\right)$ or $-x = \dfrac{x+3}{x-1}$ or $-x^2 + x = x+3$

or $x^2 = -3$ it is not defined, satisfying all real value. only one value is satisfying the equation or $x = -1$ Ans.

(12) Let f and g be functions defined by $f(x) = \dfrac{3x}{5-2x}$ and $g(x) = \dfrac{1-2x}{x+1}$ then $(fog)^{-1}(x) = \dfrac{1-x}{3x+2}$.

Solution: $-$ $f(x) = \dfrac{3x}{5-2x}$ and $g(x) = \dfrac{1-2x}{x+1}$ then $(fog)^{-1}(x)$ equal to ... ...

$(fog)^{-1}(x) = f(g(x))^{-1}(x)$ $\therefore f(g(x)) = f\left(\dfrac{1-2x}{x+1}\right) = \dfrac{\dfrac{3(1-2x)}{x+1}}{5-2\dfrac{1-2x}{x+1}} = \dfrac{1-2x}{3x+1} = h$ (say)

$\therefore h(x) = \dfrac{1-2x}{3x+1} = y$ or $3xy + y = 1 - 2x$ or $x(3y+2) = 1 - y$

or $x = \dfrac{1-y}{3y+1}$ let $h(x) = y$ or $x = h^{-1}(y)$ or $h^{-1}(y) = \dfrac{1-y}{3y+2}$ $\therefore h^{-1}(x) = \dfrac{1-x}{3x+2}$ or $f(g(x))^{-1}x = \dfrac{1-x}{3x+2}$ Ans.

(13) (i) find fofof if $f(x) = \dfrac{1-x}{2+3x}$ (ii) find fofog if $f(x) = \dfrac{\sqrt{1-x^2}}{x}$ and $g(x) = \dfrac{x^2}{1+x^2}$

(iii) find fofofog if $f(x) = \sqrt{1-x}$ and $g(x) = \dfrac{2}{1+x}$

Solution: $-$(i) $f(x) = \dfrac{1-x}{2+3x}$ then find fofof i.e $f(f(f(x)))$. or $f(f(x)) = f\left(\dfrac{1-x}{2+3x}\right) = \dfrac{1-\dfrac{1-x}{2+3x}}{2+3\left(\dfrac{1-x}{2+3x}\right)} = \dfrac{1+4x}{7+3x}$

or $f(f(f(x))) = f\left(\dfrac{1+4x}{7+3x}\right) = \dfrac{1+4\left(\dfrac{1+4x}{7+3x}\right)}{2+3\left(\dfrac{1+4x}{7+3x}\right)}$ $\therefore f(f(f(x))) = \dfrac{19x+11}{18x+17}$ Ans.

(ii) $f(x) = \dfrac{\sqrt{1-x^2}}{x}$ and $g(x) = \dfrac{x^2}{1+x^2}$ then find fofog i.e $f(f(g(x)))$.

$f(f(g(x))) = f\left(f\left(\dfrac{x^2}{1+x^2}\right)\right)$ or $f\left(\dfrac{x^2}{1+x^2}\right) = \dfrac{\sqrt{1-\dfrac{(x^2)^2}{(1+x^2)^2}}}{\dfrac{x^2}{1+x^2}} = \dfrac{\sqrt{(1+x^2)^2-(x^2)^2}}{x^2}$

or $f\left(\dfrac{x^2}{1+x^2}\right) = \dfrac{\sqrt{1+2x^2}}{x^2}$ or $f\left(\dfrac{\sqrt{1+2x^2}}{x^2}\right) = \dfrac{\sqrt{1-\dfrac{\left(\sqrt{1+2x^2}\right)^2}{x^2}}}{\dfrac{\sqrt{1+2x^2}}{x^2}} = \dfrac{\sqrt{x^4-2x^2-1}}{\sqrt{1+2x^2}}$ $\boxed{\therefore f(f(g(x))) = \sqrt{\dfrac{x^4-2x^2-1}{1+2x^2}}}$ Ans.

(iii) $f(x) = \sqrt{1-x}$ and $g(x) = \dfrac{2}{1+x}$ then find fofofog, i.e $f(f(f(g(x))))$.

or $f(g(x)) = f\left(\dfrac{2}{1+x}\right) = \sqrt{1-\dfrac{2}{1+x}} = \sqrt{\dfrac{x-1}{x+1}}$ $\therefore f\left(\sqrt{\dfrac{x-1}{x+1}}\right) = \sqrt{1-\sqrt{\dfrac{x-1}{x+1}}} = \sqrt{\dfrac{\sqrt{x+1}-\sqrt{x-1}}{\sqrt{x+1}}}$

$$\therefore f\left(\sqrt{\frac{\sqrt{x+1}-\sqrt{x-1}}{\sqrt{x+1}}}\right) = \sqrt{1-\sqrt{\frac{\sqrt{x+1}-\sqrt{x-1}}{\sqrt{x+1}}}} \qquad \boxed{\therefore f\left(f\left(f\left(g(x)\right)\right)\right) = \frac{\left(\sqrt{x+1}+\sqrt{x^2-1}-2\right)^{\frac{1}{2}}}{\left(\sqrt{x+1}+\sqrt{x^2-1}\right)^{\frac{1}{2}}}} \text{ Ans.}$$

(14)(i) find inverse of the function $y = \dfrac{5^x - 5^{-x}}{5^x + 5^{-x}}$ is _____   (ii) If $f(x) = \dfrac{\sqrt{x}}{\sqrt{x}+1}$, the domain of $f^{-1}(x)$ contains _____

(iii) The inverse of the function $y = \dfrac{1}{\sqrt{1-2\sin^2 x}}$ is _____   (iv) The inverse of the function $y = \log\left(\dfrac{x+2}{x-1}\right)$

(v) The inverse of the function $y = \tan^{-1}x + \cot^{-1}x$

Solution: $-$ (i) $y = \dfrac{5^x - 5^{-x}}{5^x + 5^{-x}} = f(x) \quad \therefore x = f^{-1}(y) \text{ or } x = \dfrac{1}{2}\log_5\left(\dfrac{y+1}{1-y}\right) \qquad \therefore \; y = \dfrac{1}{2}\log_5\left(\dfrac{y+1}{1-y}\right)$ Ans.

(ii) $f(x) = \dfrac{\sqrt{x}}{\sqrt{x}+1} = y \;\Rightarrow\; y^2 = \dfrac{x}{x+1} \;\Rightarrow\; xy^2 + y^2 = x \;\therefore\; x(y^2-1) = -y^2$

$\Rightarrow x = \dfrac{-y^2}{y^2-1} = \dfrac{y^2}{1-y^2} \quad \therefore y = f(x) \quad \therefore x = f^{-1}(y) = \dfrac{y^2}{1-y^2} \quad \therefore f^{-1}(x) = \dfrac{x^2}{1-x^2}$

Hence $f^{-1}(x)$ to be defined   or $1 - x^2 \neq 0 \;\Rightarrow\; x^2 \neq 1$ or $x \neq \pm 1$ $\therefore$ Domain $(D) = R - \{-1,1\}$ Ans.

(iii) $y = \dfrac{1}{\sqrt{1-2\sin^2 x}} = \dfrac{1}{\sqrt{\cos 2x}} \;\Rightarrow\; y^2 = \dfrac{1}{\cos 2x}$ or $x = \dfrac{1}{2}\cos^{-1}\left(\dfrac{1}{y^2}\right)$ or $y = f(x)$

$$\text{or } x = f^{-1}(y) = \dfrac{1}{2}\cos^{-1}\left(\dfrac{1}{y^2}\right) \qquad \boxed{\therefore \; y = \dfrac{1}{2}\cos^{-1}\left(\dfrac{1}{x^2}\right) \text{ Ans.}}$$

(iv) $y = \log\left(\dfrac{x+2}{x-1}\right) = f(x)$ or $e^y = \dfrac{x+2}{x-1} \;\Rightarrow\; x = \dfrac{e^y+2}{e^y-1} \;\Rightarrow\; x = \dfrac{e^y+2}{e^y-1}$

$$\text{or } y = f(x) \;\Rightarrow\; x = f^{-1}(y) = \dfrac{e^y+2}{e^y-1} \quad \therefore y = \dfrac{e^y+2}{e^y-1} \qquad \boxed{\therefore y = \dfrac{e^{x+2}}{e^x-1} \text{ Ans.}}$$

(v) $y = \tan^{-1}x + \cot^{-1}x = \dfrac{(\tan^{-1}x)^2 + 1}{\tan^{-1}x}$, let $\tan^{-1}x = z$ or $x = \tan z$

$y = \dfrac{z^2+1}{z} \;\Rightarrow\; z^2 - zy + 1 = 0 \;\Rightarrow\; z = \dfrac{y \pm \sqrt{y^2-4}}{2} = \tan^{-1}x \quad \therefore x = \tan\left(\dfrac{y \pm \sqrt{y^2-4}}{2}\right) \qquad \boxed{\therefore y = \tan\left(\dfrac{x \pm \sqrt{x^2-4}}{2}\right) \text{ Ans.}}$

(15) Determine the function $y = f(x)$ satisfying the condition $f\left(x - \dfrac{1}{x}\right) = -\dfrac{1}{x^2} + x^2 \;(x \neq 0)$.

Solution: $- \; f\left(x - \dfrac{1}{x}\right) = -\dfrac{1}{x^2} + x^2 \quad \therefore t = x - \dfrac{1}{x} \;\Rightarrow\; t^2 = x^2 - \dfrac{1}{x^2} + 2 \;\Rightarrow\; t^2 - 2 = -\dfrac{1}{x^2} + x^2$

$\therefore f(t) = t^2 - 2, \quad f(x) = x^2 - 2$ is required function which will satisfy the given condition.   Ans.

(16) Let $f(x) = \begin{cases} x^2 - 7x + 10\,, & x < 5 \\ x - 7\,, & x \geq 5 \end{cases}$ and $g(x) = \begin{cases} x - 5\,, & x < 7 \\ x^2 + 3x + 3\,, & x \geq 7 \end{cases}$ Describe the function $\dfrac{f}{g}$ and find its domain.

Solution: $-$ The critical points are $5,7$ and so we redefine the function for $x < 5, 5 \leq x < 7, x \geq 7$

$$f(x) = \begin{cases} x^2 - 7x + 10\,, & x < 5 \\ x - 7\,, & 5 \leq x < 7 \\ x - 7\,, & x \geq 7 \end{cases}, \qquad g(x) = \begin{cases} x - 5\,, & x < 5 \\ x - 5\,, & 5 \leq x < 7 \\ x^2 + 3x + 3\,, & x \geq 7 \end{cases}$$

$$\dfrac{f}{g} = \begin{cases} \dfrac{x^2 - 7x + 10}{x - 5}\,, & x < 5 \\ \dfrac{x - 7}{x - 5}\,, & 5 \leq x < 7 \\ \dfrac{x - 7}{x^2 + 3x + 3}\,, & x \geq 7 \end{cases} = \begin{cases} x - 2\,, & x < 5 \\ \dfrac{x - 7}{x - 5}\,, & 5 \leq x < 7 \\ \dfrac{x - 7}{x^2 + 3x + 3}\,, & x \geq 7 \end{cases} \qquad \text{Domain}(D) = R - \{5\} \text{ Ans.}$$

$$f + g = \begin{cases} x^2 - 7x + 10 + x - 5, & x < 5 \\ x - 7 + x - 5, & 5 \le x < 7 \\ x - 7 + x^2 + 3x + 3, & x \ge 7 \end{cases} = \begin{cases} x^2 - 6x + 5, & x < 5 \\ 2x - 12, & 5 \le x < 7 \\ x^2 + 4x - 4, & x \ge 7 \end{cases} \quad \text{Ans.}$$

(17) Let $f(x) = \dfrac{9^x}{9^x + 3}$. Prove that $f(x) + f(1 - x) = 1$, hence prove that $f\left(\dfrac{1}{1987}\right) + f\left(\dfrac{2}{1987}\right) + \cdots \cdots \cdots + f\left(\dfrac{1986}{1987}\right) = 993$

Solution: $-\ f(x) = \dfrac{9^x}{9^x + 3}$ and $f(1 - x) = \dfrac{9^{1-x}}{9^{1-x} + 3} = \dfrac{\frac{9}{9^x}}{\frac{9}{9^x} + 3} = \dfrac{\frac{9}{9^x}}{\frac{9 + 3.9^x}{9^x}} = \dfrac{9}{3(3 + 9^x)} = \dfrac{3}{3 + 9^x}$

$$f(x) + f(1 - x) = \dfrac{9^x}{9^x + 3} + \dfrac{3}{3 + 9^x} = \dfrac{3 + 9^x}{3 + 9^x} = 1 \text{ Proved.}$$

Solved Example

(1) solve the following inequalities: $-$ (a) $\left|3x - \dfrac{1}{2}\right| < 1 \quad \therefore |x| < a \quad or - a < x < a$

$$\therefore a = 1, x = 3x - \frac{1}{2} \ or -1 < 3x - \frac{1}{2} < 1 \ or -\frac{1}{2} < 3x < \frac{3}{2} \quad \boxed{or -\frac{1}{6} < x < \frac{1}{2} \quad \text{Ans.}}$$

(b) $\left|\dfrac{x}{2} - 1\right| \ge 2 \quad \therefore |x| \ge a \ \Rightarrow x \ge a \ \ or \ x \le -a \ \Rightarrow \dfrac{x}{2} - 1 \ge 2 \ \ or \ \dfrac{x}{2} - 1 \le -2 \ \Rightarrow \dfrac{x}{2} \ge 3 \ or \ \dfrac{x}{2} \le -1 \ \therefore x \ge 6 \ or \ x \le -2 \quad \text{Ans.}$

(c) $\left|\dfrac{3x}{5} + \dfrac{3}{2}\right| \le \dfrac{1}{2} \ \Rightarrow -\dfrac{1}{2} \le \dfrac{3x}{5} + \dfrac{3}{2} \le \dfrac{1}{2} \ \ or -\dfrac{1}{2} - \dfrac{3}{2} \le \dfrac{3x}{5} \le \dfrac{1}{2} - \dfrac{3}{2} \ \ or -2 \le \dfrac{3x}{5} \le -1 \ \ or -10 \le 3x \le -5 \ \boxed{\therefore -\dfrac{10}{3} \le x \le -\dfrac{5}{3} \quad \text{Ans.}}$

(d) $|-x - 1| > \dfrac{3}{2} \ \Rightarrow -x - 1 > \dfrac{3}{2} \ \ or -x - 1 < -\dfrac{3}{2} \ \Rightarrow -x > \dfrac{3}{2} + 1 \ \ or -x < -\dfrac{3}{2} + 1$

$$\therefore -x > \frac{5}{2} \ \ or -x < -\frac{1}{2} \quad \boxed{\therefore x < \frac{5}{2} \ or \ x > \frac{1}{2} \quad \text{Ans.}}$$

(e) $(2x - 1)^2 \ge 9 \ \therefore (2x - 1)^2 - 3^2 \ge 0 \ \therefore (2x - 1 + 3)(2x - 1 - 3) \ge 0 \ \therefore (2x + 2)(2x - 4) \ge 0$

$$\therefore f(x) \ge 0 \ \therefore f(x) = 0 \ \Rightarrow (2x + 2)(2x - 4) = 0 \ or \ x = -1, 2$$

$$\underline{\qquad\qquad\qquad\qquad\qquad \overset{\displaystyle f(x) \ge 0}{\qquad\qquad} \qquad\qquad\qquad\qquad\qquad}$$

| $-\infty$ | $+ve$ | -1 | $-ve$ | 2 | $+ve$ | ∞ |

$$\boxed{\therefore x \ge 2 \ or \ x \le -1 \quad \text{Ans.}}$$

(f) $4x^2 + x - 3 \le 0 \ \therefore f(x) = 4x^2 + x - 3 \ \therefore f(x) \le 0$

$\therefore f(x) = 0$ and find x. $\quad \therefore 4x^2 + x - 3 = 0 \ \Rightarrow 4x^2 + 4x - 3x - 3 = 0 \ \Rightarrow 4x(x + 1) - 3(x + 1) = 0$

$$\Rightarrow (4x - 3)(x + 1) = 0 \ \therefore x = -1, \frac{3}{4}$$

$$\underline{\qquad\qquad\qquad\qquad\qquad \overset{\displaystyle f(x) \le 0}{\qquad\qquad} \qquad\qquad\qquad\qquad\qquad}$$

| $-\infty$ | $+ve$ | -1 | $-ve$ | $\dfrac{3}{4}$ | $+ve$ | ∞ |

$$\therefore f(x) \le 0$$

$$\boxed{\therefore -1 \le x \le \frac{3}{4} \quad \text{Ans.}}$$

(2)(a) Given the function $f(x) = \dfrac{2x-1}{2x+3}$, find $f\left(\dfrac{x}{2}\right)$, $2f\left(\dfrac{x}{2}\right)$, $f\left(\dfrac{x^2}{4}\right)$, $\left[f\left(\dfrac{x}{2}\right)\right]^2$.

Solution: $-$ $f(x) = \dfrac{2x-1}{2x+3}$, $f\left(\dfrac{x}{2}\right) = \dfrac{2.\frac{x}{2}-1}{2.\frac{x}{2}+3} = \dfrac{x-1}{x+3}$ Ans. $\qquad 2f\left(\dfrac{x}{2}\right) = 2.\dfrac{x-1}{x+3}$ Ans.

$f\left(\dfrac{x^2}{4}\right) = \dfrac{2\frac{x^2}{4}-1}{2.\frac{x^2}{4}+3} = \dfrac{x^2-2}{x^2+6}$ Ans. $\qquad \left[f\left(\dfrac{x}{2}\right)\right]^2 = \left(\dfrac{x-1}{x+3}\right)^2 = \dfrac{(x-1)^2}{(x+3)^2}$ Ans.

(b) Given the function $f(x) = \log\left(\dfrac{2-x}{2+x}\right)^2$ then prove that $f(x_1) + f(x_2) = f\left(\dfrac{4+x_1.x_2}{x_1+x_2}\right)$.

Solution: $-$ $f(x) = \log\left(\dfrac{2-x}{2+x}\right)^2 = 2\log\left(\dfrac{2-x}{2+x}\right)$, $f(x_1) = 2\log\left(\dfrac{2-x_1}{2+x_1}\right)$, $f(x_2) = 2\log\left(\dfrac{2-x_2}{2+x_2}\right)$

L.H.S $= f(x_1) + f(x_2) = 2\log\left(\dfrac{2-x_1}{2+x_1}\right) + 2\log\left(\dfrac{2-x_2}{2+x_2}\right) = 2\left[\log\dfrac{(2-x_1)(2-x_2)}{(2+x_1)(2+x_2)}\right] = 2\log\left(\dfrac{4-2x_1-2x_2+x_1x_2}{4+2x_1+2x_2+x_1x_2}\right)$

R.H.S $= f\left(\dfrac{4+x_1x_2}{x_1+x_2}\right) = \log\left(\dfrac{2-\frac{4+x_1x_2}{x_1+x_2}}{2+\frac{4+x_1x_2}{x_1+x_2}}\right)^2 = \log\left(\dfrac{2x_1+2x_2-4-x_1x_2}{2x_1+2x_2+4+x_1x_2}\right)^2 = \log\left[\dfrac{-(4-2x_1-2x_2+x_1x_2)}{4+2x_1+2x_2+x_1x_2}\right]^2$

R.H.S $= \log(-1)^2.\left(\dfrac{4-2x_1-2x_2+x_1x_2}{4+2x_1+2x_2+x_1x_2}\right)^2 = 2\log\left(\dfrac{4-2x_1-2x_2+x_1x_2}{4+2x_1+2x_2+x_1x_2}\right)^2$

$$\boxed{\therefore \text{L.H.S} = \text{R.H.S Proved.}}$$

(3) (a) Given the function $f(x) = \dfrac{3^x+3^{-x}}{2}$. show that $f(x+y) + f(x-y) = 2f(x)f(y)$.

Solution: $-$ $f(x) = \dfrac{3^x+3^{-x}}{2}$, $f(x+y) = \dfrac{3^{x+y}+3^{-(x+y)}}{2}$, $f(x-y) = \dfrac{3^{x-y}+3^{-(x-y)}}{2}$

L.H.S $= f(x+y) + f(x-y) = \dfrac{3^{x+y}+3^{-(x+y)}}{2} + \dfrac{3^{x-y}+3^{-(x-y)}}{2} = \dfrac{3^{x+y}+3^{-(x+y)}+3^{x-y}+3^{-(x-y)}}{2} = \dfrac{3^{x^2-y^2}+3^{x^2-y^2}}{2} = \dfrac{2.3^{x^2-y^2}}{2} = 3^{x^2-y^2}$

R.H.S $= 2f(x)f(y) = 2\left(\dfrac{3^x+3^{-x}}{2}.\dfrac{3^y+3^{-y}}{2}\right) = \dfrac{3^x3^y+3^x3^{-y}+3^{-x}3^y+3^{-x}3^{-y}}{2} = \dfrac{3^{x+y}+3^{x-y}+3^{-(x-y)}+3^{-(x+y)}}{2} = \dfrac{3^{x^2-y^2}+3^{x^2-y^2}}{2}$

$= \dfrac{2.3^{x^2-y^2}}{2} = 3^{x^2-y^2}$

$$\boxed{\text{L.H.S} = \text{R.H.S Proved}}$$

(b) Given the function $f(x) = \dfrac{(x+2)^2}{x^2-4}$.

find $f(1), f(-1), f(2), f(-2), f(a+1), f(a)+1, f(a+2), f(a)+2, f(a-1), f(a)-1, f(a-2), f(a)-2, f(2a)$ and $f(2a+1), f(2a)+1$.

Solution: $-$ $f(x) = \dfrac{(x+2)^2}{x^2-4} = \dfrac{(x+2)^2}{x^2-2^2} = \dfrac{(x+2)^2}{(x+2)(x-2)} = \dfrac{x+2}{x-2}$

$\therefore f(1) = \dfrac{1+2}{1-2} = -3$ Ans. $\quad \therefore f(-1) = \dfrac{-1+2}{-1-2} = -\dfrac{1}{3}$ Ans. $\quad \therefore f(2) = \dfrac{2+2}{2-2} = \infty$ Ans.

$\therefore f(-2) = \dfrac{-2+2}{-2-2} = \dfrac{0}{-4} = 0$ Ans. $\quad \therefore f(a+1) = \dfrac{a+1+2}{a+1-2} = \dfrac{a+3}{a-1}$ Ans.

$\therefore f(a)+1 = \dfrac{a+2}{a-2}+1 = \dfrac{a+2+a-2}{a-2} = \dfrac{2a}{a-2}$ Ans. $\quad \therefore f(a+2) = \dfrac{a+2+2}{a+2-2} = \dfrac{a+4}{a}$ Ans.

Functional Calculus

$\therefore f(a) + 2 = \dfrac{a+2}{a-2} + 2 = \dfrac{a+2+2a-4}{a-2} = \dfrac{3a-2}{a-2}$ Ans. $\therefore f(a-1) = \dfrac{a-1+2}{a-1-2} = \dfrac{a+1}{a-3}$ Ans.

$\therefore f(a) - 1 = \dfrac{a+2}{a-2} - 1 = \dfrac{a+2-a+2}{a-2} = \dfrac{4}{a-2}$ Ans. $\therefore f(a-2) = \dfrac{a-2+2}{a-2-2} = \dfrac{a}{a-4}$ Ans.

$\therefore f(a) - 2 = \dfrac{a+2}{a-2} - 2 = \dfrac{a+2-2a+4}{a-2} = \dfrac{-a+6}{a-2}$ Ans. $\therefore f(2a) = \dfrac{2a+2}{2a-2} = \dfrac{a+1}{a-1}$ Ans.

$\therefore f(2a+1) = \dfrac{2a+1+2}{2a+1-2} = \dfrac{2a+3}{2a-1}$ Ans. $\therefore f(2a)+1 = \dfrac{2a+2}{2a-2} + 1 = \dfrac{2a+2+2a-2}{2a-2} = \dfrac{4a}{2a-2} = \dfrac{4a}{2(a-1)} = \dfrac{2a}{a-1}$ Ans.

(4)(a) Given the function $f(x) = \dfrac{x^3+1}{x+1}$. find $\dfrac{f(b)-1}{b-1}$ and $f\left(\dfrac{b+1}{2}\right)$.

(b) Given the function $f(x) = \dfrac{x^3-8}{x^2+2x+4}$. find $\dfrac{f(a+1)-f(b+1)}{a-b}$ and $f\left(\dfrac{\sqrt{3}a+b}{b}\right), f\left(\dfrac{\sqrt{2}b-a}{a}\right)$.

Solution: $-$ given $f(x) = \dfrac{x^3+1}{x+1}$ $\therefore f(x) = \dfrac{(x+1)(x^2-x+1)}{(x+1)} = x^2 - x + 1$, $f(b) = b^2 - b + 1$

$$\therefore \dfrac{f(b)-1}{b-1} = \dfrac{b^2-b+1-1}{b-1} = \dfrac{b(b-1)}{(b-1)} = b \quad \text{Ans.}$$

or $f\left(\dfrac{b+1}{2}\right) = \left(\dfrac{b+1}{2}\right)^2 - \left(\dfrac{b+1}{2}\right) + 1 = \dfrac{b^2+2b+1}{4} - \dfrac{b+1}{2} + 1 = \dfrac{b^2+2b+1-2b-2+4}{4} = \dfrac{b^2+3}{4}$ Ans.

(b) given $f(x) = \dfrac{x^3-8}{x^2+2x+4}$ $\therefore f(x) = \dfrac{(x-2)(x^2+2x+4)}{(x^2+2x+4)} = x - 2$ $\therefore f(a+1) = a+1-2 = a-1, f(b+1) = b+1-2 = b-1$

or $\dfrac{f(a+1)-f(b+1)}{a-b} = \dfrac{a-1-b+1}{a-b} = \dfrac{a-b}{a-b} = 1$ Ans.   or $f\left(\dfrac{\sqrt{3}a+b}{b}\right) = \dfrac{\sqrt{3}a+b}{b} - 2 = \dfrac{\sqrt{3}a+b-2b}{b} = \dfrac{\sqrt{3}a-b}{b}$ Ans.

$$\therefore f\left(\dfrac{\sqrt{2}b-a}{a}\right) = \dfrac{\sqrt{2}b-a}{a} - 2 = \dfrac{\sqrt{2}b-a-2a}{a} = \dfrac{\sqrt{2}b-3a}{a} \quad \text{Ans.}$$

(5)(a) Given the function $f(x) = \begin{cases} 3^x - 2^{-x} + 1, & -2 \le x < 0 \\ \sin\dfrac{x}{2} + \cos\dfrac{x}{2}, & 0 \le x < \pi \\ \dfrac{2x}{x^2-1}, & \pi \le x \le 4 \end{cases}$ find $f(-1), f(-2), f\left(\dfrac{\pi}{3}\right), f\left(\dfrac{2\pi}{3}\right), f\left(\dfrac{\pi}{2}\right), f(\pi), f(0), f(2), f(4)$.

(b) Given the function $f(x) = \begin{cases} 3x^2 + 2, & x \le 1 \\ \dfrac{3}{2x-3}, & 1 < x \le 2 \\ \dfrac{x}{2} - 3, & x > 2 \end{cases}$ find $f(\sqrt{3}), f(\sqrt{2}), f(\sqrt{2}-1), f(\sqrt{3}+1), f(\sqrt{2}+2), f(\sqrt{3}-2)$.

Solution: $-$ (2)(a) $f(x) = \begin{cases} 3^x - 2^{-x} + 1, & -2 \le x < 0 \\ \sin\dfrac{x}{2} + \cos\dfrac{x}{2}, & 0 \le x < \pi \\ \dfrac{2x}{x^2-1}, & \pi \le x \le 4 \end{cases}$

or $f(-1) = 3^x - 2^{-x} + 1 = 3^{-1} - 2^1 + 1$ $\therefore f(-1) = \dfrac{1}{3} - 2 + 1 = \dfrac{1-6+3}{3} = \dfrac{-2}{3}$ Ans.

or $f(-2) = 3^x - 2^{-x} + 1 = 3^{-2} - 2^2 + 1 = \dfrac{1}{9} - 4 + 1 = \dfrac{1-36+9}{9} = -\dfrac{26}{9}$ Ans.

or $f\left(\dfrac{\pi}{3}\right) = \sin\dfrac{x}{2} + \cos\dfrac{x}{2} = \sin\dfrac{\pi}{6} + \cos\dfrac{\pi}{6} = \dfrac{\sqrt{3}+1}{2}$ Ans.   or $f\left(\dfrac{2\pi}{3}\right) = \sin\dfrac{x}{2} + \cos\dfrac{x}{2} = \sin\dfrac{2\pi}{6} + \cos\dfrac{2\pi}{6} = \dfrac{\sqrt{3}}{2} + \dfrac{1}{2} = \dfrac{\sqrt{3}+1}{2}$ Ans.

or $f\left(\dfrac{\pi}{2}\right) = \sin\dfrac{x}{2} + \cos\dfrac{x}{2} = \sin\dfrac{\pi}{4} + \cos\dfrac{\pi}{4} = \dfrac{1}{\sqrt{2}} + \dfrac{1}{\sqrt{2}} = \dfrac{1+1}{\sqrt{2}} = \dfrac{2}{\sqrt{2}} = \dfrac{\sqrt{2} \times \sqrt{2}}{\sqrt{2}} = \sqrt{2}$ Ans.

or $f(\pi) = \dfrac{2x}{x^2 - 1} = \dfrac{2\pi}{\pi^2 - 1}$ Ans. $\quad \therefore f(0) = \sin\dfrac{x}{2} + \cos\dfrac{x}{2} = \sin\dfrac{0}{2} + \cos\dfrac{0}{2} = 1$ Ans.

or $f(2) = \sin\dfrac{x}{2} + \cos\dfrac{x}{2} = \sin\dfrac{2}{2} + \cos\dfrac{2}{2} = \sin 1 + \cos 1$ Ans. $\quad \therefore f(4) = \dfrac{2x}{x^2 - 1} = \dfrac{2 \times 4}{4^2 - 1} = \dfrac{8}{15}$ Ans.

Solution: $-(2)(b)$ same as above, Ans: $-f(\sqrt{3}) = 2\sqrt{3} + 3$, $f(\sqrt{2}) = -(6\sqrt{2} + 9)$, $f(\sqrt{2} - 1) = 11 - 6\sqrt{2}$

$f(\sqrt{3} + 1) = -\dfrac{11}{\sqrt{3} + 5}$, $f(\sqrt{2} + 2) = -\dfrac{7}{\sqrt{2} + 4}$, $f(\sqrt{3} - 2) = \dfrac{97}{23 + 12\sqrt{3}}$.

$(6)(a)$ Calculate $f(x) = \dfrac{16}{x^2} + x^2$ at the points for which $\dfrac{4}{x} + x = 2$. (b) Given the function $f(x) = \dfrac{7x^3 + 3}{x - 1}$. find $f(2x), f(x^2), 5f(x), [f(x)]^2$.

Solution: $-(6)(a)$ $f(x) = \dfrac{16}{x^2} + x^2$ at the point $\dfrac{4}{x} + x = 2$ $\therefore$ $\dfrac{4}{x} + x = 2$ squaring both sides then, $\left(\dfrac{4}{x} + x\right)^2 = 4$ $\Rightarrow$ $\dfrac{16}{x^2} + x^2 + 2.x.\dfrac{4}{x} = 4$

or $\dfrac{16}{x^2} + x^2 = 4 - 8 = -4$ $\therefore$ $f(x) = \dfrac{16}{x^2} + x^2 = -4$ Ans.

(b) $f(x) = \dfrac{7x^3 + 3}{x - 1}$ $\therefore$ $f(2x) = \dfrac{7(2x)^3 + 3}{2x - 1} = \dfrac{56x^3 + 3}{2x - 1}$ Ans. $\therefore$ $f(x^2) = \dfrac{7(x^2)^3 + 3}{x^2 - 1} = \dfrac{7x^6 + 3}{x^2 - 1}$ Ans.

$\therefore$ $5f(x) = 5\left(\dfrac{7x^3 + 3}{x - 1}\right) = \dfrac{35x^3 + 15}{x - 1}$ Ans. $\therefore$ $[f(x)]^2 = \left(\dfrac{7x^3 + 3}{x - 1}\right)^3 = \dfrac{49x^6 + 42x^3 + 9}{x^2 - 2x + 1}$ Ans.

$(7)(a)$ Let $f(x) = \begin{cases} 5^x, & -2 < x < 0 \\ 3, & 0 \le x < 1 \\ 5x - 2, & 1 \le x \le 4 \end{cases}$ find $f(-1), f(0), f(0.5), f(-0.5), f(2), f(3), f(4), f\left(\dfrac{3}{2}\right)$.

(b) $f(x) = x^3 + 5$, $g(x) = 3x$ solve the equation $g(x) = |f(x)|$ and $f(x) + 2 = |g(x)| + 1$.

Solution: $-$ (7) (a) Do yourself.

Ans: $-$ $f(-1) = \dfrac{1}{5}$, $f(0) = 3$, $f(0.5) = 3$, $f(-0.5) = \sqrt{5}$, $f(2) = 8$, $f(3) = 13$, $f(4) = 18$, $f\left(\dfrac{3}{2}\right) = \dfrac{11}{2}$.

(b) $f(x) = x^3 + 5$, $g(x) = 3x$ $\therefore$ $g(x) = |f(x)|$ $\Rightarrow 3x = |x^3 + 5|$ $\Rightarrow 3x = -(x^3 + 5)$ and $3x = x^3 + 5$

$$\boxed{\therefore \ x^3 + 3x + 5 = 0 \ \text{ or } x^3 - 3x + 5 = 0 \ \text{ Ans.}}$$

Also find $f(x) + 2 = |g(x)| + 1$ $\Rightarrow x^3 + 5 + 2 = 3x + 1$ or $x^3 + 5 + 2 = -3x + 1$

$\qquad\qquad x^3 + 7 = 3x + 1$ or $x^3 + 7 = -3x + 1$ $\qquad \therefore x^3 - 3x + 6 = 0$ or $x^3 + 3x + 6 = 0$ Ans.

$(8)(a)$ Find $f(x)$ if $f(x + 2) = x^2 + 5x - 3$ $\qquad$ (b) Find $f(x)$ if $f(2x - 1) = 4x^2 - 6x + 7$

Solution: $-$ $(8)(a)$ find $f(x)$, $f(x + 2) = x^2 + 5x - 3 = (x + 2)^2 + (x + 2) - 9$ $\quad$ put $x + 2 = x$ then $f(x) = x^2 + x - 9$ Ans.

(b) Do yourself (see above question). Ans: $-$ $f(x) = x^2 - x + 5$

(9) which of the given functions is (are) even, odd, and which of them is (are) neither even, nor odd?

(a) $f(x) = \log\left(x - \sqrt{1 + x^2}\right)$ (b) $f(x) = \log\left(\dfrac{2 + x}{2 - x}\right)$ (c) $f(x) = 2x^2 - 3x + 1$ (d) $f(x) = \dfrac{a^{2x} + 1}{a^{2x} - 1}$ (e) $f(x) = \dfrac{x(a^x - 2)}{a^x}$ (f) $f(x) = x\dfrac{e^x - 1}{e^x + 1}$

Solution: $-$ $(9)(a)$ $f(x) = \log\left(x - \sqrt{1 + x^2}\right)$, $\quad f(-x) = \log\left(-x - \sqrt{1 + x^2}\right)$

$f(x) + f(-x) = \log\left(x - \sqrt{1 + x^2}\right) + \log\left(-x - \sqrt{1 + x^2}\right) = \log\left(x - \sqrt{1 + x^2}\right)\left(-x - \sqrt{1 + x^2}\right)$

$$f(x) + f(-x) = \log\left(-x^2 - x\sqrt{1 + x^2} + x\sqrt{1 + x^2} + 1 + x^2\right) = \log 1 = 0$$

$$f(x) + f(-x) = 0 \quad \therefore f(x) = -f(-x) \quad \text{function } f(x) \text{ is odd.} \quad \text{Ans.}$$

(b) Ans: – odd   (c) Ans: – function f(x) is neither even nor odd.

(d) $f(x) = \dfrac{a^{2x} + 1}{a^{2x} - 1}$, $f(-x) = \dfrac{a^{-2x} + 1}{a^{-2x} - 1} = \dfrac{1 + a^{2x}}{1 - a^{2x}} = -\dfrac{a^{2x} + 1}{a^{2x} - 1} = -f(x)$ $f(x) + f(-x) = 0$ function f(x) is odd. Ans.

(e) $f(x) = \dfrac{x(a^x - 2)}{a^x}$, $f(-x) = -x\dfrac{a^{-x} - 2}{a^{-x}} = -x(1 - 2a^x)$ $\therefore$ function f(x) is neither even nor odd. Ans.

(f) $f(x) = x\dfrac{e^x - 1}{e^x + 1}$, $f(-x) = -x\dfrac{e^{-x} - 1}{e^{-x} + 1} = -x\dfrac{1 - e^x}{1 + e^x} = x\dfrac{e^x - 1}{e^x + 1} = f(x)$ $f(x) = f(-x)$ $\therefore f(x) - f(-x) = 0$ function f(x) is even. Ans.

(10) which of the following function is (are) even and which is (are) odd?

(a) $f(x) = 4 - \sin^2 x$ (b) $f(x) = \sqrt{-x^2 - 6x + 9} + \sqrt{-x^2 + 6x + 9}$ (c) $f(x) = \sin\dfrac{x}{2} + \cos\dfrac{x}{2}$ (d) $f(x) = \sqrt{1 + x + x^2} - \sqrt{1 - x + x^2}$

(e) $f(x) = 3 - x^2 + \cos^2 x$ (f) $f(x) = x\dfrac{1 - a^{kx}}{1 + a^{kx}}$ (g) $f(x) = \sqrt[3]{1 + x} + \sqrt[3]{1 - x}$ (h) $f(x) = 2x^2 + |x|$ (i) $f(x) = x^2 \sin x - x$

(j) $f(x) = \dfrac{(1 + 3^x)^2}{(1 - 3^x)}$ (k) $f(x) = \dfrac{1 + a^{kx}}{1 - a^{kx}}$ (l) $f(x) = \dfrac{(1 - 2^x)^2}{1 + 2^x}$

Solution: –(10)(a) Ans: – Even   (b) Ans: – Even   (c) Ans: – Neither even nor odd.   (d) Ans: – odd

(e) Ans: –Even   (f) Ans: – Even   (g) Ans: – Even   (h) Ans: – Even   (i) Ans: – odd   (j) Ans: – odd

(k) $f(x) = \dfrac{1 + a^{kx}}{1 - a^{kx}}$, $f(-x) = \dfrac{1 + a^{-kx}}{1 - a^{-kx}} = \dfrac{a^{kx} + 1}{a^{kx} - 1} = -\dfrac{a^{kx} + 1}{1 - a^{kx}} = -f(x)$ or $f(x) + f(-x) = 0$ $\therefore$ function f(x) is odd

(l) $f(x) = \dfrac{(1 - 2^x)^2}{1 + 2^x}$, $f(-x) = \dfrac{(1 - 2^{-x})^2}{1 + 2^{-x}} = \dfrac{(2^x - 1)^2}{1 + 2^x} = \dfrac{[-(1 - 2^x)]^2}{1 + 2^x} = f(x)$

$$\boxed{f(-x) - f(x) = 0 \quad \therefore \text{function } f(x) \text{ is even.}}$$

(11) Indicate the amplitude |A|, frequency ω, initial phase φ and period T of the following harmonics.

(a) $f(x) = 4\sin\dfrac{3x}{2}$ (b) $f(x) = 3\sin\left(5x + \dfrac{\pi}{3}\right)$ (c) $f(x) = 2\sin\dfrac{x}{3} + 3\cos\dfrac{x}{3}$ (d) $f(x) = \sin 2x + \cos 2x$

Solution: – (11)(a) $f(x) = 4\sin\dfrac{3x}{2}$ $\therefore f(x) = A\sin(\omega x + \varphi)$

$f(x) = 4\sin\left(\dfrac{3x}{2} + 0\right)$, $\therefore |A| = 4, \varphi = 0, \omega = \dfrac{3}{2}$ $\therefore$ Period (T) $= \dfrac{2\pi}{\omega} = \dfrac{2\pi}{\frac{3}{2}} = \dfrac{4\pi}{3}$ Ans.

(b) $f(x) = 3\sin\left(5x + \dfrac{\pi}{3}\right)$, $\therefore f(x) = A\sin(\omega x + \varphi)$ Here $|A| = 3, \omega = 5, \varphi = \dfrac{\pi}{3}$ Period (T) $= \dfrac{2\pi}{\omega} = \dfrac{2\pi}{5}$ Ans.

(c) $f(x) = 2\sin\dfrac{x}{3} + 3\cos\dfrac{x}{3}$,   Ans: – Period(T) = 6π,

(d) $f(x) = \sin 2x + \cos 2x$,   Ans: – Period(T) = π, $|A| = \sqrt{2}$ and $\varphi = \dfrac{\pi}{4}$

(12) Find the period for each of the following functions. (a) $f(x) = \tan\dfrac{3x}{2}$ (b) $f(x) = 2\cot\dfrac{x}{3}$ (c) $f(x) = \sin\dfrac{2\pi x}{3}$ (d) $f(x) = \cos 3\pi x$

Solution: – (12)(a) $f(x) = \tan\dfrac{3x}{2}$, Period of $\tan x = \pi$ $\therefore$ Period of $\tan\dfrac{3x}{2} = \dfrac{\text{period of } \tan x}{|a|}$

where $a = \dfrac{3}{2}$ $\quad \therefore$ Period of $\tan\dfrac{3x}{2} = \dfrac{\pi}{\frac{3}{2}} = \dfrac{2\pi}{3}$ Ans.

(b) $f(x) = 2\cot\dfrac{x}{3}$, $\quad$ Period of $\cot x = \pi = (T)$ $\quad \therefore$ Period of $2\cot\dfrac{x}{3} = \dfrac{T}{a} = \dfrac{\pi}{\frac{1}{3}} = 3\pi$ $\quad$ where $a = \dfrac{1}{3}$ $\quad$ Ans.

(c) $f(x) = \sin\dfrac{2\pi x}{3}$, $\quad$ Period of $\sin x = 2\pi = T$ $\quad \therefore$ Period of $\sin\dfrac{2\pi x}{3} = \dfrac{T}{a} = \dfrac{2\pi}{\frac{2\pi}{3}} = 3$ $\quad$ where $a = \dfrac{2\pi}{3}$ $\quad$ Ans.

(d) $f(x) = \cos 3\pi x$, $\quad$ Period of $\cos x = 2\pi = T$ $\quad$ Period of $\cos 3\pi x = \dfrac{T}{a} = \dfrac{2\pi}{3\pi} = \dfrac{2}{3}$ $\quad$ where $a = 3\pi$ $\quad$ Ans.

Note: $-$ $f(x) = \cos ax$ $\quad \therefore$ Period of $\cos x = 2\pi = T$ $\quad \therefore$ Period of $\cos ax = \dfrac{2\pi}{a} = \dfrac{T}{a}$.

Exercise – A1

(1) find the domain of each of the follwing function: $-$ (a) $y = \sqrt{\log_{.2}\left(\dfrac{x+1}{x-3}\right)} \times \dfrac{1}{x^2 - 16}$ $\quad$ (b) $y = \sqrt[5]{2^{2x} + 16^{\left(\frac{1}{2}\right)(x-2)} - 52 - 8^{\left(\frac{2}{3}\right)(x-2)}}$

(c) $y = \log_x\left(\dfrac{x-3}{x+1}\right)$ $\quad$ (d) $y = 3\cos^{-1}\left(\dfrac{2x+3}{4}\right) + 4\tan^{-1}\left(\dfrac{5}{x}\right) + \sin^{-1}\left(\dfrac{2x-1}{3}\right)$ $\quad$ (e) $y = \sqrt{\log_{10}\left(\dfrac{2x - x^2}{5}\right)} + \dfrac{2}{x-5}$ $\quad$ (f) $y = \sqrt{|x| - x}$

(g) $y = \sin\left(\dfrac{5 - 2x}{3}\right) + \sqrt{\dfrac{x+3}{x-5}} - \dfrac{1}{5 - x^2 + x}$ $\quad$ (h) $y = \log_x\left(\dfrac{x-5}{2}\right)$ $\quad$ (i) $y = 3\cos^{-1}\left(\dfrac{x+7}{5-2x}\right) + \sin^{-1}\left(\dfrac{5}{x-2}\right) - \dfrac{1}{\sqrt{5+x}}$

(j) $y = \dfrac{2x-3}{5-3x} - \dfrac{2}{x+\sqrt{2}} + \sqrt{x^2 - 16}$

(2)(a) If $f(x) = \log_e\left(\dfrac{1+\sqrt{x}}{1-\sqrt{x}}\right)$, then show that $f(m) + f(n) = \log_e\left[\dfrac{\left(1+\sqrt{m}\right)^2.\left(1+\sqrt{n}\right)^2}{(1-m)(1-n)}\right]$

(b) If $f(x) = \log\left(\dfrac{1 - \log x}{1 + \log x}\right)$, then show that $f(x) = f(y) + \log\left[\dfrac{(1 - \log x)(1 + \log y)}{(1 + \log x)(1 - \log y)}\right]$

(c) If $f(x) = \tan(\log x)$, then show that $f(x^2) + f(y^2) = f(x^2 y^2).[1 + f(x^2)f(y^2)]$.

(3) (a) Prove that the inverse of the function $y = \dfrac{2^x + 2^{-x}}{2^x - 2^{-x}} + 2$ is $y = \dfrac{1}{2}\log_2\left(\dfrac{x-1}{x-3}\right)$.

(b) Prove that the inverse of the function $f(x) = \dfrac{a^x + a^{-x}}{a^x - a^{-x}} + 1$ is $f(x) = \dfrac{1}{2}\log_a\left(\dfrac{y}{y-2}\right)$.

(c) Prove that the inverse of the function $f(x) = \dfrac{2.5^x - 5^{-x}}{3.5^x + 5^{-x}} + 3$ is $f(x) = \dfrac{1}{2}\log_5\left(\dfrac{2-y}{3y-11}\right)$.

(4) (a) If $y = \sin(\log x)$, then prove that $x^2 y'' + xy' + y = 0$. $\quad$ (b) If $y = \cos(\log x)$, then prove that $x^2 y'' - xy' + y = 0$.

(c) If $f(x) = \dfrac{x-1}{x}$, then prove that $f(x) + f\left(\dfrac{1}{x}\right) = f(x)f\left(\dfrac{1}{x}\right)$.

(5) (a) Let function $f: R \to R$ be defined by $f(x) = x + \cos x$ for $x \in R$, then prove that f is one to one and onto.

(b) If $f(x) = (a + x^m)^{\frac{1}{m}}$ whose $a > 0$ and m is a positive integer, then show that $f(f(x)) = (2a + x^m)^{\frac{1}{m}}$.

(c) Let $f(x) = \dfrac{\alpha x}{x-1}$, $x \neq 1$. then for what value of α is $f(f(x)) = x$.

(6) (a) Given the function $f(x) = \begin{cases} 2^{-x} - 1 & \text{for } -2 \le x < 0 \\ \cot\dfrac{x}{2} & \text{for } 0 \le x < \pi \\ \dfrac{x}{x^3 - 2} & \text{for } \pi \le x \le 8 \end{cases}$ find $f(-2), f\left(\dfrac{\pi}{6}\right), f\left(\dfrac{2\pi}{3}\right), f(6), f(8), f\left(\dfrac{\pi}{4}\right), f\left(\dfrac{3\pi}{4}\right), f(1), f\left(\dfrac{1}{2}\right), f(-1).$

(b) $f: R \to R$ is defined as under $f(x) = \begin{cases} x - 3, & x < -8 \\ x^3 + 1, & x \in [-8, 8] \\ 3x + 5, & x > 8 \end{cases}$ Evaluate $, f(-14), f(2), f(12)$ and $(fof)4$.

(c) If $f(x) = \log_2 x$ and $g(x) = 2^x$ then show that $(fog)x = (gof)x = x$.   (d) If $f(x) = \dfrac{x}{1-x}$ then show that $(fofof)x = \dfrac{x}{1-3x}$.

(e) Let $f(x) = \cot x, x \in \left(-\dfrac{\pi}{2}, \dfrac{\pi}{2}\right)$ and $g(x) = \sqrt{1 + x^2}$. Determine $(fog)x$ and $(gof)x$.

(7) (a) If $g(f(x)) = |\cos x|$ and $f(g(x)) = (\cos \sqrt{x})^2$ then find $f(x)$ and $g(x)$.

(b) If $f(x) = \begin{cases} x^2 - 2x + 3, & x < 3 \\ 2x - 1, & x \ge 3 \end{cases}$ and $g(x) = \begin{cases} x - 4, & x < 4 \\ 2x^2 + 4x + 5, & x \ge 4 \end{cases}$ Determine $f + g$ and $\dfrac{f}{g}$.

(c) If $f(x) = |x - 3|$ and $g(x) = f(f(x))$, then for $x > 3$, $g'(x)$ is equal to ... ...

(8) Find the domain of definition of the functions: — (a) $f(x) = \log_x 2$ (b) $f(x) = \sqrt{\log\left(\dfrac{9x - 4x^2}{5}\right)}$ (c) $f(x) = \log|3 - x^2|$

(d) $f(x) = \sqrt{2x - 5} - \dfrac{1}{\log_3(2 + x)}$ (e) $f(x) = \sqrt{\sin^{-1} x} + \dfrac{1}{\log x}$ (f) $f(x) = \sqrt{2 + \log_2 x}$ (g) $f(x) = \log_3\left\{\log_{\frac{1}{3}}(x^2 - 4x + 4)\right\}$

(h) $f(x) = \log_4 \log_3 \log_2(1 - 6x - 9x^2)$ (i) $f(x) = \log\{(2\log x)^2 - 12\log x + 9\}$ (j) $f(x) = \sqrt{\log_{0.2}(4x - 2x^2)}$

(k) $f(x) = \sqrt{|x| - x}$ (l) $f(x) = \dfrac{2}{2x + 3|x|}$

(9) Find the domain of definition of the functions: — (a) $f(x) = \cos^{-1}\left(\dfrac{2x}{1 + x^2}\right) + \sqrt{\sin(\cos x)}$ (b) $f(x) = \cos^{-1}\left[\log_3\left(\dfrac{1}{3}x^3\right)\right]$

(c) $f(x) = \dfrac{\sqrt{3 - x^2}}{\cos^{-1}(1 + x)}$ (d) $f(x) = \log(\sin x)$ (e) $f(x) = \log(\cos x)^2$ (f) $f(x) = \sin^{-1}\left(\dfrac{3 - |x|}{2}\right) + \log(2 - x)$

(g) $f(x) = \sqrt{4 + 2^x + 2^{2-x}} + \sqrt{\cos^{-1}(x + 2)}$

(10) (a) $y = \dfrac{\sqrt{x - 3}}{x^2 - 1}, x \ge 3$ and $x \ne \pm 1$ (b) $y = \dfrac{\cos^{-1}(3 - x)}{\sqrt{x^2 - 1}}$ (c) $y = \left\{\dfrac{3 - x^2}{x + 2}\right\}^{\frac{1}{2}}$ (d) $y = \dfrac{\sqrt{5 + 3x - x^2}}{\sqrt{\sin x - \dfrac{1}{2}}}$

(e) $y = \log_4(x^2 + 1) + \sqrt{64x - x^7}$ (f) $y = \sqrt{\dfrac{2 - |x|}{3 - |x|}}$ (g) $f(x) = \sin^{-1}\sqrt{1 - x^2} + \cos^{-1}\sqrt{1 + x^2}$

(11) Find the range of the functions: — (a) $y = \dfrac{2x}{1 - x^2}$ (b) $y = \dfrac{2}{1 - \sin 2x}$ (c) $y = \dfrac{2}{3 + x^2}$ (d) $y = \dfrac{x + 2}{x^2 - x - 6}$ (e) $y = 2 - \dfrac{3}{x^2 - 3x + 4}$

(f) $y = \log\sqrt{x^2 + 4x + 10}$ (g) $y = 3\cos x + 2\sin\left(x + \dfrac{\pi}{6}\right) + 5$ (h) $y = \cos^{-1}\left[\dfrac{3}{2} + x^2\right]$ where $[\,.\,]$ denotes the greatest integer function.

(12) Find the domain and range of the functions: — (a) $y = \dfrac{x + 1}{x^2 - 6x - 3}$ (b) $y = \dfrac{2}{\sqrt{1 + 2\cos x}}$ (c) $y = \cos\left[\log\left(\dfrac{\sqrt{1 - x^2}}{2 - x}\right)\right]$

(d) If $f(x) = 5 - x_{P_{x-1}}$, find domain and range of $f(x)$.

(e) If $f(\sin 2x) = \dfrac{(2\tan x + \sec^2 x)(1 + \sin 2x)}{4\sin^2 x}$ then determine the domain and range of $f(t)$.

(13) (a) If the function f: $[2, \infty] \to [2, \infty[$ is defined by f: $(2, \infty)$ $f^{-1}(x)$ is the $f(x) = 3^{x(x-2)}$, then $f^{-1}(x)$ is $1 \pm \sqrt{1 + \log_3 x}$.

$f^{-1}(x)$ is positive $f^{-1}(x) = 1 + \sqrt{1 + \log_3 x}$

(b) If f: $[2, \infty[\to [3, \infty[$ is given by $f(x) = x - \dfrac{2}{x}$ then $f^{-1}(x)$ is equal to $\dfrac{x \pm \sqrt{x^2 + 8}}{2}$, $f^{-1}(x)$ is positive $f^{-1}(x) = \dfrac{x + \sqrt{x^2 + 8}}{2}$

(14) Find the identical function are following: − (a) $f(x) = \dfrac{x+1}{x+1}$ and $g(x) = 1$ (b) $f(x) = 1$ and $g(x) = \sec^2 x - \tan^2 x$

(c) $f(x) = \log x^4$ and $g(x) = 4 \log x$ (d) $f(x) = \log(x+1) + \log x$ and $g(x) = \log\{x(x+1)\}$

(e) $f(x) = 2x$ and $g(x) = \sqrt{4x^2}$ (f) $f(x) = \log x - \log(x-2)$ and $g(x) = \log\left(\dfrac{x}{x+2}\right)$

(15) Classify the functions are even or odd: − (a) $f(x) = \log\left(x - \sqrt{1 + x^2}\right)$ (b) $f(x) = \cos^3 x + 2\tan^2 x$ (c) $f(x) = x^4 + |x|$

(d) $f(x) = x\dfrac{e^x - 1}{e^x + 1}$ (e) $f(x) = \dfrac{\tan x . \cot x}{\sin^2 x + \cos x}$ (f) $f(x) = \sec^2 x + \csc^2 x$

(16) (a) If f is an even function defined on the interval $(-3,3)$, then real values of x satisfying the equation

$f(x) = f\left(\dfrac{x-2}{x-3}\right)$ are $1 - \sqrt{3}, 2 - \sqrt{2}, 1 + \sqrt{3}$.

(b) show that the function $f(x) = \displaystyle\int_0^x \log_e\left(\dfrac{2+x}{2-x}\right) dx$ is an even function.

(17) Find the fundamental time period of the function: −

(a) $\sin\left(\dfrac{2x}{3}\right)$ (b) $\cos(\pi - x)$ (c) $\tan 2x$ (d) $\sin\dfrac{x}{2}$ (e) $|\sin x|$ (f) $\sin^3 x + \cos^3 x$ (g) $2\sin\dfrac{2}{3}(x + \pi)$

Solution

(1) (a) Domain (D) $= R - \{[-1,3]\}$ (b) Domain (D) $= [52, \infty[$ (c) Domain (D) $= (3, \infty)$

(d) Domain (D) $= \left[-1, \dfrac{1}{2}\right]$ (e) Domain (D) $= (-\infty, -1] \cup [5, \infty)$ (f) Domain (D) $=]-\infty, 0]$

(g) D $= [1,4] - \left\{\dfrac{1 + \sqrt{21}}{2}\right\}$ (h) D $= (5, \infty)$ (i) D $= \left[-\dfrac{2}{3}, \dfrac{2}{3}\right]$ (j) Domain (D) $= (-\infty, -4) \cup [4, \infty[$

(2) (a) $f(x) = \log_e\left(\dfrac{1 + \sqrt{x}}{1 - \sqrt{x}}\right)$, then show that $f(m) + f(n) = \log_e\left[\dfrac{(1 + \sqrt{m})^2.(1 + \sqrt{n})^2}{(1 - m)(1 - n)}\right]$

$f(x) = \log_e\left(\dfrac{1 + \sqrt{x}}{1 - \sqrt{x}}\right)$, $\quad f(m) = \log_e\left(\dfrac{1 + \sqrt{m}}{1 - \sqrt{m}}\right)$, $\quad f(n) = \log_e\left(\dfrac{1 + \sqrt{n}}{1 - \sqrt{n}}\right)$

$f(m) + f(n) = \log_e\left(\dfrac{1 + \sqrt{m}}{1 - \sqrt{m}}\right) + \log_e\left(\dfrac{1 + \sqrt{n}}{1 - \sqrt{n}}\right) = \log_e\left\{\dfrac{(1 + \sqrt{m})(1 + \sqrt{n})}{(1 - \sqrt{m})(1 - \sqrt{n})}\right\} = \log_e(1 + \sqrt{m})(1 + \sqrt{n}) - \log_e(1 - \sqrt{m})(1 - \sqrt{n})$

or $f(x) = \log_e\left(\dfrac{1 + \sqrt{x}}{1 - \sqrt{x}}\right) = \log_e\left\{\dfrac{(1 + \sqrt{x})(1 + \sqrt{x})}{(1 - \sqrt{x})(1 + \sqrt{x})}\right\} = \log_e\dfrac{(1 + \sqrt{x})^2}{(1 - x)}$

$f(m) = \log_e\dfrac{(1 + \sqrt{m})^2}{(1 - m)}$, $\quad f(n) = \log_e\dfrac{(1 + \sqrt{n})^2}{(1 - n)}$

$$f(m) + f(n) = \log_e \frac{(1+\sqrt{m})^2}{(1-m)} + \log_e \frac{(1+\sqrt{n})^2}{(1-n)} = \log_e \left\{ \frac{(1+\sqrt{m})^2 (1+\sqrt{n})^2}{(1-m)(1-n)} \right\} \quad \text{Proved.}$$

$$f(m) + f(n) = \log_e (1+\sqrt{m})^2 . (1+\sqrt{n})^2 - \log_e (1-m)(1-n) = \log_e (1+\sqrt{m})^2 + \log_e (1+n)^2 - \log_e (1-m) - \log_e (1-n)$$
$$= 2\log_e (1+\sqrt{m}) + 2\log_e (1+\sqrt{n}) - \log_e (1-m) - \log_e (1-n) \quad \text{Proved.}$$

(b) same as above question. (c) same as above question.

(3) (a) $y = \dfrac{2^x + 2^{-x}}{2^x - 2^{-x}} + 2$ is $y = \dfrac{1}{2} \log_2 \left(\dfrac{x-1}{x-3} \right)$

$$\Rightarrow y - 2 = \frac{2^{2x} + 1}{2^{2x} - 1} \quad \Rightarrow 2^{2x}.y - y - 2.2^{2x} + 2 = 2^{2x} + 1 \quad \Rightarrow 2^{2x}(y-3) = y - 1 \quad \Rightarrow 2^{2x} = \frac{y-1}{y-3}$$

$$\log_2 2^{2x} = \log_2 \left(\frac{y-1}{y-3} \right) \quad \Rightarrow 2x = \log_2 \left(\frac{y-1}{y-3} \right) \quad \therefore x = \frac{1}{2} \log_2 \left(\frac{y-1}{y-3} \right) \text{ is the inverse function.} \quad \therefore y = \frac{1}{2} \log_2 \left(\frac{x-1}{x-3} \right) \quad \text{Proved.}$$

(b) same as above question   (c) same as above question

(4) (a) $y = \sin(\log x)$ $\quad \therefore y' = \cos(\log x).\dfrac{1}{x}$ $\quad \therefore y'' = \dfrac{-x\sin(\log x).\dfrac{1}{x} - \cos(\log x).1}{x^2} = \dfrac{-[\sin(\log x) + \cos(\log x)]}{x^2}$

$$\text{L.H.S} = x^2 y'' + xy' + y = x^2 \left\{ \frac{-[\sin(\log x) + \cos(\log x)]}{x^2} \right\} + x \left\{ \cos(\log x).\frac{1}{x} \right\} + \sin(\log x)$$
$$= -\sin(\log x) - \cos(\log x) + \cos(\log x) + \sin(\log x) = 0 = \text{R.H.S} \quad \text{Proved.}$$

(b) same as above question (c) $f(x) = \dfrac{x-1}{x}$, $f\left(\dfrac{1}{x}\right) = \dfrac{\frac{1}{x} - 1}{\frac{1}{x}} = \dfrac{\frac{1-x}{x}}{\frac{1}{x}} = 1 - x$

$$\text{L.H.S} = f(x) + f\left(\frac{1}{x}\right) = \frac{x-1}{x} + 1 - x = \frac{x-1+x-x^2}{x} = \frac{-(x^2 - 2x + 1)}{x} = \frac{-(x-1)^2}{x}$$

$$\text{R.H.S} = f(x).f\left(\frac{1}{x}\right) = \frac{x-1}{x} \times 1 - x = \frac{-(x-1)^2}{x} \quad \text{or L.H.S} = \text{R.H.S} \quad \therefore \frac{-(x-1)^2}{x} = \frac{-(x-1)^2}{x} \quad \text{Proved.}$$

(5) (a) $f(x) = x + \cos x$, $\quad f(x_1) = x_1 + \cos x_1$, $\quad f(x_2) = x_2 + \cos x_2$

$$\therefore f(x_1) = f(x_2) \Rightarrow x_1 = x_2 \text{ function } f(x) \text{ is one to one function.}$$

$\Rightarrow y = x + \cos x$ it is straight line, then range$(R) = (-\infty, \infty)$ so, $y = x + \cos x$ is onto function.

$f(x) = x + \cos x$ the function $f(x)$ is one to one and onto function. Proved

(b) $f(x) = (a + x^m)^{\frac{1}{m}}$ $\Rightarrow \text{L.H.S} = f(f(x)) = f\left\{ (a+x^m)^{\frac{1}{m}} \right\} = \left[a + (a+x^m)^{\frac{1}{m}.m} \right]^{\frac{1}{m}} = [2a + x^m]^{\frac{1}{m}} = \text{R.H.S} \quad \text{Proved}$

(c) Do yourself, $a = \dfrac{x \pm (x-2)}{2} = \dfrac{x+x-2}{2}, \dfrac{x-x+2}{2}$ $\quad \therefore a = 1, (x-1)$ Ans.

(6) (a) $f(x) = \begin{cases} 2^{-x} - 1 & \text{for } -2 \le x < 0 \\ \cot \dfrac{x}{2} & \text{for } 0 \le x < \pi \\ \dfrac{x}{x^3 - 2} & \text{for } \pi \le x \le 8 \end{cases}$ find $f(-2) = 2^{-x} - 1 = 2^{-(-2)} - 1 = 2^2 - 1 = 4 - 1 = 3$ Ans.

$$f\left(\frac{\pi}{6}\right) = \cot \frac{x}{2} = \cot \frac{\pi}{12} = \cot 15^0 = \frac{\cos 15^0}{\sin 15^0} = \frac{\cos(45^0 - 30^0)}{\sin(45^0 - 30^0)} = \frac{\cos 45^0 \cos 30^0 + \sin 45^0 \sin 30^0}{\sin 45^0 \cos 30^0 - \cos 45^0 \sin 30^0}$$

$$f\left(\frac{\pi}{6}\right) = \frac{\frac{1}{\sqrt{2}}\frac{\sqrt{3}}{2} + \frac{1}{\sqrt{2}}\frac{1}{2}}{\frac{1}{\sqrt{2}}\frac{\sqrt{3}}{2} - \frac{1}{\sqrt{2}}\frac{1}{2}} = \frac{\frac{\sqrt{3}}{2\sqrt{2}} + \frac{1}{2\sqrt{2}}}{\frac{\sqrt{3}}{2\sqrt{2}} - \frac{1}{2\sqrt{2}}} = \frac{\sqrt{3} + 1}{\sqrt{3} - 1} = \frac{\sqrt{3}+1}{\sqrt{3}-1} \times \frac{\sqrt{3}+1}{\sqrt{3}+1} = \frac{(\sqrt{3}+1)^2}{3-1} = \frac{3 + 2\sqrt{3} + 1}{2} = \frac{4 + 2\sqrt{3}}{2} = 2 + \sqrt{3} \quad \text{Ans.}$$

$f\left(\dfrac{2\pi}{3}\right) = \cot\dfrac{x}{2} = \cot\dfrac{2\pi}{2.3} = \cot\dfrac{\pi}{3} = \dfrac{1}{\sqrt{3}}$ Ans. $\quad f(6) = \dfrac{x}{x^3 - 2} = \dfrac{6}{6^3 - 2} = \dfrac{6}{216 - 2} = \dfrac{6}{214} = \dfrac{3}{107}$ Ans.

$f(8) = \dfrac{x}{x^3 - 2} = \dfrac{8}{8^3 - 2} = \dfrac{8}{512 - 2} = \dfrac{8}{510} = \dfrac{4}{255}$ Ans. $\quad f\left(\dfrac{\pi}{4}\right) = \cot\dfrac{x}{2} = \cot\dfrac{\pi}{8}$ Ans. $\quad f\left(\dfrac{3\pi}{4}\right) = \cot\dfrac{x}{2} = \cot\dfrac{3\pi}{8}$ Ans.

$f(1) = \cot\dfrac{x}{2} = \cot\dfrac{1}{2}$ Ans. $\quad f\left(\dfrac{1}{2}\right) = \cot\dfrac{x}{2} = \cot\dfrac{1}{4}$ Ans. $\quad f(-1) = 2^{-(-1)} - 1 = 2^1 - 1 = 2 - 1 = 1$ Ans.

(b) $f(x) = \begin{cases} x - 3, & x < -8 \\ x^3 + 1, & x \in [-8, 8] \\ 3x + 5, & x > 8 \end{cases}$ Ans: $-$ $f(-14) = -17, f(2) = 9, f(12) = 41, (f \circ f)4 = f\big(f(4)\big) = 200,$

(c) $f(x) = \log_2 x$ and $g(x) = 2^x$ To prove, $(f \circ g)x = (g \circ f)x = x$ $\Rightarrow f\big(g(x)\big) = g\big(f(x)\big) = x$ $\Rightarrow f(2^x) = g(\log_2 x) = x$

$$\log_2 2^x = 2^{\log_2 x} = x \quad \Rightarrow x = x = x \quad \text{Proved.}$$

(d) Do yourself, same as above question (e) Do yourself, same as question no. $-(5)(c)$

(7) Ans: $-$ (a) is satisfy the equation $g\big(f(x)\big) = |\cos x|$ and $f\big(g(x)\big) = \big(\cos\sqrt{x}\big)^2$.

(b) $f(x) = \begin{cases} x^2 - 2x + 3, & x < 3 \\ 2x - 1, & x \geq 3 \end{cases}$ and $g(x) = \begin{cases} x - 4, & x < 4 \\ 2x^2 + 4x + 5, & x \geq 4 \end{cases}$

The critical points are 3,4 and so we redefine the functions for $x < 3, 3 \leq x < 4, x \geq 4$

$\therefore f(x) = \begin{cases} x^2 - 2x + 3, & x < 3 \\ 2x - 1, & 3 \leq x < 4 \\ 2x - 1, & x \geq 4 \end{cases}$ and $g(x) = \begin{cases} x - 4, & x < 3 \\ x - 4, & 3 \leq x < 4 \\ 2x^2 + 4x + 5, & x \geq 4 \end{cases}$

$\therefore \ f + g = \begin{cases} x^2 - 2x + 3 + x - 4, & x < 3 \\ 2x - 1 + x - 4, & 3 \leq x < 4 \\ 2x - 1 + 2x^2 + 4x + 5, & x \geq 4 \end{cases} = \begin{cases} x^2 - x - 1, & x < 3 \\ 3x - 5, & 3 \leq x < 4 \\ 2x^2 + 6x + 4, & x \geq 4 \end{cases}$ Ans.

$\therefore \ \dfrac{f}{g} = \begin{cases} \dfrac{x^2 - 2x + 3}{x - 4}, & x < 3 \\[2mm] \dfrac{2x - 1}{x - 4}, & 3 \leq x < 4 \\[2mm] \dfrac{2x - 1}{2x^2 + 4x + 5}, & x \geq 4 \end{cases}$ Ans.

(c) $f(x) = |x - 3|$ and $g(x) = f\big(f(x)\big)$ for $x > 3$, find $g'(x)$ is equal to $\ldots\ldots\ldots\ldots$

$\therefore g(x) = f\big(f(x)\big) = f(|x - 3|)$ for $x > 3$ $\Rightarrow g(x) = f(x - 3) = x - 3 - 3 = x - 6$ $\boxed{g'(x) = 1 \text{ Ans.}}$

(8) (a) $f(x) = \log_x 2$ $\quad \boxed{\text{Formula, } f(x) = \log_a x, f(x) \text{ to defined } f(x) = \begin{cases} x > 0, x \neq 0 \\ a > 0, a \neq 1 \end{cases}}$

function $f(x)$ to be defined $f(x) = \begin{cases} 2 > 0, & 2 \neq 0 \text{ it is true.} \\ x > 0, & x \neq 1 \end{cases} = x > 0, x \neq 1$ $\boxed{\text{Domain (D)} = (0, \infty) - \{1\} \quad \text{Ans.}}$

(b) $f(x) = \sqrt{\log\left(\dfrac{9x - 4x^2}{5}\right)}$ Ans: $-$ Domain (D) $= \varphi$ (c) $f(x) = \log|3 - x^2|$ Ans: $-$ Domain (D) $= R - \{\pm\sqrt{3}\}$.

(d) $f(x) = \sqrt{2x - 5} - \dfrac{1}{\log_3(2 + x)}$ Ans: $-$ Domain (D) $= \left[\dfrac{5}{2}, \infty\right[.$

(e) $f(x) = \sqrt{\sin^{-1} x} + \dfrac{1}{\log x}$, function $f(x)$ to be defined as $\quad f(x) = \begin{cases} \sin^{-1} x \geq 0 \\ -1 \leq x \leq 1 \end{cases}$ or $\begin{cases} \log x \neq 0 \\ x > 0 \end{cases} = \begin{cases} x \geq 0 \\ -1 \leq x \leq 1 \end{cases}$ or $\begin{cases} x \neq 1 \\ x > 0 \end{cases}$

Domain (D) $= (0, 1)$ Ans.

(f) $f(x) = \sqrt{2 + \log_2 x}$, function f(x) is defined as $f(x) = \begin{cases} 2 + \log_2 x \geq 0 \\ x > 0 \\ x \neq 0 \end{cases} = \begin{cases} \log_2 x \geq -2 = \log_2 2^{-2} \\ x > 0 \\ x \neq 0 \end{cases} = \begin{cases} x \geq \dfrac{1}{4} \\ x > 0 \\ x \neq 0 \end{cases}$

$$\text{Domain (D)} = \left[\frac{1}{4}, \infty\right[\quad \text{Ans.}$$

(g) $f(x) = \log_3\left\{\log_{\frac{1}{3}}(x^2 - 4x + 4)\right\}$ $\boxed{\text{Formula}, f(x) = \log_a x \text{ is defined as} \begin{cases} x > 0, & x \neq 0 \\ a > 0, & a \neq 1 \end{cases} \text{ and } \log_a x > \log_a y \text{ then,} \begin{cases} x > y, & a > 1 \\ x < y, & a < 1 \end{cases}}$

function $f(x) = \log_3\left\{\log_{\frac{1}{3}}(x^2 - 4x + 4)\right\}$ is defined as

$f(x) = \begin{cases} \log_{\frac{1}{3}}(x^2 - 4x + 4) > 0 = \log_{\frac{1}{3}} 1 \\ x^2 - 4x + 3 > 0 \end{cases}$ or $\begin{cases} x^2 - 4x + 4 > 1 \\ x^2 - 4x + 3 > 0 \text{ satisfy all value.} \end{cases}$ or $x^2 - 4x + 3 > 0$

or $(x - 1)(x - 3) > 0$ $\boxed{\therefore \text{ domain (D)} = (-\infty, 1) \cup (3, \infty) \text{ Ans.}}$

(h) $f(x) = \log_4 \log_3 \log_2(1 - 6x - 9x^2)$, Hence, f(x) to be defined

$f(x) = \begin{cases} \log_3 \log_2(1 - 6x - 9x^2) > 0 = \log_3 1 \\ \log_2(1 - 6x - 9x^2) > 0 = \log_2 1 \\ 1 - 6x - 9x^2 > 0 \end{cases} = \begin{cases} \log_2(1 - 6x - 9x^2) > 1 = \log_2 2 \\ 1 - 6x - 9x^2 > 1 \\ 1 - 6x - 9x^2 > 0 \end{cases} = \begin{cases} 1 - 6x - 9x^2 > 2 \\ x(3x + 2) < 0 \\ 9x^2 + 6x - 1 < 0 \end{cases}$

$$f(x) = \begin{cases} 9x^2 + 6x + 1 < 0 \\ x(3x + 2) < 0 \\ \dfrac{-1 - \sqrt{2}}{3} < x < \dfrac{-1 + \sqrt{2}}{3} \end{cases} \quad \boxed{\text{Domain (D)} = \left(\dfrac{-2}{3}, \dfrac{-1}{3}\right) \text{ Ans.}}$$

(i) $f(x) = \log\{(2\log x)^2 - 12\log x + 9\} = \log(2\log x - 3)^2 = 2\log(2\log x - 3)$

Function f(x) is defined as

$\begin{cases} 2\log x - 3 > 0 \\ 2\log x - 3 \neq 0 \\ x > 0 \text{ and } x \neq 0 \end{cases} = \begin{cases} \log x > \dfrac{3}{2} = \log_e \dfrac{3}{2} \\ \log x \neq \dfrac{3}{2} = \log_e \dfrac{3}{2} \\ x > 0 \text{ and } x \neq 0 \end{cases} = \begin{cases} x > e^{\frac{3}{2}} \\ x \neq e^{\frac{3}{2}} \\ x > 0 \text{ and } x \neq 0 \end{cases}$ $\boxed{\therefore \text{ Domain (D)} = \left(e^{\frac{3}{2}}, \infty\right) \text{ Ans.}}$

(j) $f(x) = \sqrt{\log_{0.2}(4x - 2x^2)}$ $\Rightarrow \log_{0.2}(4x - 2x^2) \geq 0$ and $4x - 2x^2 > 0$

$\Rightarrow 4x - 2x^2 \geq 1$ and $2x^2 - 4x < 0$ $\Rightarrow 2x^2 - 4x + 1 \leq 0$ and $2x(x - 2) < 0$

$\Rightarrow D_1 = \left(-\infty, \dfrac{1}{2}\right), D_2 = (-\infty, 0) \Rightarrow D = D_1 \cap D_2 = \left(-\infty, \dfrac{1}{2}\right) \cap (-\infty, 0)$ $\boxed{\therefore \text{ Domain (D)} = (-\infty, 0) \text{ Ans.}}$

(k) $f(x) = \sqrt{|x| - x}$ $\Rightarrow |x| - x \geq 0$ when, $x \geq 0$, $x - x > 0$, $0 > 0$ when, $x < 0$, $-x - x \geq 0$, $-2x \geq 0$, $2x \leq 0$, $x \leq 0$

$$\boxed{\therefore \text{ Domain (D)} =]-\infty, 0] \text{ Ans.}}$$

(l) $f(x) = \dfrac{2}{2x + 3|x|}$ $\Rightarrow 2x + 3|x| \neq 0$ when, $x > 0$, $2x + 3x \neq 0$ or $5x \neq 0$ or $x \neq 0$

when, $x < 0$, $2x - 3x \neq 0$, $x \neq 0$ $\boxed{\therefore \text{ Domain (D)} = R - \{0\} \text{ Ans.}}$

(9) (a) $f(x) = \cos^{-1}\left(\dfrac{2x}{1 + x^2}\right) + \sqrt{\sin(\cos x)}$ Ans:$-$ Domain (D) $= [-1, 1]$

(b) $f(x) = \cos^{-1}\left[\log_3\left(\dfrac{1}{3}x^3\right)\right]$ $\Rightarrow \begin{cases} -\dfrac{\pi}{2} \leq \log_3\left(\dfrac{1}{3}x^3\right) \leq \dfrac{\pi}{2} \\ \dfrac{1}{3}x^3 > 0 \end{cases} = \begin{cases} \dfrac{1}{3} \leq \dfrac{1}{3}x^3 \leq 3 \\ x^3 > 0 \end{cases} = \begin{cases} 1 \leq x^3 \leq 9 \\ x > 0 \end{cases} = \begin{cases} 1 \leq x \leq 9^{\frac{1}{3}} \\ x > 0 \end{cases}$

$$\boxed{\therefore \text{ Domain (D)} = \left[1, 3^{\frac{2}{3}}\right] \text{ Ans.}}$$

(c) $f(x) = \dfrac{\sqrt{3 - x^2}}{\cos^{-1}(1 + x)}$ function f(x) to be defined as $f(x) = \begin{cases} 3 - x^2 \geq 0 \\ -1 \leq 1 + x \leq 1 \\ \cos^{-1}(1 + x) \neq 0 = \cos^{-1} 0 \end{cases}$

$$\therefore \text{Domain (D)} = \left[-\sqrt{3}, 0\right] - \{-1\} \quad \text{Ans.}$$

(d) $f(x) = \log(\sin x)$,   Ans:− Domain (D) $= n\pi - \dfrac{\pi}{2} < x < n\pi + \dfrac{\pi}{2}$

(e) $f(x) = \log(\cos x)^2$,   Ans:− Domain (D) = R   (f) $f(x) = \sin^{-1}\left(\dfrac{3 - |x|}{2}\right) + \log(2 - x)$,   Ans:− Domain (D) $= [-5, -1] \cup [1, 2)$.

(g) $f(x) = \sqrt{4 + 2^x + 2^{2-x}} + \sqrt{\cos^{-1}(x + 2)}$,   Ans:− Domain (D) $= \left[\dfrac{\pi}{2} - 2, -1\right]$.

(10) (a) $y = \dfrac{\sqrt{x - 3}}{x^2 - 1}$, $x \geq 3$ and $x \neq \pm 1$,   Ans:− Domain (D) $= [3, \infty)$

(b) $y = \dfrac{\cos^{-1}(3 - x)}{\sqrt{x^2 - 1}}$,   Ans:−[2,4].   (c) $y = \left\{\dfrac{3 - x^2}{x + 2}\right\}^{\frac{1}{2}}$

hence y to be defined as $y = \begin{cases} \dfrac{3 - x^2}{x + 2} \geq 0 \\ x + 2 \neq 0 \end{cases} = \begin{cases} 3 - x^2 \geq 0 \\ x + 2 > 0 \\ x + 2 \neq 0 \end{cases} = \begin{cases} x \leq \pm\sqrt{3} \\ x > -2 \\ x \neq -2 \end{cases}$ $\boxed{\therefore \text{Domain (D)} = \left[-\sqrt{3}, \sqrt{3}\right] \text{ Ans.}}$

(d) $y = \dfrac{\sqrt{5 + 3x - x^2}}{\sqrt{\sin x - \dfrac{1}{2}}}$ hence y to be defined $y = \begin{cases} 5 + 3x - x^2 \geq 0 \\ \sin x - \dfrac{1}{2} > 0 \end{cases} = \begin{cases} x^2 - 3x - 5 \leq 0 \\ \sin x > \dfrac{1}{2} \end{cases}$

or $y = \begin{cases} x \in \left[\dfrac{3 - \sqrt{29}}{2}, \dfrac{3 + \sqrt{29}}{2}\right] \\ x \in \left[\dfrac{\pi}{4}, \dfrac{3\pi}{4}\right] \end{cases} = \begin{cases} \dfrac{3 - \sqrt{29}}{2} \leq x \leq \dfrac{3 + \sqrt{29}}{2} \\ \dfrac{\pi}{4} \leq x \leq \dfrac{3\pi}{4} \end{cases}$ $\boxed{\therefore \text{Domain (D)} = \left[\dfrac{\pi}{4}, \dfrac{3\pi}{4}\right] \text{ Ans.}}$

(e) $y = \log_4(x^2 + 1) + \sqrt{64x - x^7}$, hence y to be defined as

$y = \begin{cases} x^2 + 1 > 0 \\ 64x - x^7 \geq 0 \end{cases} = \begin{cases} x^2 > -1, \text{ it is satisfying all real value.} \\ x(64 - x^6) \geq 0 \end{cases} = \begin{cases} x \geq 0 \\ 64 - x^6 \geq 0 \end{cases} = \begin{cases} x \geq 0 \\ x \leq 2 \end{cases}$

$$\therefore \text{Domain (D)} = [0, 2] \quad \text{Ans.}$$

(f) $y = \sqrt{\dfrac{2 - |x|}{3 - |x|}}$,   hence y to be defined $y = \begin{cases} 2 - |x| \geq 0 \\ 3 - |x| > 0 \end{cases} = \begin{cases} -|x| \geq -2 \\ -|x| > -3 \end{cases} = \begin{cases} |x| \leq 2 \\ |x| < 3 \end{cases}$

$$\therefore \text{Domain (D)} = (-\infty, 2] \quad \text{Ans.}$$

(g) $f(x) = \sin^{-1}\sqrt{1 - x^2} + \cos^{-1}\sqrt{1 + x^2} = \dfrac{\pi}{2}$, $f(x) = \dfrac{\pi}{2}$ $\boxed{\text{Domain (D)} = R \text{ Ans.}}$

(11) (a) $y = \dfrac{2x}{1 - x^2}$,   hence y to be defined $\Rightarrow 1 - x^2 \neq 0$ or $x \neq \pm 1$ $\boxed{\therefore \text{Domain (D)} = R - \{-1, 1\} \text{ Ans.}}$

Now, $y = \dfrac{2x}{1 - x^2}$ or $y - yx^2 = 2x$ or $yx^2 + 2x - y = 0$ $\therefore x = \dfrac{-1 \pm \sqrt{1 + y^2}}{y}$

here x to be defined as $\begin{cases} 1 + y^2 \geq 0 \\ y \neq 0 \end{cases} = \begin{cases} y^2 \geq -1 \text{ satisfy all real value of y.} \\ y \neq 0 \end{cases}$ $\boxed{\therefore \text{Range} = R - \{0\} \text{ Ans.}}$

(b) Same as above question,   Ans:− Domain (D) $= R - \left\{\dfrac{\pi}{4}\right\}$,   Range $= \left[-\dfrac{\pi}{4}, \dfrac{\pi}{4}\right]$

(c) $y = \dfrac{2}{3 + x^2}$,   Ans:− Domain(D) = R,   Range $= \left(-\infty, \dfrac{2}{3}\right] - \{0\}$. see question no. −(10)(a).

(d) $y = \dfrac{x+2}{x^2-x-6} = f(x)$ here function f(x) to be defined $\Rightarrow x^2 - x - 6 \neq 0$ $\therefore (x+2)(x-) \neq 0$

$$\therefore x \neq 3, -2 \quad \boxed{\therefore \text{Domain (D)} = R - \{-2,3\} \text{ Ans.}}$$

$$\Rightarrow y = \frac{x+2}{x^2-x-6} = \frac{x+2}{(x+2)(x-3)} = \frac{1}{(x-3)} \Rightarrow xy - 3y = 1 \quad \therefore x = \frac{1+3y}{y} = f(y),$$

here function f(y) to be defined $\therefore y \neq 0$ $\boxed{\therefore \text{Range} = R - \{0\} \text{ Ans.}}$

(e) $y = 2 - \dfrac{3}{x^2-3x+4}$, Ans: − Range $= \left[\dfrac{2}{7}, \infty\right) - \{0\}$ (f) $y = \log\sqrt{x^2+4x+10}$, Ans: − Range $= \left[\dfrac{1}{2}\log_e 6, \infty\right)$ or $\left[\log_e \sqrt{6}, \infty\right)$.

(g) $y = 3\cos x + 2\sin\left(x + \dfrac{\pi}{6}\right) + 5$, Ans: − $-\sqrt{19} + 5, \sqrt{19} + 5$ or $5 - \sqrt{19}, 5 + \sqrt{19}$

(h) $y = \cos^{-1}\left[\dfrac{3}{2} + x^2\right]$ where [.] denotes the greatest integer function. $\boxed{\text{Range} = \left\{0, \dfrac{\pi}{2}\right\} \text{ Ans.}}$

(12) (a) Ans: − Domain (D) $= R - \{3 + 2\sqrt{3}, 3 - 2\sqrt{3}\}$, Range $= \left(-\infty, -\dfrac{1}{4}\right] \cup \left[\dfrac{1}{12}, \infty\right)$.

(b) Ans: − Domain (D) $= \left(\dfrac{2\pi}{3}, \infty\right)$, Range $= \left(-\infty, -\dfrac{2}{\sqrt{3}}\right] \cup \left[\dfrac{2}{\sqrt{3}}, \infty\right)$.

(c) Ans: − Domain (D) $= (-1,1)$,   (d) Ans: − Domain (D) $= [-1,0) \cup (0,1]$ or $[-1,1] - \{0\}$,

(e) Ans: − Domain (D) $= \{1,2,3\}$ consists of integer , Range $= \{1,2,3\}$.

(13) (a) Ans: − $1 + \sqrt{1 + \log_3 x}$ (b) Ans: − $\dfrac{x + \sqrt{x^2+8}}{2}$

(14) (a) Ans: − Yes, $f(0) = \dfrac{0+1}{0+1} = 1$, $g(0) = 1$ $\therefore f(0) = g(0)$

(b) Ans: − Yes, $f(0) = 1$ and $g(0) = \sec^2 0 - \tan^2 0 = 1 - 0 = 1$ $\therefore f(0) = g(0)$

(c) Ans: − No, $f(0) = \log 0$, f(x) is defined for all $x \neq 0$ and g(x) is defined only for $x > 0$.

(d) Ans: − No, f(x) is defined only for $x > 0$ *but not for* $x < -1$ *and* g(x) is defined for $x(x+1) > 0$ *i.e* $x > 0$ *and* $x < -1$.

(e) No, f(x) is defined for all x and g(x) is defined only for $x \geq 0$.

(f) No, f(x) is defined only for $x > 0$ *but not for* $x < -2$ *and* g(x) is defined for $\dfrac{x}{x+2} > 0$ *and* $x + 2 \neq 0$ *i.e* $x > 0, x \neq -2$

(15) (a) $f(x) = \log\left(x - \sqrt{1+x^2}\right)$, $f(-x) = \log\left(-x - \sqrt{1+x^2}\right) = \log\left\{\dfrac{(-x - \sqrt{1+x^2})(-x + \sqrt{1+x^2})}{(-x + \sqrt{1+x^2})}\right\}$

$f(-x) = \log\left\{\dfrac{x^2 - 1 + x^2}{-(x - \sqrt{1+x^2})}\right\} = \log\left(\dfrac{1}{x - \sqrt{1+x^2}}\right) = \log\left(x - \sqrt{1+x^2}\right)^{-1} = -\log\left(x - \sqrt{1+x^2}\right) = -f(x)$

$$\boxed{\therefore f(x) = -f(-x) \quad \therefore f(x) \text{ is odd.}}$$

(b) $f(x) = \cos^3 x + 2\tan^2 x$, Ans: − $f(x) = f(-x)$, function f(x) is even.   (c) Ans: −Even   (d) Ans: −Even   (e) Ans: −Even   (f) Ans: −Even

(16) Ans: − (a) $1 - \sqrt{3}, 2 - \sqrt{2}, 1 + \sqrt{3}$ (b) f(x) is even function.

(17) Ans: − (a) Period of $y = 3\pi$ (b) Period of $y = 2\pi$ (c) Period of $y = \dfrac{\pi}{2}$ (d) Period of $y = 4\pi$ (e) Period of $y = \pi$

(f) Period of $y = \dfrac{2\pi}{3}$ (g) Period of $y = 3\pi$

$$\boxed{\text{Formula, } y = \sin(ax+b) \text{ , Period of } y = \frac{\text{period of sinx}}{|a|}}$$

Limit

(A)Limits: — Consider the function $y = \dfrac{x^2 - 4}{x - 2}$. the value of the function at $x = 2$ is of the form $\dfrac{0}{0}$ which is meaningless. In this case

we cannot divide the numerator by denominator, since $x - 2$ is zero. Now suppose x is not actually equal to 2 but very nearly

equal to 2 then $x - 2$ is not equal to zero. Hence in this case we can divide the numerator by denominator.

$$\therefore \frac{x^2 - 4}{x - 2} = \frac{(x - 2)(x + 2)}{(x - 2)} = x + 2$$

Thus we see that when x has fixed value 2, the value of y is meaningless but when x tends to 2 (i. e $x \to 2$),

y tends to 4 and we say that the limit of y is 4 when x tends to 2. This we writes as

$$\lim_{x \to 2} \frac{x^2 - 4}{x - 2} = \lim_{x \to 2} \frac{(x - 2)(x + 2)}{(x - 2)} = \lim_{x \to 2}(x + 2) = 2 + 2 = 4 \quad \text{Ans.}$$

Definition of limit: —The number A is said to be the limit of $f(x)$ at $x = a$ if for any arbitrarily chosen positive number ϵ however

small but not zero. there exists a corresponding number, δ greater then zero such that

$$|f(x) - A| < \epsilon \text{ , for all values of } x \text{ for which } 0 < |x - a| < \delta$$

where $|x - a|$ means the absolute value of $x - a$ without any regard to sign.

Right hand limits: —If x approaches a from the right, that is from larger value of x then a, the limit of f is called right hand limit of $f(x)$

and written as $\lim_{x \to a+0} f(x)$ or $\lim_{x \to a+0} f(a + 0)$

put, $x \to a + h$ in $f(x)$, where $h \to 0$ we have $\boxed{f(a + 0) = \lim_{h \to 0} f(a + h)}$ Right hand limit

Left hand limit: —If x approaches from the left, that is from smaller value of x than a , the limit of f is called left hand limit

and is written as $\lim_{x \to a-0} f(x)$ or $\lim_{x \to a-0} f(a - 0)$

put, $x \to a - h$, where $h \to 0$ we have $\boxed{f(a - 0) = \lim_{h \to 0} f(a - h)}$ left hand limit

Algebra of limits: — # $\lim_{x \to a}[f(x) \pm g(x)] = \lim_{x \to a} f(x) \pm \lim_{x \to a} g(x)$ # $\lim_{x \to a}[f(x) . g(x)] = \lim_{x \to a} f(x) . \lim_{x \to a} g(x)$

$\lim_{x \to a} \dfrac{f(x)}{g(x)} = \dfrac{\lim_{x \to a} f(x)}{\lim_{x \to a} g(x)} = \lim_{x \to a} f(x) \div \lim_{x \to a} g(x)$ but $\lim_{x \to a} g(x) \neq 0$

Indeterminate forms and L'Hospital rule: — If a function $f(x)$ takes the form $\dfrac{0}{0}$ or $\dfrac{\infty}{\infty}$ at $x = a_i$ then we say that $f(x)$ is

indeterminate at $x = a$ other indeterminate forms are $\infty - \infty, 0 \times \infty, 1^\infty, 0^0, \infty^0$

L'Hospital's Rule: —If $\phi(x)$, and $\psi(x)$ are functions of x such that $\phi(a) = 0, \psi(a) = 0$, then $\lim_{x \to a} \dfrac{\phi(x)}{\psi(x)} = \lim_{x \to a} \dfrac{\phi'(x)}{\psi'(x)}$.

Note: —Applying L'Hospital rule differentiate $\dfrac{\phi(x)}{\psi(x)}$, N^r and D^r separately.

Some important expansions: —

(i) $\sin x = x - \dfrac{x^3}{3!} + \dfrac{x^5}{5!} - \cdots \ldots \ldots \ldots \ldots \ldots \ldots \ldots \ldots \ldots \ldots (\text{odd} + -)$

$\sin hx = x + \dfrac{x^3}{3!} + \dfrac{x^5}{5!} + \cdots \ldots \ldots \ldots \ldots \ldots \ldots \ldots \ldots \ldots (\text{odd} + +)$

(ii) $\cos x = 1 - \dfrac{x^2}{2!} + \dfrac{x^4}{4!} - \cdots$ ………………………………………… (even + −)

$\cos hx = 1 + \dfrac{x^2}{2!} + \dfrac{x^4}{4!} + \cdots$ ……………………………………. (even + +)

(iii) $\tan x = x + \dfrac{1}{3}x^3 + \dfrac{2}{15}x^5 + \cdots$ …………………………………………

(iv) $\tan^{-1} x = x - \dfrac{1}{3}x^3 + \dfrac{1}{5}x^5 - \cdots$ ……………………………………

(v) $e^x = 1 + x + \dfrac{x^2}{2!} + \dfrac{x^3}{3!} + \cdots$ ………………………………… (All + +)

$e^{-x} = 1 - x + \dfrac{x^2}{2!} - \dfrac{x^3}{3!} + \cdots$ …………………………………. (Alternate + −)

(vi) If $|x| < 1$, then $\log_e(1 + x) = x - \dfrac{1}{2}x^2 + \dfrac{1}{3}x^3 - \dfrac{1}{4}x^4 + \cdots$ ………………………………………………

(vii) $\log_e(1 - x) = -x - \dfrac{1}{2}x^2 - \dfrac{1}{3}x^3 - \dfrac{1}{4}x^4 - \cdots$ …………………………… (All−)

Some important limits should be remembered: −

(1)(a) $\lim\limits_{x \to 0} \dfrac{\sin x}{x} = 1$ (b) $\lim\limits_{x \to 0} \cos x = 1$ (c) $\lim\limits_{x \to 0} \dfrac{\tan x}{x} = 1$ (d) $\lim\limits_{x \to 0} \dfrac{\tan^{-1} x}{x} = 1$ (e) $\lim\limits_{x \to 0} \dfrac{\sin^{-1} x}{x} = 1$

(2) $\lim\limits_{x \to 0}(1 + x)^{\frac{1}{x}} = e$ put $x = \dfrac{1}{x}, x \to \infty$ or $\lim\limits_{x \to \infty}\left(1 + \dfrac{1}{x}\right)^x = e$ or $\lim\limits_{x \to 0}(1 + mx)^{\frac{1}{x}} = e^m$

or $\lim\limits_{x \to \infty}\left(1 + \dfrac{m}{x}\right)^x = e^m$ or $\lim\limits_{x \to 0}\left(1 + \dfrac{x}{m}\right)^{\frac{1}{x}} = e^{\frac{1}{m}}$ or $\lim\limits_{x \to \infty}\left(1 + \dfrac{1}{mx}\right)^x = e^{\frac{1}{m}}$

(3) $\lim\limits_{x \to 0} \dfrac{a^x - 1}{x} = \log a$ form $\left(\dfrac{0}{0}\right)$

Proof: − $\lim\limits_{x \to 0} \dfrac{a^x - 1}{x}$ using L'Hospital rule.

Differentiate with respect to x, $\dfrac{\lim\limits_{x \to 0} a^x - 1}{\lim\limits_{x \to 0} x} = \dfrac{\lim\limits_{x \to 0} a^x \log a}{\lim\limits_{x \to 0} 1} = \dfrac{\log a}{1} = \log a$ Proved. $\lim\limits_{x \to 0} \dfrac{e^x - 1}{x} = \log e = 1$ (above formula)

Example: − (a) $\lim\limits_{x \to 0} \dfrac{5^x - 1}{x} = \log 5$ Ans.

(b) $\lim\limits_{x \to 0} \dfrac{e^{3x} - 1}{3x} = \log 2$ Ans. Hint − $x \to 0$, $3x \to 0$ limit does not change. $\lim\limits_{3x \to 0} \dfrac{e^{3x} - 1}{3x} = \log 2$ Ans.

(4) $\lim\limits_{x \to 0} \dfrac{(1 + x)^n - 1}{x} = n$, Hint: − $\lim\limits_{x \to 0} \dfrac{1 + nx + \dfrac{n(n-1)}{2}x^2 + \cdots \ldots \ldots - 1}{x} = n$ proved. (using L'Hospital rule)

(5) $\lim\limits_{x \to a} \dfrac{x^n - a^n}{x - a} = na^{n-1}$, Hint: − $\lim\limits_{x \to a} \dfrac{nx^{n-1} - na^{n-1}.0}{1 - 0} = na^{n-1}$ proved. (using L'Hospital rule)

(6) $\lim\limits_{x \to 0} \dfrac{\log_a(1 + x)}{x} = \log_a e$, $(a > 0, a \neq 0)$. (7) $\lim\limits_{x \to \infty} \dfrac{\log x}{x^m} = 0$, $(m > 0)$.

(8) (a) $\lim\limits_{x \to \infty} \dfrac{1 + x}{x} = 1$ (b) $\lim\limits_{x \to 0} \dfrac{x(x - 1)}{(x + 1)^n - 1} = -\dfrac{1}{n}$, $n \neq 0$ (c) $\lim\limits_{x \to 0} \dfrac{(1 + \sin x)^n - 1}{x} = n, n \neq 0$

Evaluation of limits: − (1) Factorisation or substitution: − $\lim\limits_{x \to a} \dfrac{x^2 - a^2}{x - a} = \lim\limits_{x \to a} \dfrac{(x - a)(x + a)}{(x - a)} = \lim\limits_{x \to a}(x + a) = a + a = 2a$ Ans.

(2) L'Hospital rule: $-\lim\limits_{x\to 0} x\log x$ form $(0\times\infty) = \lim\limits_{x\to 0}\dfrac{\log x}{\dfrac{1}{x}}$ form $\left(\dfrac{\infty}{\infty}\right)$, using L'Hospital rule.

Differentiate N^r and D^r separately, $\lim\limits_{x\to 0}\dfrac{\dfrac{1}{x}}{-\dfrac{1}{x^2}} = \lim\limits_{x\to 0}-\dfrac{1}{x}\times\dfrac{x^2}{1} = -\lim\limits_{x\to 0} x = 0$ Ans.

(3) Expansion rule: $-\lim\limits_{x\to 0}\dfrac{\log(1+x)}{x}$, form $\left(\dfrac{0}{0}\right)$

$$\lim_{x\to 0}\dfrac{-\left\{x+\dfrac{x^2}{2}+\dfrac{x^3}{3}+\cdots\ldots\ldots\right\}}{x} = \lim_{x\to 0}\dfrac{-x\left\{1+\dfrac{x}{2}+\dfrac{x^2}{3}+\cdots\ldots\ldots\right\}}{x} = \lim_{x\to 0}-\left(1+\dfrac{x}{2}+\dfrac{x^2}{3}+\cdots\ldots\ldots\right) = -1 \quad \text{Ans.}$$

(4) Rationalisation: $-\lim\limits_{x\to 2}\dfrac{x-2}{\sqrt{x-1}-\sqrt{3-x}}$

or $\lim\limits_{x\to 2}\dfrac{x-2}{\sqrt{x-1}-\sqrt{3-x}}\times\dfrac{\sqrt{x-1}+\sqrt{3-x}}{\sqrt{x-1}+\sqrt{3-x}} = \lim\limits_{x\to 2}\dfrac{(x-2)(\sqrt{x-1}+\sqrt{3-x})}{x-1-3+x} = \lim\limits_{x\to 2}\dfrac{(x-2)(\sqrt{x-1}+\sqrt{3-x})}{2x-4}$

$= \lim\limits_{x\to 2}\dfrac{(x-2)(\sqrt{x-1}+\sqrt{3-x})}{2(x-2)} = \lim\limits_{x\to 2}\dfrac{(\sqrt{x-1}+\sqrt{3-x})}{2} = \dfrac{2}{2} = 1$ Ans.

Continuity

Continuity: $- f(x)$ is said to be continuous at $x = a$ if $\lim\limits_{x\to a_+} f(x) = \lim\limits_{x\to a_-} f(x) = f(x)$ or $\lim\limits_{h\to a} f(a+h) = \lim\limits_{h\to 0} f(a-h) = f(a)$

The function is not defined at $x = a$, $f(a)$ does not exist or $\lim\limits_{x\to a_+} f(x) \neq \lim\limits_{x\to a_-} f(x)$ then we say that function is discontinuous at $x = a$.

Functions continuous on the open interval: $-$ (a, b) or $]a, b[$ or $\{a < x < b\}$, $f(x)$ will be continuous

if it is continuous at every point of the interval.

Function continuous on the closed interval: $-$ $[a, b]$ or $a \leq x \leq b$,

if will satisfy the following three condition: $-$ (i) $f(x)$ is continuous at each point of (a, b)

(ii) $\lim\limits_{x\to a_+} f(x) = f(a)$ (iii) $\lim\limits_{x\to b_-} f(x) = f(b)$

A continuous fnction in the closed interval $a \leq x \leq b$ has the following properties: $-$

(a) If $f(x)$ and $f(b)$ are of opposite signs, then there exists at least one and in general odd solutions of the equation $f(x) = 0$ for any x

in the open interval (a, b).

(b) If λ is any real number between $f(a)$ and $f(b)$, then there exists at least one solution of the equation $f(x) = \lambda$ in the open interval (a, b).

Type of discontinuities: $-$ (i) Discontinuity of first kind: $-$ The point of $x = a$ will be a point of discontinuity.

If both right hand and left hand limit at $x = a$ exist but not equal. $\lim\limits_{x\to a_+} f(x) \neq \lim\limits_{x\to a_-} f(x)$ or $\lim\limits_{h\to 0} f(a-h) \neq \lim\limits_{h\to 0} f(a+h)$

(ii) Discontinuity of second kind: $-$ The point at $x = a$ will be a point of discontinuity.

If either or both the RHL and LHL do not exist or if either or both the limits $\lim\limits_{x\to a_+} f(x) = \lim\limits_{x\to a_-} f(x)$ are infinite.

(iii) Removable discontinuity: $-$ If $\lim\limits_{x\to a_+} f(x) = \lim\limits_{x\to a_-} f(x)$ limit exists but it is not equal to $f(a)$.

Differentiability

f(x) is said to be differentiable at x = a if $\lim\limits_{h\to 0_+}\dfrac{f(a+h)-f(a)}{h}=\lim\limits_{h\to 0_-}\dfrac{f(a-h)-f(a)}{-h}=f(a)$

An important point: —A function which is differentiable at a point x = a must also be continuous at that point. If a function is continuous at a point x = a , it is not necessarily differentiable at x = a.

Differentiability implies continuity always: — Since the function is differentiable

$$\lim\limits_{h\to 0_+}\dfrac{f(a+h)-f(a)}{h}=\lim\limits_{h\to 0_-}\dfrac{f(a-h)-f(a)}{-h}=f(a)$$

$\lim\limits_{h\to 0_+}\dfrac{f(a+h)-f(a)}{h}$ exists and is finite $-------$ (A)

or $\lim\limits_{h\to 0}f(a+h)=\lim\limits_{h\to 0}[f(a+h)-f(a)]+f(a)=\lim\limits_{h\to 0}h\left[\dfrac{f(a+h)-f(a)}{h}\right]+f(a)$

$\lim\limits_{h\to 0}h\ .\lim\limits_{h\to 0}\dfrac{f(a+h)-f(a)}{h}+f(a)=0+f(a)=f(a)$

similarly, $\lim\limits_{h\to 0}f(a-h)=f(a)$ since, $\lim\limits_{h\to 0}f(a+h)=\lim\limits_{h\to 0}f(a-h)=f(a),$ the function is continuous.

Continuity does not necessarily imply differentiability: —

Example: — consider $f(x)=\begin{cases} x\cos x, & x\neq 0\\ 0, & x=0\end{cases}$ differentiability at x = 0.

R. H. L $[R'(0)]=\lim\limits_{h\to 0_+}\dfrac{f(0+h)-f(0)}{h}=\lim\limits_{h\to 0_+}\dfrac{f(h)-f(0)}{h}=\lim\limits_{h\to 0}\dfrac{h\cos h}{h}=\lim\limits_{h\to 0}\cos h=1$

L. H. L $=\lim\limits_{h\to 0_-}\dfrac{f(0-h)-f(0)}{-h}=\lim\limits_{h\to 0_-}\dfrac{f(-h)-f(0)}{-h}=\lim\limits_{h\to 0}\dfrac{-h\cos h}{-h}=\lim\limits_{h\to 0}\cos h=1$

$f(x)=f(0)=0$

R. H. L = L. H. L ≠ f(0) the function is not differentiable at x = 0.

Continuity: — $\lim\limits_{x\to a_+}f(x)=\lim\limits_{x\to a_-}f(x)=f(x)$

R. H. L $=\lim\limits_{x\to 0_+}f(x)=\lim\limits_{x\to 0_+}x\cos x=0$, L. H. L $=\lim\limits_{x\to 0_-}f(x)=\lim\limits_{x\to 0_-}x\cos x=0$, v $=f(x)=f(0)=0$

R. H. L = L. H. L = vf(x), the function is continuous at x = 0.

Differentiability of f(x) on closed interval [a, b] : —

(i) It should be differentiable at every point on the open interval (a, b) or]a, b[.

(ii) Right hand derivative R. H. L (R') at a and left hand derivative L. H. L (L')at b exist finitely.

If both the above are satisfied the we say that f(x)is differentiable in the closed interval [a, b].

Solved example

(1) Evaluate: — (a) $\lim\limits_{h\to 0}\dfrac{\sqrt{x+2h}-\sqrt{x}}{2h}$ (b) $\lim\limits_{x\to 2}\dfrac{x-2}{x^3-8}$ (c) $\lim\limits_{x\to \pi}\dfrac{\sin x-\cos x}{\cos x}$ (d) $\lim\limits_{x\to 1}\dfrac{x^2-3x+2}{2x^2-7x+5}$

Solution: — (a) $\lim\limits_{h\to 0}\dfrac{\sqrt{x+2h}-\sqrt{x}}{2h}$, (Rationalize the N^r)

$$\lim_{h \to 0} \frac{\sqrt{x+2h} - \sqrt{x}}{2h} \times \frac{\sqrt{x+2h} + \sqrt{x}}{\sqrt{x+2h} + \sqrt{x}} = \lim_{h \to 0} \frac{x+2h-x}{2h(\sqrt{x+2h} + \sqrt{x})} = \lim_{h \to 0} \frac{1}{\sqrt{x+2h} + \sqrt{x}} = \frac{1}{2\sqrt{x}} \quad \text{Ans.}$$

(b) $\displaystyle \lim_{x \to 2} \frac{x-2}{x^3 - 8} = \lim_{x \to 2} \frac{x-2}{x^3 - 2^3} = \lim_{x \to 2} \frac{(x-2)}{(x^2 + 2x + 4)(x-2)} = \lim_{x \to 2} \frac{1}{(x^2 + 2x + 4)} = \frac{1}{4+4+4} = \frac{1}{12}$ Ans.

IInd method: $-\displaystyle \lim_{x \to 2} \frac{x-2}{x^3 - 8}$, form $\left(\frac{0}{0}\right)$ using L'Hospital rule. $\displaystyle \lim_{x \to 2} \frac{x-2}{x^3 - 8} = \lim_{x \to 2} \frac{\frac{d}{dx}(x-2)}{\frac{d}{dx}(x^3 - 8)} = \lim_{x \to 2} \frac{1}{3x^2} = \frac{1}{3 \times 4} = \frac{1}{12}$ Ans.

(c) $\displaystyle \lim_{x \to \pi} \frac{\sin x - \cos x}{\cos x}$ put value $x = \pi$ then $\displaystyle \lim_{x \to \pi} \frac{\sin x - \cos x}{\cos x} = \lim_{x \to \pi} \frac{\sin \pi - \cos \pi}{\cos \pi} = \frac{0 - (-1)}{-1} = -1$ Ans.

(d) $\displaystyle \lim_{x \to 1} \frac{x^2 - 3x + 2}{2x^2 - 7x + 5}$, form $\left(\frac{0}{0}\right)$ using L'Hospital rule, $\displaystyle \lim_{x \to 1} \frac{x^2 - 3x + 2}{2x^2 - 7x + 5} = \lim_{x \to 1} \frac{2x-3}{4x-7} = \frac{2-3}{4-7} = -\frac{1}{3}$ Ans.

(2) use the formula $\displaystyle \lim_{x \to 0} \frac{a^x - 1}{x} = \log a$ to find (a) $\displaystyle \lim_{x \to 0} \frac{(1+x)^{\frac{1}{2}} - 1}{2^x - 1}$ (b) $\displaystyle \lim_{x \to 0} \frac{7^x - 5^x}{5^x - 3^x}$ (c) $\displaystyle \lim_{x \to 1} \frac{\frac{3^x}{3} + \frac{5^x}{5} - 2}{(x-1)}$ (d) $\displaystyle \lim_{x \to 0} \frac{81^x - 27^x - 3^x + 1^x}{x^2}$

(e) $\displaystyle \lim_{x \to 0} \frac{(3^x - 1)(5^x - 1)}{\sqrt{2} - \sqrt{1 + \cos x}}$ (f) $\displaystyle \lim_{x \to 0} \frac{5^x - 1}{(1-x)^{\frac{1}{2}} - 1}$ (g) $\displaystyle \lim_{x \to 0} \frac{7^x + 3^x - 2}{\sin x}$ (h) $\displaystyle \lim_{x \to 0} \frac{x^2 3^x - x^2}{(1 - \cos x)x}$

Solution: $-$(a) $\displaystyle \lim_{x \to 0} \frac{(1+x)^{\frac{1}{2}} - 1}{2^x - 1} = \lim_{x \to 0} \frac{\sqrt{1+x} - 1}{2^x - 1} \times \frac{\sqrt{1+x} + 1}{\sqrt{1+x} + 1} = \lim_{x \to 0} \frac{1+x-1}{(2^x - 1)(\sqrt{1+x} + 1)} = \lim_{x \to 0} \frac{x}{(2^x - 1)(\sqrt{1+x} + 1)}$

$\displaystyle = \lim_{x \to 0} \frac{1}{\frac{(2^x - 1)}{x} \cdot (\sqrt{1+x} + 1)} = \frac{1}{2 \log 2} = \frac{1}{\log 4}$ Ans.

(b) $\displaystyle \lim_{x \to 0} \frac{7^x - 5^x}{5^x - 3^x} = \lim_{x \to 0} \frac{(7^x - 1) - (5^x - 1)}{(5^x - 1) - (3^x - 1)} = \lim_{x \to 0} \frac{\frac{(7^x - 1)}{x} - \frac{(5^x - 1)}{x}}{\frac{(5^x - 1)}{x} - \frac{(3^x - 1)}{x}} = \frac{\log 7 - \log 5}{\log 5 - \log 3} = \frac{\log\left(\frac{7}{5}\right)}{\log\left(\frac{5}{3}\right)}$ Ans.

(c) $\displaystyle \lim_{x \to 1} \frac{\frac{3^x}{3} + \frac{5^x}{5} - 2}{(x-1)} = \lim_{x \to 1} \frac{3^{x-1} + 5^{x-1} - 2}{(x-1)}$, $x \to 1$, $x - 1 \to 0$

$\displaystyle \lim_{(x-1) \to 0} \frac{\left(3^{(x-1)} - 1\right) + \left(5^{(x-1)} - 1\right)}{(x-1)} = \lim_{(x-1) \to 0} \left\{ \frac{\left(3^{(x-1)} - 1\right)}{(x-1)} + \frac{\left(5^{(x-1)} - 1\right)}{(x-1)} \right\} = \log 3 + \log 5 = \log 3.5 = \log 15$ Ans.

(d) $\displaystyle \lim_{x \to 0} \frac{81^x - 27^x - 3^x + 1^x}{x^2} = \lim_{x \to 0} \frac{(27^x - 1)(3^x - 1)}{x^2} = \lim_{x \to 0} \frac{(27^x - 1)}{x} \cdot \lim_{x \to 0} \frac{(3^x - 1)}{x} = \log 27 . \log 3 = \log 3^3 . \log 3 = 3 \log 3 . \log 3$

$= 3(\log 3)^2$ Ans.

(e) $\displaystyle \lim_{x \to 0} \frac{(3^x - 1)(5^x - 1)}{\sqrt{2} - \sqrt{1 + \cos x}} = \lim_{x \to 0} \frac{(3^x - 1)(5^x - 1)}{\sqrt{2} - \sqrt{2 \cos^2 \frac{x}{2}}}$ use formula, $1 - \cos x = 2 \sin^2 \frac{x}{2}$, $1 + \cos x = 2 \cos^2 \frac{x}{2}$

$\displaystyle \lim_{x \to 0} \frac{(3^x - 1)(5^x - 1)}{\sqrt{2}\left[1 - \cos \frac{x}{2}\right]} = \lim_{x \to 0} \frac{(3^x - 1)(5^x - 1)}{\sqrt{2} . 2 \sin^2 \frac{x}{4}} = \lim_{x \to 0} \frac{\frac{(3^x - 1)}{x} \frac{(5^x - 1)}{x}}{2\sqrt{2}\left(\frac{\sin \frac{x}{4}}{\frac{x}{4}}\right)^2 \times \frac{1}{16}}$ formula, $1 - \cos \frac{x}{2} = 2 \sin^2 \frac{x}{4}$

or $\displaystyle \frac{\log 3 . \log 5}{\frac{\sqrt{2}}{8}} = 8 \frac{\log 3 . \log 5}{\sqrt{2}} = 4 . \sqrt{2} . \sqrt{2} \frac{\log 3 . \log 5}{\sqrt{2}} = 4\sqrt{2} . \log 3 . \log 5$ Ans.

(f) $\displaystyle \lim_{x \to 0} \frac{5^x - 1}{(1-x)^{\frac{1}{2}} - 1} = \lim_{x \to 0} \frac{5^x - 1}{(1-x)^{\frac{1}{2}} - 1} \times \frac{(1-x)^{\frac{1}{2}} + 1}{(1-x)^{\frac{1}{2}} + 1} = \lim_{x \to 0} \frac{(5^x - 1)\left[(1-x)^{\frac{1}{2}} + 1\right]}{1 - x - 1} = -\lim_{x \to 0} \frac{(5^x - 1)}{x} . \left[(1-x)^{\frac{1}{2}} + 1\right] = -\log 5 . 2$

$= -2 \log 5 = \log 5^{-2} = \log\left(\frac{1}{25}\right)$ Ans.

(g) $\lim\limits_{x\to 0}\dfrac{7^x+3^x-2}{\sin x}=\lim\limits_{x\to 0}\dfrac{(7^x-1)+(3^x-1)}{\sin x}=\lim\limits_{x\to 0}\dfrac{\dfrac{(7^x-1)}{x}+\dfrac{(3^x-1)}{x}}{\dfrac{\sin x}{x}}=\dfrac{\log 7+\log 3}{1}=\log 7.3=\log 21$ Ans.

(h) $\lim\limits_{x\to 0}\dfrac{x^2 3^x-x^2}{(1-\cos x)x}=\lim\limits_{x\to 0}\dfrac{x^2(3^x-1)}{\left(2\sin^2\frac{x}{2}\right)x}=\lim\limits_{x\to 0}\dfrac{x(3^x-1)}{2\left(\dfrac{\sin\frac{x}{2}}{\frac{x}{2}}\right)^2\times\frac{x^2}{4}}=\lim\limits_{x\to 0}\dfrac{(3^x-1)}{x}\times\dfrac{2}{\left(\dfrac{\sin\frac{x}{2}}{\frac{x}{2}}\right)^2}=2\log 3=\log 3^2=\log 9$ or $2\log 3$ Ans.

(3) using the formula $\lim\limits_{x\to 0}(1+x)^{\frac{1}{x}}=e=\lim\limits_{x\to\infty}\left(1+\dfrac{1}{x}\right)^x$, to find the limit (a) $\lim\limits_{x\to 0}\left(\dfrac{1+5x}{1+7x}\right)^{\frac{1}{x}}$

(b) $\lim\limits_{x\to\infty}\left(\dfrac{1+\frac{5}{x^2}}{1+\frac{3}{x^2}}\right)^{x^2}$ (c) $\lim\limits_{x\to\infty}\left(\dfrac{x^2+4x+5}{x^2+2x+1}\right)^x$ (d) $\lim\limits_{x\to 0}\left(\dfrac{x^2+3x+7}{x^2+x+2}\right)^{\frac{1}{x}}$

Solution:$-$ (a) $\lim\limits_{x\to 0}\left(\dfrac{1+5x}{1+7x}\right)^{\frac{1}{x}}=\lim\limits_{x\to 0}\dfrac{(1+5x)^{\frac{1}{5x}\times 5}}{(1+7x)^{\frac{1}{7x}\times 7}}$ put $5x=y$ and $7x=z$ $\boxed{\text{formula }\lim\limits_{x\to 0}(1+mx)^{\frac{1}{x}}=e^m}$

or $\lim\limits_{x\to 0}\dfrac{(1+y)^{\frac{1}{y}\times 5}}{(1+z)^{\frac{1}{z}\times 7}}$ $\boxed{\text{use formula,}\quad\lim\limits_{x\to 0}(1+x)^{\frac{1}{x}}=e}$ or $\lim\limits_{x\to 0}\left(\dfrac{1+5x}{1+7x}\right)^{\frac{1}{x}}=\dfrac{e^5}{e^7}=e^{5-7}=e^{-2}=\dfrac{1}{e^2}$ Ans.

(b) $\lim\limits_{x\to\infty}\left(\dfrac{1+\frac{5}{x^2}}{1+\frac{3}{x^2}}\right)^{x^2}$ use formula , $e=\lim\limits_{x\to\infty}\left(1+\dfrac{1}{x}\right)^x$ or $\lim\limits_{x\to\infty}\left(1+\dfrac{m}{x}\right)^x=e^m$ or $\lim\limits_{x\to\infty}\dfrac{\left(1+\frac{5}{x^2}\right)^{x^2}}{\left(1+\frac{3}{x^2}\right)^{x^2}}=\dfrac{e^5}{e^3}=e^{5-3}=e^2$ Ans.

(c) $\lim\limits_{x\to\infty}\left(\dfrac{x^2+4x+5}{x^2+2x+1}\right)^x=\lim\limits_{x\to\infty}\left(1+\dfrac{2x+4}{x^2+2x+1}\right)^x=\lim\limits_{x\to\infty}\left(1+\dfrac{2x+4}{x^2+2x+1}\right)^{\frac{2x+4}{x^2+2x+1}\times\frac{x^2+2x+1}{2x+4}\times x}=\lim\limits_{x\to\infty}\left(1+\dfrac{2x+4}{x^2+2x+1}\right)^{\frac{x^2+2x+1}{2x+4}\cdot\frac{2x^2+4x}{x^2+2x+1}}$

$=\lim\limits_{x\to\infty}\left(1+\dfrac{2x+4}{x^2+2x+1}\right)^{\frac{x^2+2x+1}{2x+4}\cdot\lim\limits_{x\to\infty}\frac{2x^2+4x}{x^2+2x+1}}=e^{\lim\limits_{x\to\infty}\frac{2x^2\left(1+\frac{2}{x}\right)}{x^2\left(1+\frac{2}{x}+\frac{1}{x^2}\right)}}=e^2$ Ans.

(d) $\lim\limits_{x\to 0}\left(\dfrac{x^2+3x+7}{x^2+x+2}\right)^{\frac{1}{x}}=\lim\limits_{x\to 0}\left(1+\dfrac{2x+5}{x^2+x+2}\right)^{\frac{1}{x}}=e^\infty$ Ans. (Do yourself)

IInd Method: $-$(3) (a) $\lim\limits_{x\to 0}\left(\dfrac{1+5x}{1+7x}\right)^{\frac{1}{x}}$ Let $y=\left(\dfrac{1+5x}{1+7x}\right)^{\frac{1}{x}}$ or $\log y=\log\left(\dfrac{1+5x}{1+7x}\right)^{\frac{1}{x}}=\dfrac{1}{x}\log\left(\dfrac{1+5x}{1+7x}\right)$

We Take a limit both of side, $\lim\limits_{x\to 0}\log y=\lim\limits_{x\to 0}\dfrac{1}{x}\log\left(\dfrac{1+5x}{1+7x}\right)=\dfrac{\lim\limits_{x\to 0}\log\left(\dfrac{1+5x}{1+7x}\right)}{\lim\limits_{x\to 0}x}$,using L'Hospital rule

or $\log y=\lim\limits_{x\to 0}\dfrac{1}{\frac{1+5x}{1+7x}}\times\dfrac{(1+7x)5-(1+5x)7}{(1+7x)^2}=\lim\limits_{x\to 0}\dfrac{5+35x-7-35x}{(1+5x)(1+7x)}=\lim\limits_{x\to 0}\dfrac{5-7}{(1+5x)(1+7x)}=\lim\limits_{x\to 0}\dfrac{-2}{(1+5x)(1+7x)}=-2$

or $\log y=-2$ or $y=e^{-2}=\dfrac{1}{e^2}$ Ans.

(b) and (c) Do yourself, same as above question

(4) using the formula $\lim\limits_{x\to 0}\dfrac{e^x-1}{x}=1$ to find (a) $\lim\limits_{x\to 0}\dfrac{e^{x^2}-1+\sin^2 x}{x^2}$ (b) $\lim\limits_{x\to 1}\dfrac{e^{(x-1)}-1+\sin(x-1)}{(x-1)}$

Solution:$-$ (a) $\lim\limits_{x\to 0}\dfrac{e^{x^2}-1+\sin^2 x}{x^2}=\lim\limits_{x\to 0}\left(\dfrac{e^{x^2}-1}{x^2}+\dfrac{\sin^2 x}{x^2}\right)=\lim\limits_{x\to 0}\left(\dfrac{e^{x^2}-1}{x^2}\right)+\lim\limits_{x\to 0}\left(\dfrac{\sin x}{x}\right)^2=1+1=2$ Ans.

(b) $\lim\limits_{x\to 1}\dfrac{e^{(x-1)}-1+\sin(x-1)}{(x-1)}=\lim\limits_{x-1\to 0}\dfrac{e^{(x-1)}-1+\sin(x-1)}{(x-1)}=\lim\limits_{x-1\to 0}\dfrac{e^{(x-1)}-1}{(x-1)}+\lim\limits_{x-1\to 0}\dfrac{\sin(x-1)}{(x-1)}=1+1=2$ Ans.

(5) using the formula $\lim\limits_{x\to 0}\dfrac{\log(1+x)}{x}=1$, to find (a) $\lim\limits_{x\to 0}\dfrac{\log(1-x)-\log(1+3x)}{x}$ (b) $\lim\limits_{x\to 0}\dfrac{\log(1-5x)}{\log(1+2x)}$ (c) $\lim\limits_{x\to 0}\dfrac{\log\left(\frac{1+x}{1-x}\right)}{x}$

(d) $\lim\limits_{x\to 0}\dfrac{\log(x^2+2x+1)}{\log(7x+1)}$ (e) $\lim\limits_{x\to -1}\dfrac{\log(x+2)}{\log(x^2+4x+4)}$

Solution:$-$ (a) $\lim\limits_{x\to 0}\dfrac{\log(1-x)-\log(1+3x)}{x}=\lim\limits_{x\to 0}\left[\dfrac{\log(1-x)}{x}-\dfrac{\log(1+3x)}{x}\right]=\lim\limits_{x\to 0}\left[\dfrac{\log(1-x)}{x}\right]-\lim\limits_{3x\to 0}\left[\dfrac{\log(1+3x)}{3x}\right].3=-1-3$

$\qquad\qquad\qquad =-4$ Ans.

(b) $\lim\limits_{x\to 0}\dfrac{\log(1-5x)}{\log(1+2x)}=\lim\limits_{x\to 0}\dfrac{\dfrac{\log[1+(-5)x]}{-5x}\times -5}{\dfrac{\log[1+2x]}{2x}\times 2}=\dfrac{\lim\limits_{x\to 0}\dfrac{\log[1+(-5)x]}{-5x}\times -5}{\lim\limits_{x\to 0}\dfrac{\log[1+2x]}{2x}\times 2}=\dfrac{1\times -5}{1\times 2}=-\dfrac{5}{2}$ Ans.

(c) $\lim\limits_{x\to 0}\dfrac{\log\left(\frac{1+x}{1-x}\right)}{x}=\lim\limits_{x\to 0}\dfrac{\log(1+x)-\log(1-x)}{x}=\lim\limits_{x\to 0}\dfrac{\log(1+x)}{x}-\lim\limits_{x\to 0}\dfrac{\log(1-x)}{x}=1-\lim\limits_{x\to 0}\dfrac{\log[1+(-x)]}{-x}\times -1=1-(-1)=1+1$

$\qquad\qquad\qquad =2$ Ans.

(d) $\lim\limits_{x\to 0}\dfrac{\log(x^2+2x+1)}{\log(7x+1)}=\lim\limits_{x\to 0}\dfrac{\log(1+x)^2}{\log(7x+1)}=\lim\limits_{x\to 0}\dfrac{\dfrac{2\log(1+x)}{x}}{\dfrac{\log(1+7x)}{7x}\times 7}=\dfrac{2\times 1}{1\times 7}=\dfrac{2}{7}$ Ans.

(e) $\lim\limits_{x\to -1}\dfrac{\log(x+2)}{\log(x^2+4x+4)}=\lim\limits_{x+1\to 0}\dfrac{\log(x+1+1)}{\log(x+2)^2}=\lim\limits_{x+1\to 0}\dfrac{\log[1+(x+1)]}{2\log(x+2)}=\lim\limits_{x+1\to 0}\dfrac{\dfrac{\log[1+(x+1)]}{(x+1)}}{2\dfrac{\log[1+(x+1)]}{(x+1)}}=\dfrac{1}{2}$ Ans.

IInd Method:$-$ $\lim\limits_{x\to -1}\dfrac{\log(x+2)}{\log(x^2+4x+4)}=\lim\limits_{x+1\to 0}\dfrac{\log(x+2)}{\log(x+2)^2}=\lim\limits_{x+1\to 0}\dfrac{\log(x+2)}{2\log(x+2)}=\lim\limits_{x+1\to 0}\dfrac{1}{2}=\dfrac{1}{2}$ Ans.

(6) using the formula $\lim\limits_{x\to 0}\dfrac{(1+x)^n-1}{x}=n$, to find following limits (a) $\lim\limits_{x\to 0}\dfrac{\sqrt{1+2x}-\sqrt{1+3x}}{x}$

(b) $\lim\limits_{x\to 0}\dfrac{\sqrt{1+3x}-4x^2-4x-1}{x}$ (c) $\lim\limits_{x\to 0}\dfrac{x^2+2x+1-(1+2x)^{\frac{3}{2}}}{x}$ (d) $\lim\limits_{x\to -2}\dfrac{(x+3)^2-(x+3)^{\frac{5}{2}}}{(x+2)}$

Solution:$-$ (a) $\lim\limits_{x\to 0}\dfrac{\sqrt{1+2x}-\sqrt{1+3x}}{x}=\lim\limits_{x\to 0}\dfrac{\left(\sqrt{1+2x}-1\right)-\left(\sqrt{1+3x}-1\right)}{x}=\lim\limits_{x\to 0}\dfrac{(1+2x)^{\frac{1}{2}}-1}{x}-\lim\limits_{x\to 0}\dfrac{(1+3x)^{\frac{1}{2}}-1}{x}$

$\qquad =\lim\limits_{x\to 0}\dfrac{(1+2x)^{\frac{1}{2}}-1}{2x}\times 2-\lim\limits_{x\to 0}\dfrac{(1+3x)^{\frac{1}{2}}-1}{3x}\times 3=2\dfrac{1}{2}-3\dfrac{1}{2}=1-\dfrac{3}{2}=-\dfrac{1}{2}$ Ans.

IInd Method:$-$ $\lim\limits_{x\to 0}\dfrac{\sqrt{1+2x}-\sqrt{1+3x}}{x}$ without use the formula or $\lim\limits_{x\to 0}\dfrac{\sqrt{1+2x}-\sqrt{1+3x}}{x}$, $\left[\left(\dfrac{0}{0}\right)\text{form}\right]$ using L'Hospital rule

or $\lim\limits_{x\to 0}\dfrac{\sqrt{1+2x}-\sqrt{1+3x}}{x}=\lim\limits_{x\to 0}\dfrac{\dfrac{1}{2\sqrt{1+2x}}.2-\dfrac{1}{2\sqrt{1+3x}}.3}{1}$ take a limit $=1-\dfrac{3}{2}=-\dfrac{1}{2}$ Ans.

IIIrd Method:$-$ $\lim\limits_{x\to 0}\dfrac{\sqrt{1+2x}-\sqrt{1+3x}}{x}=\lim\limits_{x\to 0}\dfrac{\sqrt{1+2x}-\sqrt{1+3x}}{x}\times\dfrac{\sqrt{1+2x}+\sqrt{1+3x}}{\sqrt{1+2x}+\sqrt{1+3x}}=\lim\limits_{x\to 0}\dfrac{1+2x-1-3x}{x\left(\sqrt{1+2x}+\sqrt{1+3x}\right)}$

$\qquad =\lim\limits_{x\to 0}\dfrac{-x}{x\left(\sqrt{1+2x}+\sqrt{1+3x}\right)}=\lim\limits_{x\to 0}\dfrac{-1}{\left(\sqrt{1+2x}+\sqrt{1+3x}\right)}$ take a limit $=-\dfrac{1}{2}$ Ans.

(b) $\lim\limits_{x\to 0}\dfrac{\sqrt{1+3x}-4x^2-4x-1}{x}=\lim\limits_{x\to 0}\dfrac{\sqrt{1+3x}-(2x+1)^2}{x}=\lim\limits_{x\to 0}\dfrac{\left[(1+3x)^{\frac{1}{2}}-1\right]-[(2x+1)^2-1]}{x}$

$\qquad =\lim\limits_{x\to 0}\dfrac{\left[(1+3x)^{\frac{1}{2}}-1\right]}{3x}\times 3-\lim\limits_{x\to 0}\dfrac{[(2x+1)^2-1]}{2x}\times 2=1\times 3-1\times 2=3-2=1$ Ans.

Note: $-$ without use the formula, check indeterminate form and use L'Hospital rule

(c) $\lim\limits_{x\to 0}\dfrac{x^2+2x+1-(1+2x)^{\frac{3}{2}}}{x} = \lim\limits_{x\to 0}\dfrac{[(x+1)^2-1]-[(1+2x)^{\frac{3}{2}}-1]}{x} = \lim\limits_{x\to 0}\dfrac{[(x+1)^2-1]}{x} - \lim\limits_{x\to 0}\dfrac{[(1+2x)^{\frac{3}{2}}-1]}{2x}\times 2 = 1-2.\dfrac{3}{2} = 1-3$

$\qquad\qquad\qquad\qquad = -2$ Ans.

IInd Method: $-$ check indeterminate form and use L'Hospital rule.

(d) $\lim\limits_{x\to -2}\dfrac{(x+3)^2-(x+3)^{\frac{5}{2}}}{(x+2)} = \lim\limits_{x+2\to 0}\dfrac{\{[(x+2)+1]^2-1\}-\{[(x+2)+1]^{\frac{5}{2}}-1\}}{(x+2)} = \lim\limits_{x+2\to 0}\dfrac{\{[(x+2)+1]^2-1\}}{(x+2)} - \lim\limits_{x+2\to 0}\dfrac{\{[(x+2)+1]^{\frac{5}{2}}-1\}}{(x+2)}$

$\qquad\qquad\qquad\qquad = 2-\dfrac{5}{2} = -\dfrac{1}{2}$ Ans.

IInd Method: $-$ check indeterminate form and use L'Hospital rule.

(7) using the formula $\lim\limits_{x\to a}\dfrac{x^n-a^n}{x-a} = na^{n-1}$, to find following limits (a) $\lim\limits_{x\to 2}\dfrac{\sqrt{x}-\sqrt{2}}{x-2}$ (b) $\lim\limits_{x\to 1}\dfrac{x-1}{x^3-1}$ (c) $\lim\limits_{x\to 2}\dfrac{x^{\frac{3}{2}}-2^{\frac{3}{2}}}{x^{\frac{1}{3}}-2^{\frac{1}{3}}}$

(d) $\lim\limits_{x\to 1}\dfrac{\sqrt{2}-\sqrt{1+x}}{2^{\frac{3}{2}}-(1+x)^{\frac{3}{2}}}$ (e) $\lim\limits_{x\to 1}\dfrac{x^p-1}{x^q-1}$

Solution: $-$ (a) $\lim\limits_{x\to 2}\dfrac{\sqrt{x}-\sqrt{2}}{x-2} = \lim\limits_{x\to 2}\dfrac{x^{\frac{1}{2}}-2^{\frac{1}{2}}}{x-2}$ use formula $\lim\limits_{x\to a}\dfrac{x^n-a^n}{x-a} = na^{n-1}$ or $\lim\limits_{x\to 2}\dfrac{x^{\frac{1}{2}}-2^{\frac{1}{2}}}{x-2} = \dfrac{1}{2}\times 2^{\frac{1}{2}-1} = \dfrac{1}{2}\times 2^{-\frac{1}{2}} = \dfrac{1}{2}\times\dfrac{1}{\sqrt{2}} = \dfrac{1}{2\sqrt{2}}$ Ans.

(b) $\lim\limits_{x\to 1}\dfrac{x-1}{x^3-1} = \lim\limits_{x\to 1}\dfrac{1}{\frac{x^3-1}{x-1}} = \dfrac{1}{\lim\limits_{x\to 1}\frac{x^3-1^3}{x-1}} = \dfrac{1}{3.1^{3-1}} = \dfrac{1}{3\times 1^2} = \dfrac{1}{3}$ Ans.

(c) $\lim\limits_{x\to 2}\dfrac{x^{\frac{3}{2}}-2^{\frac{3}{2}}}{x^{\frac{1}{3}}-2^{\frac{1}{3}}} = \lim\limits_{x\to 2}\dfrac{\frac{x^{\frac{3}{2}}-2^{\frac{3}{2}}}{x-2}}{\frac{x^{\frac{1}{3}}-2^{\frac{1}{3}}}{x-2}} = \dfrac{\frac{3}{2}\times 2^{\frac{3}{2}-1}}{\frac{1}{3}\times 2^{\frac{1}{3}-1}} = \dfrac{\frac{3}{2}\times 2^{\frac{1}{2}}}{\frac{1}{3}\times 2^{-\frac{2}{3}}} = \dfrac{\frac{3\times\sqrt{2}}{2}}{\frac{1}{3\times 2^{\frac{2}{3}}}} = \dfrac{3\sqrt{2}}{2}\times\dfrac{3\times 2^{\frac{2}{3}}}{1} = \dfrac{9\times 2^{\frac{1}{2}}.2^{\frac{2}{3}}}{2} = \dfrac{9}{2}\times 2^{\frac{7}{6}}$ Ans.

(d) $\lim\limits_{x\to 1}\dfrac{\sqrt{2}-\sqrt{1+x}}{2^{\frac{3}{2}}-(1+x)^{\frac{3}{2}}} = \lim\limits_{x\to 1}\dfrac{(1+x)^{\frac{1}{2}}-2^{\frac{1}{2}}}{(1+x)^{\frac{3}{2}}-2^{\frac{3}{2}}}$ $\therefore$ Let $1+x = t$ $\therefore$ $x\to 1$ then $t\to 2$

$\qquad\qquad\qquad$ or $\lim\limits_{t\to 2}\dfrac{t^{\frac{1}{2}}-2^{\frac{1}{2}}}{t^{\frac{3}{2}}-2^{\frac{3}{2}}} = \lim\limits_{t\to 2}\dfrac{\left(\frac{t^{\frac{1}{2}}-2^{\frac{1}{2}}}{t-2}\right)}{\left(\frac{t^{\frac{3}{2}}-2^{\frac{3}{2}}}{t-2}\right)} = \dfrac{\frac{1}{2}.2^{\frac{1}{2}-1}}{\frac{3}{2}.2^{\frac{3}{2}-1}} = \dfrac{2^{-\frac{1}{2}}}{3.2^{\frac{1}{2}}} = \dfrac{1}{3.\sqrt{2}.\sqrt{2}} = \dfrac{1}{3.2} = \dfrac{1}{6}$ Ans.

(e) $\lim\limits_{x\to 1}\dfrac{x^p-1}{x^q-1} = \lim\limits_{x\to 1}\dfrac{\frac{x^p-1^p}{x-1}}{\frac{x^q-1^q}{x-1}} = \dfrac{p\times 1^{p-1}}{q\times 1^{q-1}} = \dfrac{p}{q}$ Ans.

(8) (a) A function f is defined as $f(x) = \begin{cases}\dfrac{x^2-3x+2}{x^2-1}, & \text{for } x\neq 1 \\ 2, & \text{for } x = 1\end{cases}$ Test the continuity at the function at $x = 1$.

(b) If $f(x) = \begin{cases}5-x, & x < 0 \\ 3a-x, & x\geq 0\end{cases}$ be continuous at $x = 0$ then show that $a = \dfrac{5}{3}$.

Solution: $-$ (a) $f(x) = \begin{cases}\dfrac{x^2-3x+2}{x^2-1}, & x\neq 1 \\ 2, & x = 1\end{cases}$

R.H.L $= \lim\limits_{x\to 1_+} f(x) = \lim\limits_{x\to 1_+}\dfrac{x^2-3x+2}{x^2-1} = \lim\limits_{x\to 1_+}\dfrac{(x-1)(x-2)}{(x-1)(x+1)} = \lim\limits_{x\to 1_+}\dfrac{(x-2)}{(x+1)} = -\dfrac{1}{2}$

L.H.L $= \lim\limits_{x\to 1_-} f(x) = \lim\limits_{x\to 1_-}\dfrac{x^2-3x+2}{x^2-1} = \lim\limits_{x\to 1_-}\dfrac{(x-1)(x-2)}{(x-1)(x+1)} = \lim\limits_{x\to 1_-}\dfrac{(x-2)}{(x+1)} = -\dfrac{1}{2}$

we have $f(x) = f(1) = 2$ then, $\lim\limits_{x \to 1_+} f(x) = \lim\limits_{x \to 1_-} f(x) \neq f(x)$ function $f(x)$ is not continuous at $x = 1$.

$\lim\limits_{x \to 1_+} f(x) \neq f(x)$ and $\lim\limits_{x \to 1_-} f(x) \neq f(x)$ it is discontinuous function at $x = 1$.

Note: $-$ If $\lim\limits_{x \to 1_+} f(x) = \lim\limits_{x \to 1_-} f(x) = f(x)$ the function continuous at $x = 1$.

but $\lim\limits_{x \to 1_+} f(x) \neq \lim\limits_{x \to 1_-} f(x) \neq f(x)$ or $\lim\limits_{x \to 1_+} f(x) = \lim\limits_{x \to 1_-} f(x) \neq f(x)$ the function discontinuous at $x = 1$.

(b) If $f(x) = \begin{cases} 5 - x, & x < 0 \\ 3a - x, & x \geq 0 \end{cases}$ be continuous at $x = 0$.

$\lim\limits_{x \to 0_+} f(x) = \lim\limits_{x \to 0_-} f(x) = f(x) \ldots \ldots \ldots \ldots \ldots \ldots \ldots$ (A)

R.H.L $= \lim\limits_{x \to 0_+} f(x) = \lim\limits_{x \to 0_+} (3a - x) = 3a$, $\qquad$ L.H.L $= \lim\limits_{x \to 0_-} f(x) = \lim\limits_{x \to 0_-} (5 - x) = 5$

$V = \lim\limits_{x \to 0} f(x) = \lim\limits_{x \to 0} (3a - x) = 3a$ $\quad$ put value L.H.L, R.H.L and $f(x)$ in equation (A) we get,

$$\boxed{\text{R.H.L} = \text{L.H.L} = f(x) \text{ or } 3a = 5 = 3a \text{ or } 3a = 5 \quad \therefore a = \frac{5}{3} \text{ Proved.}}$$

(9) (a) If $f(x) = x\left[\sqrt{1 - x} - \sqrt{x}\right]$, then $f(x)$ is continuous and differentiable at $x = 0$.

(b) Discuss the continuity and differentiability of $f(x) = \begin{cases} |x - 2|, & x \geq 1 \\ \dfrac{x^2}{3} - \dfrac{3x}{4} + \dfrac{11}{4}, & x < 1 \end{cases}$ at $x = 1, 2$.

Solution: $-$ (a) $f(x) = x\left[\sqrt{1 - x} - \sqrt{x}\right]$ $\qquad$ for continuity, $R = L = V$ or $f(x)$ at $x = 0$

$$\lim\limits_{x \to 0_+} f(x) = \lim\limits_{x \to 0_-} f(x) = f(x)$$

R.H.L $= \lim\limits_{x \to 0_+} f(x) = \lim\limits_{x \to 0_+} x\left[\sqrt{1 - x} - \sqrt{x}\right] = 0$ and L.H.L $= \lim\limits_{x \to 0_-} f(x) = \lim\limits_{x \to 0_-} x\left[\sqrt{1 - x} - \sqrt{x}\right] = 0$, $f(x) = 0$

$$\lim\limits_{x \to 0_+} f(x) = \lim\limits_{x \to 0_-} f(x) = f(x) = 0 \text{ or } R = L = V = 0$$

The function $f(x)$ to be continuous at $x = 0$.

For differentiable, $\qquad \lim\limits_{h \to 0_+} \dfrac{f(0 + h) - f(0)}{h} = \lim\limits_{h \to 0_-} \dfrac{f(0 - h) - f(0)}{-h} = f(0)$ $\quad$ or $R' = L' = V$

$f(x) = x\left[\sqrt{1 - x} - \sqrt{x}\right]$

R.H.L $= \lim\limits_{h \to 0_+} \dfrac{f(0 + h) - f(0)}{h} = \lim\limits_{h \to 0_+} \dfrac{h\left[\sqrt{1 - h} - \sqrt{h}\right]}{h} = \lim\limits_{h \to 0_+} \left(\sqrt{1 - h} - \sqrt{h}\right) = 1$

L.H.L $= \lim\limits_{h \to 0_-} \dfrac{f(0 - h) - f(0)}{-h} = \lim\limits_{h \to 0_-} \dfrac{-h\left[\sqrt{1 + h} - \sqrt{-h}\right]}{-h} = \lim\limits_{h \to 0_-} \left[\sqrt{1 + h} - \sqrt{-h}\right] = 1$

$V = f(0) = 0$ $\quad \therefore$ R.H.L $=$ L.H.L $\neq Vf(x)$, The function $f(x)$ is not differentiable at $x = 0$.

(b) $f(x) = \begin{cases} |x - 2|, & x \geq 1 \\ \dfrac{x^2}{3} - \dfrac{3x}{4} + \dfrac{11}{4}, & x < 1 \end{cases} = \begin{cases} x - 2, & x \geq 2 \\ 2 - x, & 1 \leq x < 2 \\ \dfrac{x^2}{3} - \dfrac{3x}{4} + \dfrac{11}{4}, & x < 1 \end{cases}$

For continuity at $x = 1$, $\qquad \lim\limits_{x \to 1_+} f(x) = \lim\limits_{x \to 1_-} f(x) = f(x)$ and at $x = 2$, $\qquad \lim\limits_{x \to 2_+} f(x) = \lim\limits_{x \to 2_-} f(x) = f(x)$

CaseI: $-$ $f(x)$ to be continuous at $x = 1$.

R.H.L $= \lim\limits_{x \to 1_+} f(x) = \lim\limits_{x \to 1_+} (2 - x) = 1$, $\;$ L.H.L $= \lim\limits_{x \to 1_-} f(x) = \lim\limits_{x \to 1_-} \left(\dfrac{x^2}{3} - \dfrac{3x}{4} + \dfrac{11}{4}\right) = \dfrac{1}{3} - \dfrac{3}{4} + \dfrac{11}{4} = \dfrac{7}{3}$

$\therefore$ R.H.L $\neq$ L.H.L the function f(x) is not continuous at x = 1.

Differentiability at x = 1, $\lim\limits_{h\to 0_+} \dfrac{f(1+h)-f(1)}{h} = \lim\limits_{h\to 0_-} \dfrac{f(1-h)-f(1)}{-h} = f(1)$

R.H.L $= \lim\limits_{h\to 0_+} \dfrac{f(1+h)-f(1)}{h} = \lim\limits_{h\to 0_+} \dfrac{2-(1+h)-1}{h} = \lim\limits_{h\to 0_+} \dfrac{1-1-h}{h} = \lim\limits_{h\to 0_+} \dfrac{-h}{h} = -1$

L.H.L $= \lim\limits_{h\to 0_-} \dfrac{f(1-h)-f(1)}{-h}$, $f(x) = \dfrac{x^2}{3} - \dfrac{3x}{4} + \dfrac{11}{4}$, x < 1

or $f(1-h) = \dfrac{(1-h)^2}{3} - \dfrac{3(1-h)}{4} + \dfrac{11}{4} = \dfrac{1-2h+h^2}{3} - \dfrac{3(1-h)}{4} + \dfrac{11}{4} = \dfrac{4-8h+4h^2-9+9h+33}{12} = \dfrac{4h^2+h+28}{12}$

L.H.L $= \lim\limits_{h\to 0_-} \dfrac{f(1-h)-f(1)}{-h} = \lim\limits_{h\to 0_-} \dfrac{\frac{4h^2+h+28}{12} - \frac{7}{3}}{-h} = \lim\limits_{h\to 0_-} \dfrac{\frac{4h^2+h+28-28}{12}}{-h} = \lim\limits_{h\to 0_-} \dfrac{h(4h+1)}{-12h} = \lim\limits_{h\to 0_-} \dfrac{(4h+1)}{-12} = -\dfrac{1}{12}$

$\therefore$ R.H.L $\neq$ L.H.L, then the function f(x) is not differentiable at x = 1.

CaseII:$-$ f(x) to be continuous at x = 2, R.H.L = L.H.L = f(x) or f(2)

R.H.L $= \lim\limits_{x\to 2_+} f(x) = \lim\limits_{x\to 2_+} (x-2) = 0$ and L.H.L $= \lim\limits_{x\to 2_-} f(x) = \lim\limits_{x\to 2_-} (2-x) = 0$

$f(x) = f(2) = x - 2 = 0$ $\therefore$ $\lim\limits_{x\to 2_+} f(x) = \lim\limits_{x\to 2_-} f(x) = f(x)$ or f(2)

$\therefore$ R.H.L = L.H.L = f(x) or f(2) $\therefore$ function f(x) to be continuous at x = 2.

Differentiable at x = 2

$\lim\limits_{h\to 0_+} \dfrac{f(2+h)-f(2)}{h} = \lim\limits_{h\to 0_-} \dfrac{f(2-h)-f(2)}{-h} = f(2)$ or R.H.L = L.H.L = f(2)

R.H.L $= \lim\limits_{h\to 0_+} \dfrac{f(2+h)-f(2)}{h}$, $f(x) = x - 2, x \geq 2$ or $f(2+h) = 2 + h - 2 = h, f(2) = 0$

R.H.L $= \lim\limits_{h\to 0_+} \dfrac{f(2+h)-f(2)}{h} = \lim\limits_{h\to 0_+} \dfrac{h-0}{h} = \lim\limits_{h\to 0_+} \dfrac{h}{h} = 1$

L.H.L $= \lim\limits_{h\to 0_-} \dfrac{f(2-h)-f(2)}{-h}$, $f(x) = 2 - x, 1 \leq x < 2$ or $f(2-h) = 2 - 2 + h = h, f(2) = 0$

L.H.L $= \lim\limits_{h\to 0_-} \dfrac{f(2-h)-f(2)}{-h} = \lim\limits_{h\to 0_-} \dfrac{h-0}{-h} = \lim\limits_{h\to 0_-} \dfrac{h}{-h} = -1$

$\therefore$ R.H.L $\neq$ L.H.L, the function f(x) is not differentiable at x = 2.

(10) (a) Let $f(x) = \begin{cases} x^4 + x^3 - 16x^2 + 20x + 15, & x \neq 2 \\ m, & x = 2 \end{cases}$ If f(x) is continuous for all x, then m is equal to. ... ?

(b) Find the values of a and b so that the function $f(x) = \begin{cases} x - a\sqrt{2}\cos x , 0 \leq x < \dfrac{\pi}{4} \\ 3x\tan x + b, \dfrac{\pi}{4} \leq x \leq \dfrac{\pi}{2} \\ a\sin x - b\cos 2x, \dfrac{\pi}{2} < x \leq \pi \end{cases}$ is continuous for $0 \leq x \leq \pi$.

Solution:$-$ (a) $f(x) = \begin{cases} x^4 + x^3 - 16x^2 + 20x + 15, & x \neq 2 \\ m, & x = 2 \end{cases}$ to be continuous at x = 2.

For continuity, $\lim\limits_{x\to 2_+} f(x) = \lim\limits_{x\to 2_-} f(x) = f(x)$

R.H.L $= \lim\limits_{x\to 2_+} f(x) = \lim\limits_{x\to 2_+} (x^4 + x^3 - 16x^2 + 20x + 15) = 16 + 8 - 64 + 40 + 15 = 79 - 64 = 15$

similarly , L.H.L $= \lim\limits_{x\to 2_-} f(x) = 15$, f(2) = m $\therefore$ $\lim\limits_{x\to 2_+} f(x) = \lim\limits_{x\to 2_-} f(x) = f(x)$ or 15 = 15 = m Ans.

(b) $f(x) = \begin{cases} x - a\sqrt{2}\cos x, & 0 \le x < \dfrac{\pi}{4} \\ 3x\tan x + b, & \dfrac{\pi}{4} \le x \le \dfrac{\pi}{2} \\ a\sin x - b\cos 2x, & \dfrac{\pi}{2} < x \le \pi \end{cases}$ is continuous for $0 \le x \le \pi$.

At $x = \dfrac{\pi}{4}$ we have, $\quad \lim\limits_{x \to \frac{\pi}{4}+} f(x) = \lim\limits_{x \to \frac{\pi}{4}-} f(x) = f\left(\dfrac{\pi}{4}\right)$

R.H.L $= \lim\limits_{x \to \frac{\pi}{4}+} f(x) = \lim\limits_{x \to \frac{\pi}{4}+} (3x\tan x + b) = 3\dfrac{\pi}{4} \times \tan\dfrac{\pi}{4} + b = 3\dfrac{\pi}{4} \times 1 + b = \dfrac{3\pi}{4} + b$

L.H.L $= \lim\limits_{x \to \frac{\pi}{4}-} f(x) = \lim\limits_{x \to \frac{\pi}{4}-} \left(x - a\sqrt{2}\cos x\right) = \dfrac{\pi}{4} - a.\sqrt{2}\cos\dfrac{\pi}{4} = \dfrac{\pi}{4} - a.\sqrt{2} \times \dfrac{1}{\sqrt{2}} = \dfrac{\pi}{4} - a$

$f(x) = f\left(\dfrac{\pi}{4}\right) = 3x\tan x + b = \dfrac{3\pi}{4} + b$

For continuity, $\quad \lim\limits_{x \to \frac{\pi}{4}+} f(x) = \lim\limits_{x \to \frac{\pi}{4}-} f(x) = f\left(\dfrac{\pi}{4}\right)$ or $\dfrac{3\pi}{4} + b = \dfrac{\pi}{4} - a = \dfrac{3\pi}{4} + b$

$\therefore \dfrac{3\pi}{4} + b = \dfrac{\pi}{4} - a$ or $a + b = \dfrac{\pi}{4} - \dfrac{3\pi}{4} = \dfrac{\pi - 3\pi}{4} = \dfrac{-2\pi}{4} = -\dfrac{\pi}{2} \quad \therefore a + b = -\dfrac{\pi}{2}$ .... (A)

At $x = \dfrac{\pi}{2}$ we have , $\quad \lim\limits_{x \to \frac{\pi}{2}+} f(x) = \lim\limits_{x \to \frac{\pi}{2}-} f(x) = f\left(\dfrac{\pi}{2}\right)$

R.H.L $= \lim\limits_{x \to \frac{\pi}{2}+} f(x) = \lim\limits_{x \to \frac{\pi}{2}+} (a\sin x - b\cos x) = a\sin\dfrac{\pi}{2} - b\cos\dfrac{\pi}{2} = a - 0 = a$

L.H.L $= \lim\limits_{x \to \frac{\pi}{2}-} f(x) = \lim\limits_{x \to \frac{\pi}{2}-} (3x\tan x + b) = 3\dfrac{\pi}{2} \times \tan\dfrac{\pi}{2} + b = \infty + b = \infty$

$f(x) = f\left(\dfrac{\pi}{2}\right) = 3x\tan x + b = \infty$

$\therefore \lim\limits_{x \to \frac{\pi}{2}+} f(x) = \lim\limits_{x \to \frac{\pi}{2}-} f(x) = f\left(\dfrac{\pi}{2}\right)$ or $a = \infty = \infty \quad \therefore a = \infty$ .... (B)

solving equation (A) and (B), we get $\quad \therefore a + b = -\dfrac{\pi}{2}, \quad a = \infty \quad \therefore \infty + b = -\dfrac{\pi}{2}$ or $b = \infty$

$$\boxed{\therefore \ a = \infty, b = \infty \quad \text{Ans.}}$$

(11) (a) If $f(x) = \dfrac{(2^x - 1)^2}{\sin x.\log(1 - x)}, x \ne 0$ is continuous at $x = 0$ then find $f(0)$.

(b) If $f(x) = \begin{cases} 5x - 7, & 0 \le x \le 1 \\ 3x + a, & 1 < x \le 2 \end{cases}$ to be continuous at $x = 1$, then prove that $a = -5$.

Solution:$-$ (a) $f(x) = \dfrac{(2^x - 1)^2}{\sin x.\log(1 - x)}, x \ne 0 \qquad$ For continuity, $\lim\limits_{x \to 0+} f(x) = \lim\limits_{x \to 0-} f(x) = f(0)$

R.H.L $= \lim\limits_{x \to 0+} f(x) = \lim\limits_{x \to 0+} \dfrac{(2^x - 1)^2}{\sin x.\log(1 - x)} = \lim\limits_{x \to 0+} \dfrac{\left(\dfrac{2^x - 1}{x}\right)^2 \times x^2}{\left(\dfrac{\sin x}{x}\right) \times x.\left[-\left(x + \dfrac{x^2}{2} + \dfrac{x^3}{3} + \cdots \ldots\right)\right]} = (\log 2)^2 \lim\limits_{x \to 0+} \dfrac{x}{-x\left[1 + \dfrac{x}{2} + \dfrac{x^2}{3}\ldots\ldots\ldots\right]}$

$= (\log 2)^2 \times (-1) = -(\log 2)^2$

similarly, L.H.L $= \lim\limits_{x \to 0-} f(x) = -(\log 2)^2$, $f(0) = 0 \quad \therefore \lim\limits_{x \to 0+} f(x) = \lim\limits_{x \to 0-} f(x) = f(0)$ or $-(\log 2)^2 = -(\log 2)^2 = f(0)$

$$\boxed{\therefore \ f(0) = -(\log 2)^2 \ \text{Ans.}}$$

(b) $f(x) = \begin{cases} 5x - 7, & 0 \le x \le 1 \\ 3x + a, & 1 < x \le 2 \end{cases}$ to be continuous at $x = 1$.

for continuity, $\lim_{x\to 1_+} f(x) = \lim_{x\to 1_-} f(x) = f(x) = f(1)$ or R.H.L = L.H.L = f(1)

R.H.L $= \lim_{x\to 1_+} f(x) = \lim_{x\to 1_+} 3x + a = a + 3$, L.H.L $= \lim_{x\to 1_-} f(x) = \lim_{x\to 1_-} 5x - 7 = -2$, $f(1) = 5x - 7 = -2$

for continuity, $\lim_{x\to 1_+} f(x) = \lim_{x\to 1_-} f(x) = f(x) = f(1)$ $\therefore$ $a + 3 = -2 = -2$ or $a + 3 = -2$ or $a = -2 - 3 = -5$ $\therefore$ $a = -5$ Proved.

(12) (a) Test the continuity and differentiability of the function defined as under at x = 2 and x = 3. function

$$f(x) = \begin{cases} -x, & x < 2 \\ 3 + x, & 2 \le x \le 3 \\ -5 + 2x - x^2, & x > 3 \end{cases}$$

(b) Find the derivative of $f(x) = \begin{cases} \dfrac{x - 2}{x^2 + 3x - 10}, & x \ne 2 \\ \dfrac{1}{7}, & x = 2 \end{cases}$ at x = 2.

Solution: $-$ (a) $f(x) = \begin{cases} -x, & x < 2 \\ 3 + x, & 2 \le x \le 3 \\ -5 + 2x - x^2, & x > 3 \end{cases}$

For continuity at x = 2, $\lim_{x\to 2_+} f(x) = \lim_{x\to 2_-} f(x) = f(2)$

R.H.L $= \lim_{x\to 2_+} f(x) = \lim_{x\to 2_+} (3 + x) = 5$, L.H.L $= \lim_{x\to 2_-} f(x) = \lim_{x\to 2_-} (-x) = -2$ or $f(2) = 3 + x = 5$ $\therefore$ R.H.L $\ne$ L.H.L

f(x) is not continuous at x = 2, f(x) is discontinuous at x = 2.

For continuity at x = 3, $\lim_{x\to 3_+} f(x) = \lim_{x\to 3_-} f(x) = f(3)$

R.H.L $= \lim_{x\to 3_+} f(x) = \lim_{x\to 3_+} (-5 + 2x - x^2) = -5 + 6 - 9 = -8$

L.H.L $= \lim_{x\to 3_-} f(x) = \lim_{x\to 3_-} (3 + x) = 3 + 3 = 6$, $f(3) = 3 + x = 6$

$\therefore$ R.H.L $\ne$ L.H.L, f(x) is not continuous at x = 3. $\therefore$ f(x) is discontinuous at x = 3.

Differentiability at x = 2, Ist method: $-$ $\lim_{h\to 0_+} \dfrac{f(2 + h) - f(2)}{h} = \lim_{h\to 0_-} \dfrac{f(2 - h) - f(2)}{-h} = f(2)$

IInd method: $-$ $\lim_{x\to 2_+} f'(x) = \lim_{x\to 2_-} f'(x) = f'(2)$

using Ist method formula: $-$ $\lim_{h\to 0_+} \dfrac{f(2 + h) - f(2)}{h} = \lim_{h\to 0_-} \dfrac{f(2 - h) - f(2)}{-h} = f(2)$

R.H.L $= \lim_{h\to 0_+} \dfrac{f(2 + h) - f(2)}{h} = \lim_{h\to 0_+} \dfrac{3 + 2 + h - 5}{h} = \lim_{h\to 0_+} \dfrac{h}{h} = 1$ where $f(2 + h) = 3 + 2 + h = 5 + h$

L.H.L $= \lim_{h\to 0_-} \dfrac{f(2 - h) - f(2)}{-h} = \lim_{h\to 0_-} \dfrac{-2 + h + 2}{-h} = \lim_{h\to 0_-} \dfrac{h}{-h} = -1$, $f(2) = 3 + x = 3 + 2 = 5$

$\therefore$ R.H.L $\ne$ L.H.L . f(x) is not differentiable at x = 2.

Differentiable at x = 3, $\lim_{h\to 0_+} \dfrac{f(3 + h) - f(3)}{h} = \lim_{h\to 0_-} \dfrac{f(3 - h) - f(3)}{-h} = f(3)$

R.H.L $= \lim_{h\to 0_+} \dfrac{f(3 + h) - f(3)}{h}$ $\therefore$ $f(3 + h) = -5 + 2(3 + h) - (3 + h)^2 = -5 + 6 + 2h - 9 - 6h - h^2 = -4h - h^2 - 8$

$\therefore$ $f(3) = -5 + 2x - x^2 = -5 + 6 - 9 = -8$

R.H.L $= \lim_{h\to 0_+} \dfrac{f(3 + h) - f(3)}{h} = \lim_{h\to 0_+} \dfrac{-4h - h^2 - 8 + 8}{h} = \lim_{h\to 0_+} \dfrac{-h(4 - h)}{h} = \lim_{h\to 0_+} -(4 - h) = -4$

$$L.H.L = \lim_{h \to 0_-} \frac{f(3-h) - f(3)}{-h} = \lim_{h \to 0_-} \frac{3+3-h-6}{-h} = \lim_{h \to 0_-} \frac{-h}{-h} = 1$$

$\therefore$ R.H.L $\neq$ L.H.L the function $f(x)$ is not differentiable at $x = 3$.

using IInd method formula, $f(x) = \begin{cases} -x, & x < 2 \\ 3 + x, & 2 \leq x \leq 3 \\ -5 + 2x - x^2, & x > 3 \end{cases}$ Differentiate, $f'(x) = \begin{cases} -1, & x < 2 \\ 1, & 2 \leq x \leq 3 \\ 2 - 2x, & x > 3 \end{cases}$

IInd method formula, $\lim_{x \to 2_+} f'(x) = \lim_{x \to 2_-} f'(x) = f'(2)$ and $\lim_{x \to 3_+} f'(x) = \lim_{x \to 3_-} f'(x) = f'(3)$

(b) $f(x) = \begin{cases} \dfrac{x-2}{x^2 + 3x - 10}, & x \neq 2 \\ \dfrac{1}{7}, & x = 2 \end{cases} = \begin{cases} \dfrac{(x-2)}{(x-2)(x+5)}, & x \neq 2 \\ \dfrac{1}{7}, & x = 2 \end{cases}$

Derivative at $x = 2$, $\lim_{h \to 0_+} \dfrac{f(2+h) - f(2)}{h} = \lim_{h \to 0_-} \dfrac{f(2-h) - f(2)}{-h} = $ finite

$$R.H.L = \lim_{h \to 0_+} \frac{f(2+h) - f(2)}{h} = \lim_{h \to 0_+} \frac{\frac{1}{2+h+5} - \frac{1}{7}}{h} = \lim_{h \to 0_+} \frac{\frac{7-h-7}{7(h+7)}}{h} = \lim_{h \to 0_+} \frac{-h}{7h(h+7)} = -\frac{1}{49}$$

similarly, R.H.L $=$ L.H.L $=$ finite $\therefore$ finite $= -\dfrac{1}{49}$ Ans.

(13) (a) If $G(x) = \sqrt{16 - x^2}$, then $\lim_{x \to 2} \dfrac{G(x) - G(2)}{x-2}$ equal to … … … … ….??

(b) If $f(16) = 16, f'(16) = 8$ then $\lim_{x \to 16} \dfrac{\sqrt{f(x)} - 4}{\sqrt{x} - 4}$ equal to … … … … … ….??

Solution: $-$ (a) $G(x) = \sqrt{16 - x^2}$, $G(2) = \sqrt{16 - 4} = \sqrt{12}$

or $\lim_{x \to 2} \dfrac{G(x) - G(2)}{x-2} = \lim_{x \to 2} \dfrac{\sqrt{16 - x^2} - \sqrt{12}}{x-2}$ $\left(\dfrac{0}{0} \text{ form}\right)$ using L'Hospital rule.

or $\lim_{x \to 2} \dfrac{\frac{1}{2\sqrt{16 - x^2}} \times (-2x) - 0}{1} = \lim_{x \to 2} \dfrac{-x}{\sqrt{16 - x^2}} = \dfrac{-2}{\sqrt{16 - 4}} = \dfrac{-2}{\sqrt{12}} = \dfrac{-2}{2\sqrt{3}} = -\dfrac{1}{\sqrt{3}}$ Ans.

(b) Given, $f(16) = 16, f'(16) = 8$

$\Rightarrow \lim_{x \to 16} \dfrac{\sqrt{f(x)} - 4}{\sqrt{x} - 4}$ $\left(\dfrac{0}{0} \text{ form}\right)$ using L'Hospital rule.

or $\lim_{x \to 16} \dfrac{\sqrt{f(x)} - 4}{\sqrt{x} - 4} = \lim_{x \to 16} \dfrac{\frac{1}{2\sqrt{f(x)}} \times f'(x)}{\frac{1}{2\sqrt{x}}} = \dfrac{\frac{f'(16)}{2\sqrt{f(16)}}}{\frac{1}{2\sqrt{16}}} = \dfrac{8 \times f'(16)}{2\sqrt{f(16)}} = \dfrac{8 \times 8}{2\sqrt{16}} = \dfrac{64}{8} = 8$ Ans.

Exercise $-$ A2

(1) Evaluate the following limits: $-$ (a) $\lim_{x \to 0} \dfrac{\sin x}{x}$ (b) $\lim_{x \to 0} \dfrac{\tan x}{x}$

(2) Evaluate: $-$ (a) $\lim_{x \to \infty} \dfrac{(3+x)^{30}.(6+x)^5}{(3-x)^{35}} = -1$ (b) $\lim_{x \to \infty} \dfrac{3.\sqrt{x} + 5.\sqrt[3]{x} + 7.\sqrt[5]{x}}{\sqrt{3x - 1} + \sqrt[3]{5x - 2}} = $ … … … … …..

(3) Evaluate: $-$ (a) $\lim_{x \to \frac{\pi}{4}} \dfrac{1 - \cot x}{1 - \sqrt{2}\cos x}$ (b) $\lim_{x \to 0} \dfrac{\sqrt{1 - \sin x} + \sqrt{1 + \sin x}}{x}$

(4) Evaluate: − (a) $\lim\limits_{x\to\infty}\left(\dfrac{x+7}{x+3}\right)^{x+2}$ (b) $\lim\limits_{x\to\infty}\left(\dfrac{x+2}{x-5}\right)^{2x+3}$

(5) Evaluate: − (a) $\lim\limits_{x\to 0}\left(\dfrac{1+7x^2}{1+3x^2}\right)^{\frac{1}{x^2}}$ (b) $\lim\limits_{x\to 0}\left(\dfrac{4x^2+4x+1}{9x^2+6x+1}\right)^{\frac{1}{x}}$

(6) Evaluate: − (a) $\lim\limits_{x\to 0}\dfrac{x-\sin x}{x\cos x}$ (b) $\lim\limits_{x\to\infty}\dfrac{\sqrt{x^3+1}-\sqrt[3]{x^6+1}}{\sqrt[5]{x^5+1}+\sqrt[6]{x^6+1}}$

(7) (a) Let $f(x)=\begin{cases}x-2, & x\le 2\\ 5+ax^2, & x>2\end{cases}$ to be continuous at $x=2$, then find the value of a.

(b) The value of $f(0)$ for which $f(x)=\dfrac{3-\sqrt{x-4}}{\sin 2x}$ is continuous is ... ...??

(8) (a) $\lim\limits_{x\to 2}(2-x)\tan\dfrac{\pi x}{4}$ (b) $\lim\limits_{x\to\frac{\pi}{4}}\dfrac{\sin\left(\frac{\pi}{4}-x\right)}{\sqrt{2}\cos x-1}$

(9) (a) $\lim\limits_{x\to 0}\dfrac{x\sin x+\log(1+x^2)}{x^2}$ (b) $\lim\limits_{x\to\frac{\pi}{4}}\dfrac{\tan x-\cot x}{x-\frac{x}{4}}$

(10) (a) $\lim\limits_{x\to 0}\dfrac{\log\left(1+\frac{x}{m}\right)+\log(1-nx)-\log\left(1+\frac{x}{p}\right)}{x}$ (b) $\lim\limits_{x\to 0}\dfrac{(5^x-1)^2}{\sin\frac{x}{2}\cdot\log\left(1-\frac{x^2}{4}\right)}$

(11) (a) $\lim\limits_{x\to\tan^{-1}5}\left(\dfrac{\tan^2 x-3\tan x+10}{\tan^2 x-7\tan x+10}\right)$ (b) $\lim\limits_{x\to 0}\dfrac{e^{2x}-e^{-2x}}{x}$

(12) (a) Discuss the limits and continuity of the function $f(x)=\begin{cases}\dfrac{a^{[x]-x}}{[x]-x}, & x\ne 0\\ \log a, & x<0\end{cases}$ at $x=0$.

(b) Discuss the continuity of the function $f(x)=\begin{cases}\dfrac{x-3}{1+e^{\frac{1}{(x-3)}}}, & x\ne 3\\ 1, & x=3\end{cases}$.

(13) (a) Find the value of A so that the function $f(x)=\begin{cases}\dfrac{2^{x+3}-64}{4^x-64}, & x\ne 3\\ A, & x=3\end{cases}$.

(b) Find the value of A , B such that $\lim\limits_{x\to 0}\dfrac{A\cos x-B\sin x}{x^2}=1$.

(14) (a) Determine the value of x for which the following function fails to be continuous or differentiable.

$$f(x)=\begin{cases}2-x, & x<2\\ (2-x)(3-x), & 2\le x\le 3\\ 4-x, & x>3\end{cases}.$$

(b) The function $f(x)=\begin{cases}|x-2| & \text{for } x\ge 1\\ \dfrac{x^2}{5}+\dfrac{3x}{2}+\dfrac{13}{10} & \text{for } x<1\end{cases}$ Discuss the continuity and differentiability.

(15) (a) $f(x)=\begin{cases}ax^2+b, |x|<2\\ \dfrac{1}{|x|}, & |x|\ge 2\end{cases}$ The function is continuous and differentiable, then prove that

(b) Discuss the continuity and differentiability of the function $f(x)=\begin{cases}\dfrac{x}{1-|x|}, & |x|\ge 1\\ \dfrac{x}{1+|x|}, & |x|<1\end{cases}$.

(16) (a) Prove that the function $f(x) = \cos[\pi|x|]$ is continuous at $x = 0$ but is also differentiable them.

(b) $f(x) = \begin{cases} \dfrac{\left(e^{-\frac{1}{x}} + e^{\frac{1}{x}}\right)}{\left(e^{-\frac{1}{x}} - e^{\frac{1}{x}}\right)}, & x \neq 0 \\ 0, & x = 0 \end{cases}$ prove that $f(x)$ is not differentiable at $x = 0$.

(17) (a) Find the value of a and b so that the function $f(x) = \begin{cases} x - b\sqrt{2}\sin x, & 0 \leq x < \dfrac{\pi}{4} \\ -3x \cot x - a, & \dfrac{\pi}{4} \leq x \leq \dfrac{\pi}{2} \\ b\cos 2x + a \sin x, & \dfrac{\pi}{2} < x \leq \pi \end{cases}$ is continuous for $0 \leq x \leq \pi$.

(b) $f(x) = \begin{cases} A \tan^{-1} \dfrac{1}{x-3}, & 0 \leq x < 3 \\ \dfrac{\pi}{2}, & x = 3 \\ B \tan^{-1} \dfrac{2}{x-3}, & 3 < x < 5 \\ \sin^{-1}(6-x) + A\dfrac{\pi}{3}, & 5 \leq x \leq 7 \end{cases}$, Determine the value of A and B if $f(x)$ is continuous in the interval $[0,7]$.

(18) (a) Discuss the continuity and differentiability of the function $f(x)$ in $(0,5)$ where $f(x) = \begin{cases} |3x-5|[x], & 0 \leq x \leq 3 \\ \dfrac{x^2}{3}, & 3 < x \leq 5 \end{cases}$

(b) $f(x) = \begin{cases} (x-3).3^{-\left[\frac{1}{|x-3|} + \frac{1}{(x-3)}\right]}, & x \neq 3 \\ 0, & x = 3 \end{cases}$ prove that the function is not differentiable at $x = 3$.

(19) (a) If $f(3) = 5$ and $f'(3) = 2$ then find $\lim\limits_{x \to 3} \dfrac{xf(3) - 2f(x)}{x - 2}$.

(b) If $f(a) = 3$, $f'(a) = 1$, $g(a) = -2$, $g'(a) = -1$ then the value of $\lim\limits_{x \to a} \dfrac{g(x)f(x) - g(a)f(x)}{x - a} = -1$

(20) (a) If $P(x) = -\sqrt{36 - x^2}$ then $\lim\limits_{x \to 2} \dfrac{P(x) - P(2)}{x - 2} = \dfrac{1}{2\sqrt{2}}$. (b) If $f(25) = 25, f'(25) = 5$ then $\lim\limits_{x \to 25} \dfrac{\sqrt{f(x)} - 5}{\sqrt{x} - 5} = 5$.

(c) If $P(x) = \sqrt{64 - x^2}$ then $\lim\limits_{x \to 3} \dfrac{P(x) - P(3)}{x - 3} = -\dfrac{3}{\sqrt{55}}$.

Answer

(1) (a) Ans: 1 (b) Ans: 1

(2) (a) $\lim\limits_{x \to \infty} \dfrac{(3+x)^{30}.(6+x)^5}{(3-x)^{35}} = \lim\limits_{x \to \infty} \dfrac{x^{30}\left(\frac{3}{x}+1\right)^{30}.x^5\left(\frac{6}{x}+1\right)^5}{x^{35}\left(\frac{3}{x}-1\right)^{35}} = \lim\limits_{x \to \infty} \dfrac{\left(\frac{3}{x}+1\right)^{30}.\left(\frac{6}{x}+1\right)^5}{\left(\frac{3}{x}-1\right)^{35}} = -1$ Ans.

(b) Ans: $\sqrt{3}$. Hint,   put $x = \dfrac{1}{y}$, $x \to \infty, y \to 0$ (3) (a) Ans: 2   (b) Ans: -1 (4) (a) Ans: e^4 (b) Ans: e^{14}

(5) (a) Ans: $\dfrac{e^7}{e^3} = e^{7-3} = e^4$.   (b) Ans: e^{-2} (6) (a) Ans: 0   (b) Ans: ∞

(7) (a) Ans: $a = -\dfrac{5}{4}$ (b) Ans: (8) (a) Ans: $\dfrac{4}{\pi}$ (b) Ans: 1 (9) (a) Ans: 2   (b) Ans: 1

(10) (a) Ans: $\dfrac{p - n - m}{mp}$ (b) Ans: $(2\log 5)^3$ (11) (a) Ans: $\dfrac{7}{3}$ (b) Ans: 4 (13) (a) Ans: $A = 0$ (b) Ans:

(17) (a) Ans: $a = \pi, b = 2\pi$ (b) Ans: $A = 1, \dfrac{9}{4}$ and $B = 1, -\dfrac{2}{3}$

(19) (a) Ans: 1   (b) Ans: -1 (20) (a) Ans: $\dfrac{1}{2\sqrt{2}}$ (b) Ans: 5   (c) Ans: $\dfrac{-3}{\sqrt{55}}$

Exercise – A3

(1) Find $\lim\limits_{x\to\infty} f(x)$ if, (a) $f(x) = \dfrac{4x^2 + 3x + 2}{1 + x^2}$ (b) $f(x) = \dfrac{3 - x^2}{x^2 - 2x + 3}$

(c) $f(x) = \dfrac{5x^3 + 2x^2 - x + 3}{3x^3 - 4x + 10}$ (d) $f(x) = \dfrac{1^2 + 2^2 + \cdots\ldots\ldots + x^2}{4x^3 - x + 1}$ (e) $f(x) = \dfrac{1 + 2 + 3 \ldots\ldots\ldots + x}{x^2 + 1}$

(2) Find the following limit: – (a) $\lim\limits_{n\to\infty} \left(\dfrac{2n^3 - n + 3}{3n^3 + 2n + 1}\right)^2$ (b) $\lim\limits_{n\to\infty} \sqrt[n]{1 + n}$ (c) $\lim\limits_{x\to\infty} \sqrt[x]{x^5 + 2}$ (d) $\lim\limits_{x\to\infty} \sqrt[x]{x^3}$

(3) Find the following limit: – (a) $\lim\limits_{n\to 0} \dfrac{1 + 2 + 3 + \cdots\ldots\ldots + n}{n}$ (b) $\lim\limits_{n\to 0} \dfrac{1 + 2 + 3 + \cdots\ldots\ldots\ldots + n}{n} - 1$

(c) $\lim\limits_{n\to 0} \dfrac{1^2 + 2^2 + 3^2 + \cdots\ldots\ldots + n^2}{n}$ (d) $\lim\limits_{n\to 0} \dfrac{1^2 + 2^2 + 3^2 + \cdots\ldots\ldots + n^2}{n^2}$

(4) Find the following limit: – (a) $\lim\limits_{x\to 0} \dfrac{(x^3 - 1)^5 + 1}{x}$ (b) $\lim\limits_{x\to 1}(x^7 + x^5 + x^3 + x)$ (c) $\lim\limits_{x\to 0} \dfrac{2^x + 3^x - 2}{x}$ (d) $\lim\limits_{x\to 0} \left(\dfrac{2\sin 2x + \cos 2x - 1}{x}\right)$

(e) $\lim\limits_{x\to\frac{\pi}{4}} \dfrac{1 - \tan x}{1 - \sqrt{2}\sin x}$ (f) $\lim\limits_{x\to 2} \dfrac{\log(x - 1)}{x - 2}$ (solve without use L'Hospital rule)   (g) $\lim\limits_{x\to\infty} \left(\dfrac{1 - x^2}{x^2}\right)$

(h) $\lim\limits_{x\to\infty} x\left(\dfrac{\pi}{2} - \tan^{-1} x\right)$ (i) $\lim\limits_{x\to 0} \dfrac{5^x - 3^x}{4^x - 2^x}$ (j) $\lim\limits_{x\to 2} \dfrac{e^{\sqrt{x}} - e^{\sqrt{2}}}{x - 2}$ (k) $\lim\limits_{x\to 0} \dfrac{(x.3^x - x)}{\dfrac{1 - \cos 2x}{2}}$

(5) Find the following limit: – (a) $\lim\limits_{x\to 0} \dfrac{27^x + 9^x + 3^x - 3}{x}$ (b) $\lim\limits_{x\to 0} \dfrac{81^x - 27^x + 9^x - 3^x}{\left(\sqrt{24 + \cos x} - 3\right)\sin x}$ (c) $\lim\limits_{x\to\frac{\pi}{4}} \dfrac{\sqrt{2}\left(x - \dfrac{\pi}{4}\right)}{(\sin x - \cos x)}$ (d) $\lim\limits_{x\to 0} \dfrac{x}{\sin x + \tan x}$

(e) $\lim\limits_{x\to 1} \dfrac{\sqrt{x + 2} - \sqrt{2x + 1}}{x - 1}$ (f) $\lim\limits_{x\to 0}(1 + \sin x)^{\frac{1}{\sin x}}$ (g) $\lim\limits_{x\to 2} \dfrac{x - 2}{2x^2 - 5x + 2}$ (h) $\lim\limits_{x\to 0} \dfrac{\sqrt{1 + \tan x} - \sqrt{1 - \sin x}}{x}$ (i) $\lim\limits_{x\to\frac{\pi}{2}} \dfrac{\sqrt{1 - \sin x}}{\cos x}$

(j) $\lim\limits_{x\to 2} \dfrac{(1 + x)^3 - 27}{x - 2}$ (k) $\lim\limits_{x\to 3} \dfrac{\left(1 + \sqrt{1 + x}\right)^2 - 9}{x - 3}$

(6) Find the following limit: – (a) $\lim\limits_{x\to\frac{\pi}{4}} \dfrac{1 - \sin 2x}{x - \dfrac{\pi}{4}}$ (b) $\lim\limits_{x\to a} \dfrac{x^{\frac{1}{3}} - a^{\frac{1}{3}}}{x^{\frac{1}{2}} - a^{\frac{1}{2}}}$ (c) $\lim\limits_{x\to\frac{\pi}{2}} \dfrac{\log(\cos x + 1)}{\sin 2x}$ (d) $\lim\limits_{x\to 0} \dfrac{1 + \sin x - \cos x}{\cos x - \sin x - 1}$

(e) $\lim\limits_{x\to 0} \dfrac{\sqrt{9 + \tan 2x} - 3}{\log(1 + \sin 2x)}$ (f) $\lim\limits_{x\to 1} \dfrac{\sqrt[4]{15 + x} - 2}{3(x - 1)}$ (g) $\lim\limits_{x\to 0} \dfrac{\sqrt[3]{1 + x} + \sqrt[4]{1 + x} - 2}{3^x + 2^x - 2}$ (h) $\lim\limits_{x\to\frac{\pi}{2}} \dfrac{\sqrt{1 + \cos x} - \sqrt{1 - \cos x}}{\cos x}$

(i) $\lim\limits_{x\to a} \dfrac{\sqrt{x + 3a} - 2.\sqrt{a}}{3.\sqrt{x + a} - 3.\sqrt{2a}}$, $(a \neq 0)$ (j) $\lim\limits_{x\to 0} \dfrac{x\sin x + \log(1 + x^2)}{x^2}$

(7) Find the following limit: – (a) $\lim\limits_{x\to\infty} \dfrac{\sqrt{x^2 + 1} + \sqrt[4]{x^4 + 1}}{\sqrt[3]{x^3 + 1} + \sqrt[5]{x^5 + 1}}$ (b) $\lim\limits_{x\to\infty} \dfrac{\sqrt[3]{x^2 + 1} - \sqrt[4]{x^4 + 1}}{\sqrt[3]{x^3 + 1} - \sqrt[5]{x^4 + 1}}$ (c) $\lim\limits_{x\to\infty} \dfrac{x + \cos x}{x - \sin^2 x}$ (d) $\lim\limits_{x\to 3} \dfrac{x^2 - 4x + 3}{x^2 - 5x + 6}$

(e) $\lim\limits_{x\to 1} \dfrac{(3x - 1)\left(\sqrt{x^2} - 1\right)}{(3x^3 + x^2 - 3x - 1)}$ (f) $\lim\limits_{x\to 0} \dfrac{(x.3^x - x)}{\sin^2 x}$ (g) $\lim\limits_{x\to\pi} \dfrac{(x - \pi)\sin x}{1 + \cos x}$ (h) $\lim\limits_{x\to 0} \dfrac{\tan 2x.(1 - \tan^2 x)}{x + \sin x}$ (i) $\lim\limits_{x\to 0} \dfrac{x\cos x + \tan x}{\tan x}$

(j) $\lim\limits_{x\to 0} \dfrac{2x - \sin x}{3x - \tan x}$ (k) $\lim\limits_{x\to 0} \dfrac{\tan x + x}{x^2}$

(8) Evaluate: — (a) $\lim\limits_{x\to0}\dfrac{\cot(\pi x)}{\dfrac{\pi}{2x}}$ (b) $\lim\limits_{\theta\to\pi}\dfrac{\sqrt{1+\cos\theta}-\sin\theta}{\cos\left(\dfrac{\theta}{2}\right)}$ (c) $\lim\limits_{\theta\to\frac{\pi}{2}}\dfrac{(\sin\theta)^{\sin\theta}-1}{1-\sin\theta}$

(d) If $f'(a)$ is exists then show that $\lim\limits_{x\to a}\dfrac{(2x-a)f(x)-af(2x-a)}{x-a}=2f(a)-af'(a)$.

(e) If $f(a)=3, f'(a)=2, g(2a)=-1$ then the value of $\lim\limits_{x\to a}\dfrac{f(a).g(x+a)-f(2a-x).g(2a)}{x-a}=-1$.

(9) (a) If $F(x)=\sqrt{24-x^3}$, then $\lim\limits_{x\to2}\dfrac{F(x)-F(2)}{x-2}$ equal to ... ... ?

(b) If $F(5)=5, F'(5)=3$, then $\lim\limits_{x\to5}\dfrac{\sqrt{F(x)+3}-\sqrt{8}}{\sqrt{x}-\sqrt{5}}$ equal to ... ... ?

(10) Evaluate: — (a) $\lim\limits_{x\to0}\dfrac{\tan x-x-\dfrac{x^3}{3}}{x^5}$ (b) $\lim\limits_{x\to\frac{\pi}{2}}\sec x.\log_e(\cosec x)$ (c) $\lim\limits_{x\to1}\dfrac{(1+x)-2}{(1-x)}$ (d) $\lim\limits_{x\to0}\dfrac{\log\sin\left(\dfrac{\pi}{2}+ax\right)}{\tan bx}$ (e) $\lim\limits_{x\to0}\dfrac{\dfrac{\pi}{2x}}{\cosec\left(\dfrac{\pi}{2}-x\right)}$

(11) (a) Find the value of constant a and b such that $\lim\limits_{x\to0}\dfrac{a\tan x+x(b+\cos x)}{x^3}=1$

(b) Find the value of constant a, b and c such that $\lim\limits_{x\to0}\dfrac{axe^x-b\log(1+x)+cxe^{-x}}{x^2\sin x}=2$

(c) $\lim\limits_{x\to0}\left(\dfrac{1-x}{1+x}+ax-b\right)=0$, then prove that $a=0, b=-1$ (d) $\lim\limits_{x\to0}\left(\dfrac{ax^2}{x+1}-bx+1\right)=1$, then prove that $a=0, b=0$

(12) (a) $\lim\limits_{x\to0}\dfrac{\cos 2x+a\cos x}{x^2}=$ finite, find a and the limit.   (b) $\lim\limits_{x\to0}\dfrac{a\tan x+b\sin x-2}{x^3}=$ finite, find a and b and the limit.

(c) $\lim\limits_{x\to0}\dfrac{ae^x+b\log(1+x)-ce^{-x}}{x\tan x}=$ finite, find a, b and c.   (d) $\lim\limits_{x\to0}\dfrac{ax\cos x+b\log(1+x)-cxe^x}{x^2\sin x}=2$, find a, b and c.

(13) (a) $\lim\limits_{x\to0}\left\{\sin\left(\dfrac{\pi}{2}+x\right)\right\}^{\frac{1}{x}}=1$ (b) $\lim\limits_{x\to\infty}\left(\dfrac{x+3}{x+2}\right)^{x+4}=e$ (c) $\lim\limits_{x\to\infty}\left(\dfrac{x+1}{x-2}\right)^{x+3}=e^3$ (d) $\lim\limits_{x\to0}\left(\dfrac{1+2x^2}{1+3x^2}\right)^{\frac{1}{x^2}}=\dfrac{1}{e}$

(e) $\lim\limits_{x\to0}\left\{\dfrac{2}{3}\left(\dfrac{3+x}{2+x}\right)\right\}^{\frac{1}{x}}=e^{-\frac{1}{6}}=\dfrac{1}{e^{\frac{1}{6}}}$ (f) $\lim\limits_{x\to0}\dfrac{\log(1+x)}{x}-2$ (g) $\lim\limits_{x\to\infty}\left(\dfrac{x-5}{x+3}\right)^x$ (h) $\lim\limits_{x\to0}\left(\dfrac{\sin 2x}{2x}\right)^{\frac{1}{x^2}}$ (i) $\lim\limits_{x\to0}\left(\dfrac{\tan 3x}{3x}\right)^{\frac{1}{x^2}}$

(j) $\lim\limits_{x\to0}\left(\dfrac{\tan x}{x}\right)^{\frac{1}{x}}$ (k) $\lim\limits_{x\to0}\left(\dfrac{\sin x}{x}\right)^{\frac{1}{x}}$ (l) $\lim\limits_{x\to0}\left(\dfrac{\tan x}{x}\right)^{\frac{1}{x^3}}$ (m) $\lim\limits_{x\to0}\left(\dfrac{\sin 3x}{x}\right)^{\frac{1}{x^3}}$

(14) Evaluate: — (a) $\lim\limits_{x\to0}\dfrac{\sin 2x}{3x}$ (b) $\lim\limits_{x\to\infty}\dfrac{3x^2-2}{\sqrt{x^4+2x^3-x^2}}$ (c) $\lim\limits_{x\to0}\dfrac{\log(1+\sin x)}{\sin x}$ (solve without L'Hospital rule)

(d) $\lim\limits_{x\to\infty}\dfrac{\sqrt{2x^2-3}}{\sqrt{3x^2-1}}$ (e) $\lim\limits_{x\to2}\dfrac{e^{\frac{x}{2}}-1}{(x-2)}$ (f) $\lim\limits_{x\to1}\dfrac{\sqrt{3}-\sqrt{4-x}}{2-\sqrt{3+x}}$

(15) Find the following limits (without L'Hospital rule): — (a) $\lim\limits_{x\to\frac{\pi}{2}}\dfrac{\sqrt{\sin x}+\sqrt{\cos x}-1}{\sqrt{\cos x}-\sqrt{\sin x}+1}$ (b) $\lim\limits_{x\to\frac{\pi}{4}}\dfrac{\tan x-1}{x-\dfrac{\pi}{4}}$

(c) $\lim\limits_{x\to3}\dfrac{x-\sqrt{6+x}}{x-3}$ (d) $\lim\limits_{x\to0}\dfrac{\dfrac{3^x}{5^x}-1}{x}$ (e) $\lim\limits_{x\to\infty}\dfrac{3x^3+2x^2-5x+1}{2x^3+x^2+2x+1}$ (f) $\lim\limits_{x\to0}\dfrac{\sqrt{1-\cos 2x}}{\sqrt{2}x}$

(16) Evaluate: — (a) $\lim\limits_{x\to0}\dfrac{\log(1-e^x)-\log 2}{x}$ (b) $\lim\limits_{x\to0}\dfrac{2\left[e^{(1-\cos x)}-\cos x\right]}{\sin x}$ (c) $\lim\limits_{x\to1}\dfrac{5^{(x^2-2x+1)}-3^{(x-1)}}{(x-1)}$

(d) $\lim\limits_{x \to \frac{\pi}{4}} \dfrac{(\sin x - \cos x)}{4x - \pi}$ (without L'Hospital rule).   (e) $\lim\limits_{x \to \infty} \dfrac{x^{32}(1-x)^8}{(2+x)^{40}}$

(f) $\lim\limits_{x \to 0} \dfrac{x \sin 2x + 2x \sin x}{(1 - \cos 2x)}$ (without L'Hospital rul).   (g) $\lim\limits_{x \to 0} \dfrac{e^{\tan x} - e^{\sin x}}{5^{\sin x} - 4^{\tan x}}$ (h) $\lim\limits_{x \to \frac{\pi}{2}} \dfrac{\sqrt{2}\sin\left(\frac{x}{2}\right) - 1}{\sqrt{3} - \tan\left(\frac{2x}{3}\right)}$

(17) Evaluate: – (a) $\lim\limits_{x \to \frac{\pi}{4}}(\sec x - 2\cos x)(1 - \tan x)$ (b) $\lim\limits_{x \to \frac{\pi}{6}} \dfrac{\operatorname{cosec} x - 2\sin x - 1}{\sqrt{3}\sec x - 2}$ (c) $\lim\limits_{x \to \infty} \dfrac{\sqrt{x^2 - 3} - \sqrt[3]{2 + 3x^3}}{x + \sqrt{1 + x^2}}$

(d) $\lim\limits_{x \to 3} \dfrac{\sin(x^2 - 2x - 3) - \cos(2x - 6) + 1}{(x - 3)(x + 1)}$ (e) $\lim\limits_{x \to 1} \sqrt{x - 1}.(x^2 - 1)$ (f) $\lim\limits_{x \to 0} \dfrac{(1 - \sin x)^2 - 1}{\sin x}$ (g) $\lim\limits_{x \to 2}|x - 2|$

(18) Evaluate: – (a) $\lim\limits_{x \to 1} x - [x]$ (b) $\lim\limits_{x \to 2} e^{|x - 2|}$ (c) $\lim\limits_{x \to 0} \dfrac{x - |x|}{x}$ (d) $\lim\limits_{x \to 0} \dfrac{\sin|x|}{x}$ (e) $\lim\limits_{x \to 0} \dfrac{\tan|x|}{x}$

(f) $\lim\limits_{\theta \to 0} \tan 2\theta . \cot \theta$ (g) $\lim\limits_{x \to 0} \dfrac{\sqrt{1 + \sin x} - \sqrt{\cos x}}{\sqrt{1 - \cos x}}$ (h) $\lim\limits_{x \to 0} \dfrac{\sec x - 1}{\tan x}$ (solve without use L'Hospital rule)

(19) Evaluate: – (a) $\lim\limits_{x \to -1} \dfrac{\sqrt{x + 2} - 1}{x + 1}$ (b) $\lim\limits_{x \to -2} \dfrac{\sqrt[3]{x + 3} - 1}{x + 2}$ [use formula and solve question no. (a)and (b)]   (c) $\lim\limits_{x \to -\pi} \dfrac{\sin(x + \pi)}{1 + \cos x}$

(d) $\lim\limits_{x \to -\frac{\pi}{2}} \dfrac{\cos x}{\left(x + \frac{\pi}{2}\right)}$ (e) $\lim\limits_{x \to -\frac{\pi}{4}} \dfrac{\left(x + \frac{\pi}{4}\right)(1 - \tan x)}{1 + \tan x}$ (f) $\lim\limits_{x \to 0}(1 - \sin x)^{\frac{1}{\sin x}}$ (g) $\lim\limits_{x \to \frac{\pi}{2}}\left(1 + \dfrac{2}{\tan x}\right)^{\tan x}$ (h) $\lim\limits_{x \to 0}\left(1 + \dfrac{\sin x}{3}\right)^{\operatorname{cosec} x}$

(20) Evaluate: – (a) $\lim\limits_{x \to \infty}\left(1 - \dfrac{3}{\sin x}\right)^{\sin x}$ (b) $\lim\limits_{x \to 0} \dfrac{\sin(x + \pi)}{x \cos x}$ (c) $\lim\limits_{x \to -\frac{\pi}{2}} \dfrac{e^{2x} - e^{-\pi}}{e^{\left(x + \frac{\pi}{2}\right)} - 1}$ (d) $\lim\limits_{x \to -5} \dfrac{\sqrt{26 - x^2} - 1}{1 - \sqrt{x + 6}}$

(e) $\lim\limits_{x \to \log 2} \dfrac{\log x - \log 2}{2x - \log 4}$ (f) $\lim\limits_{x \to \infty} \dfrac{\sqrt{x + \tan x}}{\sqrt{x - \sin x}}$ (g) $\lim\limits_{x \to \infty} \dfrac{\tan x - \frac{\pi}{2}}{\cot x}$ (h) $\lim\limits_{x \to e} \dfrac{\log(x^2 + 2x + 1) - \log(x^2 + x)}{\log(2x + 3)}$

(21) Evaluate: – (a) $\lim\limits_{x \to 0} \dfrac{9^x + 7^x - 5^x - 3^x}{8^x + 6^x - 4^x - 2^x}$ (b) $\lim\limits_{x \to a - 1} \dfrac{x^2 + a^2 + 2x + 1}{x - a + 1}$ $\left(\text{use formula}, \lim\limits_{x \to a} \dfrac{x^n - a^n}{x - a} = na^{n-1}\right)$

(c) $\lim\limits_{x \to \frac{\pi}{2}} \dfrac{\cos^2 x}{1 - \sin x}$ $\left(\text{use formula } \lim\limits_{x \to a} \dfrac{x^n - a^n}{x - a} = na^{n-1}\right)$ (d) $\lim\limits_{x \to 0} x^{\frac{1}{|x|}}$ (e) $\lim\limits_{x \to 0}|x|^x$ (f) $\lim\limits_{x \to 0} e^{\frac{1}{|x|}}$ (g) $\lim\limits_{x \to 0} xe^{|x|}$ (h) $\lim\limits_{x \to 1} e^{\frac{1}{|x - 1|}}$

(i) $\lim\limits_{x \to 2}|x - 2|^x$ (j) $\lim\limits_{x \to 3} \sin|x - 3|$

(22) Evaluate: – (a) $\lim\limits_{x \to \frac{\pi}{4}}\left(x - \dfrac{\pi}{4}\right)\cot\left(x - \dfrac{\pi}{4}\right)$ (b) $\lim\limits_{x \to 0} \dfrac{e^{ax} - e^{-2ax}}{\log(1 + x)}$ (c) $\lim\limits_{x \to 1}\left(\dfrac{1}{x^2 - 1} - \dfrac{2}{x^4 - 1}\right)$ (d) $\lim\limits_{x \to 2}\left(\dfrac{1}{x - 1} - \dfrac{5}{x^2 + 1}\right)$

(e) $\lim\limits_{x \to 2}\left(\dfrac{x^3 + x^2 + x + 1}{x^2 + 2x + 1}\right)^{\frac{1 - \cos[2(x - 2)]}{(x - 2)^2}} = L$, prove that $L = \dfrac{25}{9}$ (f) $\lim\limits_{x \to -1} \dfrac{(3x + 2)\sqrt{1 + x}}{3x^2 + x - 2}$ (g) $\lim\limits_{x \to \frac{\pi}{2}} \dfrac{1}{\cos x} . \log_e(\sin x)$

(h) $\lim\limits_{x \to \infty} x\left[\tan^{-1}\dfrac{x}{x + 1} - \dfrac{\pi}{4}\right]$

(23) Evaluate: – (a) $\lim\limits_{x \to 0} \dfrac{\tan(\pi \cos^2 x)}{x^2}$ (b) $\lim\limits_{x \to \infty}\left(\dfrac{x + 5}{x + 1}\right)^x$ (c) $\lim\limits_{x \to \infty}\left(\dfrac{x - 1}{x + 2}\right)^{3x}$ (d) $\lim\limits_{x \to 0}(1 - 2\tan x)^{\frac{1}{\tan x}}$

(e) $\lim\limits_{x \to \frac{\pi}{2}}\left(1 + \dfrac{\cos x}{3}\right)^{\frac{1}{\cos x}}$ (f) $\lim\limits_{x \to 1} \dfrac{\sin \pi x}{x - 1} + \lim\limits_{x \to \infty}\left(\dfrac{x + 1}{x - 3}\right)^{2x}$ (g) $\lim\limits_{x \to a} \dfrac{x^{\frac{3}{2}} - a^{\frac{3}{2}}}{x^{\frac{1}{2}} - a^{\frac{1}{2}}}$ (h) $\lim\limits_{x \to 3a} \dfrac{\sqrt{x} - \sqrt{3a} + \sqrt{x - 3a}}{\sqrt{2x^2 - 18a^2}}$ (i) $\lim\limits_{x \to 0} \dfrac{\sqrt{9 + \tan 3x} - 3}{\log(1 + \sin 2x)}$

(24) (a) $\lim\limits_{x \to \infty}\left(\dfrac{x^2 + x + 1}{x + 1} - ax - b\right) = 4$ (b) If $\lim\limits_{x \to 0}[1 + x\log(1 + b^2)]^{\frac{1}{x}} = 2b\sin^2\theta , b > 0 \text{ } and \text{ } \theta \in (-\pi, \pi]$, then the value of θ.

(c) Let $L = \lim\limits_{x \to 0} \dfrac{a - \sqrt{a^2 - x^2} - \dfrac{x^2}{4}}{x^4}$, $a > 0$. If L is $finite$.

Answer

(1) (a) $f(x) = \dfrac{4x^2 + 3x + 2}{1 + x^2}$, $\displaystyle\lim_{x\to\infty} \dfrac{4x^2 + 3x + 2}{1 + x^2} = \lim_{x\to\infty} \dfrac{x^2\left(4 + \frac{3}{x} + \frac{2}{x^2}\right)}{x^2\left(1 + \frac{1}{x^2}\right)} = 4$ Ans.   (b) Ans: 1   (c) Ans: $\dfrac{5}{3}$

(d) $f(x) = \dfrac{1^2 + 2^2 + \cdots \ldots \ldots + x^2}{4x^3 - x + 1}$ or $\displaystyle\lim_{x\to\infty} \dfrac{1^2 + 2^2 + \cdots \ldots \ldots + x^2}{4x^3 - x + 1}$ $\left[\text{formula, } 1^2 + 2^2 + \cdots + x^2 = \dfrac{x(x+1)(2x+1)}{6}\right]$

$\displaystyle\lim_{x\to\infty} \dfrac{1^2 + 2^2 + \cdots \ldots \ldots + x^2}{4x^3 - x + 1} = \lim_{x\to\infty} \dfrac{x(x+1)(2x+1)}{(4x^3 - x + 1)6} = \dfrac{1}{6}\lim_{x\to\infty} \dfrac{x^3\left(1 + \frac{1}{x}\right)\left(2 + \frac{1}{x}\right)}{x^3\left(4 - \frac{1}{x^2} + \frac{1}{x^3}\right)} = \dfrac{1}{6} \times \dfrac{2}{4} = \dfrac{1}{12}$ Ans.

(e) $f(x) = \dfrac{1 + 2 + 3 \ldots \ldots \ldots + x}{x^2 + 1}$ or $\displaystyle\lim_{x\to\infty} \dfrac{1 + 2 + 3 \ldots \ldots \ldots + x}{x^2 + 1}$ $\left[\text{formula, } 1 + 2 + 3 \ldots \ldots \ldots + x = \dfrac{x(x+1)}{2}\right]$

$\displaystyle\lim_{x\to\infty} \dfrac{1 + 2 + 3 \ldots \ldots \ldots + x}{x^2 + 1} = \lim_{x\to\infty} \dfrac{x(x+1)}{x^2\left(1 + \frac{1}{x^2}\right).2} = \lim_{x\to\infty} \dfrac{x^2\left(1 + \frac{1}{x}\right)}{x^2\left(1 + \frac{1}{x^2}\right).2} = \dfrac{1}{2}$ Ans.

(2) (a) $\displaystyle\lim_{n\to\infty} \left(\dfrac{2n^3 - n + 3}{3n^3 + 2n + 1}\right)^2 = \lim_{n\to\infty} \left[\dfrac{n^3\left(2 - \frac{1}{n^2} + \frac{3}{n^3}\right)}{n^3\left(3 + \frac{2}{n^2} + \frac{1}{n^3}\right)}\right]^2 = \left(\dfrac{2}{3}\right)^2 = \dfrac{4}{9}$ Ans.   (b) Ans: 1   (c) Ans: 1   (d) Ans: 1

(3) (a) $\displaystyle\lim_{n\to 0} \dfrac{1 + 2 + 3 + \cdots \ldots \ldots + n}{n} = \lim_{n\to 0} \dfrac{n(n+1)}{2n} = \dfrac{1}{2}$ Ans.   (b) Ans: $-\dfrac{1}{2}$ (c) Ans: $\dfrac{1}{6}$ (d)

(4) (a) $\displaystyle\lim_{x\to 0} \dfrac{(x^3 - 1)^5 + 1}{x}$ use L'Hospital rule. Ans: 0   (b) Ans: 4

(c) $\displaystyle\lim_{x\to 0} \dfrac{2^x + 3^x - 2}{x} = \lim_{x\to 0} \dfrac{(2^x - 1) + (3^x - 1)}{x} = \lim_{x\to 0} \left[\dfrac{(2^x - 1)}{x} + \dfrac{(3^x - 1)}{x}\right] = \lim_{x\to 0} \dfrac{(2^x - 1)}{x} + \lim_{x\to 0} \dfrac{(3^x - 1)}{x} = \log 2 + \log 3 = \log 6$ Ans.

(d) $\displaystyle\lim_{x\to 0} \left(\dfrac{2\sin 2x + \cos 2x - 1}{x}\right) = \lim_{x\to 0} \dfrac{2\sin 2x - 2\sin^2 x}{x} = \lim_{x\to 0} \left(\dfrac{2\sin 2x}{x} - \dfrac{2\sin^2 x}{x}\right) = \lim_{x\to 0} 4\left(\dfrac{\sin 2x}{2x}\right) - \lim_{x\to 0} \dfrac{2\sin^2 x}{x} = 4 - 0 = 4$ Ans.

(e) $\displaystyle\lim_{x\to \frac{\pi}{4}} \dfrac{1 - \tan x}{1 - \sqrt{2}\sin x} = 2$ Ans.   (f) $\displaystyle\lim_{x\to 2} \dfrac{\log(x - 1)}{x - 2} = 1$ Ans. $\left(\text{hint: } -\text{use formula } \lim_{x\to 0} \dfrac{\log(1 + x)}{x} = 1\right)$

(g) Ans: -1 (h) Ans: $-\left[1 + \dfrac{\pi}{2}\right]$

(i) $\displaystyle\lim_{x\to 0} \dfrac{5^x - 3^x}{4^x - 2^x} = \lim_{x\to 0} \dfrac{(5^x - 1) - (3^x - 1)}{(4^x - 1) - (2^x - 1)} = \lim_{x\to 0} \dfrac{\frac{(5^x - 1) - (3^x - 1)}{x}}{\frac{(4^x - 1) - (2^x - 1)}{x}} = \lim_{x\to 0} \dfrac{\left[\frac{5^x - 1}{x} - \frac{3^x - 1}{x}\right]}{\left[\frac{4^x - 1}{x} - \frac{2^x - 1}{x}\right]},$

use formula $\displaystyle\lim_{x\to 0} \dfrac{a^x - 1}{x} = \log a$ then limit $= \dfrac{\log 5 - \log 3}{\log 4 - \log 2} = \dfrac{\log \frac{5}{3}}{\log \frac{4}{2}} = \dfrac{\log \frac{5}{3}}{\log 2} = \log_2\left(\dfrac{5}{3}\right)$ Ans.

Note: $-$ Direct formula, $\displaystyle\lim_{x\to 0} \dfrac{a^x - b^x}{c^x - d^x} = \dfrac{\log\left(\frac{a}{b}\right)}{\log\left(\frac{c}{d}\right)}$

(j) $\displaystyle\lim_{x\to 2} \dfrac{e^{\sqrt{x}} - e^{\sqrt{2}}}{x - 2} = \dfrac{1}{2\sqrt{2}} e^{\sqrt{2}}$ Ans. (use L'Hospital rule)

(k) $\displaystyle\lim_{x\to 0} \dfrac{(x.3^x - x)}{\frac{1 - \cos 2x}{2}} = \lim_{x\to 0} \dfrac{x(3^x - 1)}{\sin^2 x} = \lim_{x\to 0} \dfrac{\frac{(3^x - 1)}{x}}{\left(\frac{\sin x}{x}\right)^2} = \log 3$ Ans.

$$\left(\lim_{x\to 0}\frac{a^x-1}{x}=\log a \ , \quad \frac{1-\cos 2x}{2}=\frac{1-(\cos^2 x-\sin^2 x)}{2}=\frac{1-\cos^2 x+\sin^2 x}{2}=\frac{2\sin^2 x}{2}=\sin^2 x\right)$$

(5) (a) $\lim_{x\to 0}\dfrac{27^x+9^x+3^x-3}{x}=\lim_{x\to 0}\left\{\dfrac{(27^x-1)}{x}+\dfrac{(9^x-1)}{x}+\dfrac{(3^x-1)}{x}\right\}=\lim_{x\to 0}\dfrac{(27^x-1)}{x}+\lim_{x\to 0}\dfrac{(9^x-1)}{x}+\lim_{x\to 0}\dfrac{(3^x-1)}{x}$

$\qquad =\log 27+\log 9+\log 3=\log 3^3+\log 3^2+\log 3=3\log 3+2\log 3+\log 3=6\log 3=\log 3^6=\log 729$ Ans.

(b) $\lim_{x\to 0}\dfrac{81^x-27^x+9^x-3^x}{\left(\sqrt{24+\cos x}-3\right)\sin x}=\log 3$ Ans.

(c) $\lim_{x\to\frac{\pi}{4}}\dfrac{\sqrt{2}\left(x-\frac{\pi}{4}\right)}{(\sin x-\cos x)}=\lim_{x\to\frac{\pi}{4}}\dfrac{\sqrt{2}\left(x-\frac{\pi}{4}\right)}{\sqrt{2}\left(\sin x.\cos\frac{\pi}{4}-\cos x.\sin\frac{\pi}{4}\right)}=\lim_{x\to\frac{\pi}{4}}\dfrac{\left(x-\frac{\pi}{4}\right)}{\sin\left(x-\frac{\pi}{4}\right)}=\lim_{x-\frac{\pi}{4}\to 0}\dfrac{1}{\dfrac{\sin\left(x-\frac{\pi}{4}\right)}{\left(x-\frac{\pi}{4}\right)}}=1$ Ans.

(d) $\lim_{x\to 0}\dfrac{x}{\sin x+\tan x}=\lim_{x\to 0}\dfrac{1}{\dfrac{\sin x+\tan x}{x}}=\lim_{x\to 0}\dfrac{1}{\left(\dfrac{\sin x}{x}+\dfrac{\tan x}{x}\right)}=\dfrac{1}{1+1}=\dfrac{1}{2}$ Ans.   (e) $\lim_{x\to 1}\dfrac{\sqrt{x+2}-\sqrt{2x+1}}{x-1}=-\dfrac{1}{2\sqrt{3}}$ Ans.

(f) $\lim_{x\to 0}(1+\sin x)^{\frac{1}{\sin x}}=y$ (say)

$\Rightarrow \log y=\lim_{x\to 0}\log(1+\sin x)^{\frac{1}{\sin x}}=\lim_{x\to 0}\dfrac{\log(1+\sin x)}{\sin x}=\lim_{\sin x\to 0}\dfrac{\log(1+\sin x)}{\sin x}$

use formula, $\lim_{x\to 0}\dfrac{\log(1+x)}{x}=1$ $\therefore \log y=\lim_{\sin x\to 0}\dfrac{\log(1+\sin x)}{\sin x}=1=\log_e e$ $\therefore y=e$

$$\therefore \lim_{x\to 0}(1+\sin x)^{\frac{1}{\sin x}}=y=e$$ Ans.

(g) $\lim_{x\to 2}\dfrac{x-2}{2x^2-5x+2}=\dfrac{1}{3}$ Ans.   (h) $\lim_{x\to 0}\dfrac{\sqrt{1+\tan x}-\sqrt{1-\sin x}}{x}=1$ Ans.

(i) $\lim_{x\to\frac{\pi}{2}}\dfrac{\sqrt{1-\sin x}}{\cos x}=\lim_{x\to\frac{\pi}{2}}\dfrac{\sqrt{1-\sin x}}{\cos x}\times\dfrac{\sqrt{1+\sin x}}{\sqrt{1+\sin x}}=\lim_{x\to\frac{\pi}{2}}\dfrac{\sqrt{(1-\sin x)(1+\sin x)}}{\cos x\sqrt{1+\sin x}}=\lim_{x\to\frac{\pi}{2}}\dfrac{\sqrt{\cos^2 x}}{\cos x\sqrt{1+\sin x}}=\lim_{x\to\frac{\pi}{2}}\dfrac{\cos x}{\cos x\sqrt{1+\sin x}}=\lim_{x\to\frac{\pi}{2}}\dfrac{1}{\sqrt{1+\sin x}}$

$\qquad =\dfrac{1}{\sqrt{2}}$ Ans.

IInd method :$-$ $\lim_{x\to\frac{\pi}{2}}\dfrac{\sqrt{1-\sin x}}{\cos x}=\lim_{x\to\frac{\pi}{2}}\dfrac{\sqrt{1-\sin x}}{\sqrt{1-\sin^2 x}}=\lim_{x\to\frac{\pi}{2}}\dfrac{\sqrt{1-\sin x}}{\sqrt{(1-\sin x)(1+\sin x)}}=\lim_{x\to\frac{\pi}{2}}\dfrac{1}{\sqrt{(1+\sin x)}}=\dfrac{1}{\sqrt{2}}$ Ans.

(j) $\lim_{x\to 2}\dfrac{(1+x)^3-27}{x-2}=27$ Ans. (without use L'Hospital rule)   (k) $\lim_{x\to 3}\dfrac{\left(1+\sqrt{1+x}\right)^2-9}{x-3}=\dfrac{3}{2}$ Ans.

(6) (a) $\lim_{x\to\frac{\pi}{4}}\dfrac{1-\sin 2x}{x-\frac{\pi}{4}}=0$ Ans.   (b) $\lim_{x\to a}\dfrac{x^{\frac{1}{3}}-a^{\frac{1}{3}}}{x^{\frac{1}{2}}-a^{\frac{1}{2}}}=\lim_{x\to a}\dfrac{\dfrac{x^{\frac{1}{3}}-a^{\frac{1}{3}}}{x-a}}{\dfrac{x^{\frac{1}{2}}-a^{\frac{1}{2}}}{x-a}}=\dfrac{\frac{1}{3}a^{\frac{1}{3}-1}}{\frac{1}{2}a^{\frac{1}{2}-1}}=\dfrac{2}{3}a^{-\frac{1}{6}}=\dfrac{2}{3a^{\frac{1}{6}}}$ Ans. $\left(\text{use formula},\lim_{x\to a}\dfrac{x^n-a^n}{x-a}=na^{n-1}\right)$

(c) $\lim_{x\to\frac{\pi}{2}}\dfrac{\log(\cos x+1)}{\sin 2x}=\dfrac{1}{2}$ Ans.   (solve without L'Hospital rule)   (d) $\lim_{x\to 0}\dfrac{1+\sin x-\cos x}{\cos x-\sin x-1}=-1$ Ans.

(e) $\lim_{x\to 0}\dfrac{\sqrt{9+\tan 2x}-3}{\log(1+\sin 2x)}=\dfrac{1}{3}$ Ans. (without use L'Hospital rule)   (f) $\lim_{x\to 1}\dfrac{\sqrt[4]{15+x}-2}{3(x-1)}=\dfrac{1}{12}$ Ans.

(g) $\lim_{x\to 0}\dfrac{\sqrt[3]{1+x}+\sqrt[4]{1+x}-2}{3^x+2^x-2}=\dfrac{7}{12\log 6}$ Ans. $\left(\text{formula},\lim_{x\to 0}\dfrac{(1+x)^n-1}{x}=n,\ \lim_{x\to 0}\dfrac{a^x-1}{x}=\log a\ ,\log m+\log n=\log m.n\right)$

(h) $\lim_{x\to\frac{\pi}{2}}\dfrac{\sqrt{1+\cos x}-\sqrt{1-\cos x}}{\cos x}=1$ Ans.   (i) $\lim_{x\to a}\dfrac{\sqrt{x+3a}-2.\sqrt{a}}{3.\sqrt{x+a}-3.\sqrt{2a}}=\dfrac{1}{3\sqrt{2}}$ $(a\neq 0)$

(j) $\lim\limits_{x\to 0}\dfrac{x\sin x+\log(1+x^2)}{x^2}=2$ Ans. (solve the question without use L'Hospital rule)

(7) (a) $\lim\limits_{x\to\infty}\dfrac{\sqrt{x^2+1}+\sqrt[4]{x^4+1}}{\sqrt[3]{x^3+1}+\sqrt[5]{x^5+1}}=1$ Ans.   (b) $\lim\limits_{x\to\infty}\dfrac{\sqrt[3]{x^2+1}-\sqrt[4]{x^4+1}}{\sqrt[3]{x^3+1}-\sqrt[5]{x^4+1}}=-1$ Ans.

(c) $\lim\limits_{x\to\infty}\dfrac{x+\cos x}{x-\sin^2 x}=1$ Ans.   (d) $\lim\limits_{x\to 3}\dfrac{x^2-4x+3}{x^2-5x+6}=2$ Ans.   (e) $\lim\limits_{x\to 1}\dfrac{(3x-1)(\sqrt{x^2}-1)}{(3x^3+x^2-3x-1)}=\lim\limits_{x\to 1}\dfrac{(3x-1)(x-1)}{(x-1)(x+1)(3x+1)}=\dfrac{1}{4}$ Ans.

(f) $\lim\limits_{x\to 0}\dfrac{(x.3^x-x)}{\sin^2 x}=\log 3$ Ans.   (g) $\lim\limits_{x\to\pi}\dfrac{(x-\pi)\sin x}{1+\cos x}=-2$ Ans.

(h) $\lim\limits_{x\to 0}\dfrac{\tan 2x.(1-\tan^2 x)}{x+\sin x}=\lim\limits_{x\to 0}\dfrac{2\tan x.(1-\tan^2 x)}{(1-\tan^2 x)(x+\sin x)}=\lim\limits_{x\to 0}\dfrac{2\tan x}{(x+\sin x)}=\lim\limits_{x\to 0}\dfrac{2\tan x}{x\left(1+\dfrac{\sin x}{x}\right)}=\lim\limits_{x\to 0}\left(\dfrac{\tan x}{x}\right)\dfrac{2}{\left(1+\dfrac{\sin x}{x}\right)}=\dfrac{2}{1+1}=\dfrac{2}{2}$

$=1$ Ans.

(i) $\lim\limits_{x\to 0}\dfrac{x\cos x+\tan x}{\tan x}=2$ Ans.   (j) $\lim\limits_{x\to 0}\dfrac{2x-\sin x}{3x-\tan x}=\dfrac{1}{2}$ Ans.   (k) $\lim\limits_{x\to 0}\dfrac{\tan x+x}{x^2}=\infty$ Ans.

(8) (a) $\lim\limits_{x\to 0}\dfrac{\cot(\pi x)}{\dfrac{\pi}{2x}}=\dfrac{2}{\pi^2}$ Ans.

(b) $\lim\limits_{\theta\to\pi}\dfrac{\sqrt{1+\cos\theta}-\sin\theta}{\cos\left(\dfrac{\theta}{2}\right)}=\lim\limits_{\theta\to\pi}\dfrac{\sqrt{2\cos^2\dfrac{\theta}{2}}-2\sin\dfrac{\theta}{2}\cos\dfrac{\theta}{2}}{\cos\left(\dfrac{\theta}{2}\right)}=\lim\limits_{\theta\to\pi}\dfrac{\cos\dfrac{\theta}{2}\left(\sqrt{2}-2\sin\dfrac{\theta}{2}\right)}{\cos\left(\dfrac{\theta}{2}\right)}=\lim\limits_{\theta\to\pi}\left(\sqrt{2}-2\sin\dfrac{\theta}{2}\right)=(\sqrt{2}-2)$ Ans.

(c) $\lim\limits_{\theta\to\frac{\pi}{2}}\dfrac{(\sin\theta)^{\sin\theta}-1}{1-\sin\theta}=-1$ Ans. $\left(\text{hint:}-\text{ put }\sin\theta=t,\ \theta\to\dfrac{\pi}{2}\text{ then }t\to 1\text{ and use L'Hospital rule}\right)$

(d) If $f'(a)$ is exists then show that $\lim\limits_{x\to a}\dfrac{(2x-a)f(x)-af(2x-a)}{x-a}=2f(a)-af'(a)$.

L.H.S $=\lim\limits_{x\to a}\dfrac{(2x-a)f(x)-af(2x-a)}{x-a}=\lim\limits_{x\to a}\dfrac{(2x-a)f'(x)+f(x).2-af'(2x-a).2}{1}=af'(x)+2f(a)-2af'(a)$

$=2f(a)-af'(a)$ proved. (use L'Hospital rule)

(e) Do yourself. (same as above question, use L'Hospital rule)

(9) (a) $F(x)=\sqrt{24-x^3}$, $F'(x)=\dfrac{1}{2.\sqrt{24-x^3}}\times -3x^2$, $\lim\limits_{x\to 2}\dfrac{F(x)-F(2)}{x-2}$

use L'Hospital rule, $\lim\limits_{x\to 2}\dfrac{F(x)-F(2)}{x-2}=\lim\limits_{x\to 2}\dfrac{F'(x)}{1}=F'(2)=\dfrac{1}{2.\sqrt{24-8}}\times -3(2)^2=\dfrac{-12}{8}=-\dfrac{3}{2}$ Ans.

(b) $F(5)=5,F'(5)=3$ then find $\lim\limits_{x\to 5}\dfrac{\sqrt{F(x)+3}-\sqrt{8}}{\sqrt{x}-\sqrt{5}}=\lim\limits_{x\to 5}\dfrac{\left(\sqrt{F(x)+3}-\sqrt{8}\right)\left(\sqrt{F(x)+3}+\sqrt{8}\right)(\sqrt{x}+\sqrt{5})}{(\sqrt{x}-\sqrt{5})(\sqrt{x}+\sqrt{5})\left(\sqrt{F(x)+3}+\sqrt{8}\right)}$

$=\lim\limits_{x\to 5}\dfrac{[F(x)+3-8](\sqrt{x}+\sqrt{5})}{(x-5)\left(\sqrt{F(x)+3}+\sqrt{8}\right)}=\lim\limits_{x\to 5}\dfrac{F(x)-5}{x-5}\times\lim\limits_{x\to 5}\dfrac{\sqrt{x}+\sqrt{5}}{\sqrt{F(x)+3}+\sqrt{8}}=\dfrac{2\sqrt{5}}{2\sqrt{8}}\times\lim\limits_{x\to 5}F'(x)=F'(5).\dfrac{\sqrt{5}}{\sqrt{8}}=\dfrac{3\sqrt{5}}{2\sqrt{2}}$

$=\dfrac{3\sqrt{5}}{2\sqrt{2}}\times\dfrac{\sqrt{2}}{\sqrt{2}}=\dfrac{3}{4}\sqrt{10}$ Ans.

(10) (a) $\lim\limits_{x\to 0}\dfrac{\tan x-x-\dfrac{x^3}{3}}{x^5}=\lim\limits_{x\to 0}\dfrac{\left[x+\dfrac{1}{3}x^3+\dfrac{2}{15}x^5+\cdots\ldots\ldots\right]-x-\dfrac{x^3}{3}}{x^5}=\lim\limits_{x\to 0}\dfrac{\dfrac{2}{15}x^5}{x^5}=\dfrac{2}{15}$ Ans.

(b) $\lim\limits_{x\to\frac{\pi}{2}}\sec x.\log_e(\text{cosec}x)=\lim\limits_{x\to\frac{\pi}{2}}\dfrac{\log_e(\text{cosec}x)}{\cos x}=\lim\limits_{x\to\frac{\pi}{2}}\dfrac{-\cot x.\text{cosec}x}{\text{cosec}x.-\sin x}=\lim\limits_{x\to\frac{\pi}{2}}\dfrac{\cot x}{\sin x}=0$ Ans.

(c) $\lim\limits_{x \to 1} \dfrac{(1+x)-2}{(1-x)} = -1$ Ans. (Do yourself) (d) $\lim\limits_{x \to 0} \dfrac{\log \sin\left(\frac{\pi}{2}+ax\right)}{\tan bx} = 0$ (use L'Hospital rule) (e) $\lim\limits_{x \to 0} \dfrac{\frac{\pi}{2x}}{\operatorname{cosec}\left(\frac{\pi}{2}-x\right)} = \dfrac{\pi^2}{4}$ Ans.

(11) (a) $\lim\limits_{x \to 0} \dfrac{a\tan x + x(b+\cos x)}{x^3} = 1$

$\Rightarrow \lim\limits_{x \to 0} \dfrac{a\left(x+\frac{1}{3}x^3+\frac{2}{15}x^5+\cdots\ldots\right)+x\left(b+1-\frac{x^2}{2!}+\frac{x^4}{4!}-\cdots\ldots\right)}{x^3} = 1$

$\Rightarrow \lim\limits_{x \to 0} ax + \frac{1}{3}ax^3 + \frac{2}{15}ax^5 + \cdots \ldots \ldots \ldots \ldots \ldots + bx + x - \frac{x^3}{2!} + \frac{x^5}{4!} - \cdots \ldots \ldots \ldots = x^3$

$\Rightarrow x(a+b+1) + x^3\left(\frac{1}{3}a - \frac{1}{2}\right) + \cdots \ldots \ldots \ldots \ldots \ldots = x^3 + 0.x$

$\therefore\ a+b+1 = 0$ or $\frac{1}{3}a - \frac{1}{2} = 1$ $\therefore\ \frac{1}{3}a = 1 + \frac{1}{2} = \frac{3}{2}$ $\therefore a = \frac{9}{2}$

$\therefore\ a+b+1 = 0$ $\therefore\ b = -1 - a = -1 - \frac{9}{2} = -\frac{11}{2}$ $\therefore\ a = \frac{9}{2}, b = -\frac{11}{2}$ Ans.

(b) $\lim\limits_{x \to 0} \dfrac{axe^x - b\log(1+x) + cxe^{-x}}{x^2 \sin x} = 2$, $a = 3, b = 12$ and $c = 9$ Ans. (same as above question)

(c) $\lim\limits_{x \to 0}\left(\dfrac{1-x}{1+x} + ax - b\right) = 0$, then prove that $a = 0, b = -1$ $\therefore\ \lim\limits_{x \to 0}\left(\dfrac{1-x}{1+x} + ax - b\right) = 0$ or $\lim\limits_{x \to 0} \dfrac{1-x+ax+ax^2-b-bx}{1+x} = 0$

or $\lim\limits_{x \to 0} ax^2 + x(a-b-1) + (1-b) = 0.x^2 + 0.x$ $\Rightarrow a = 0$ or $a - b - 1 = 0$ $\therefore b = -1$

$\boxed{\therefore\ a = 0 \text{ and } b = -1 \quad \text{proved.}}$

(d) $\lim\limits_{x \to 0}\left(\dfrac{ax^2}{x+1} - bx + 1\right) = 1$, then prove that $a = 0, b = 0$. (solve same as above question)

(12) (a) $\lim\limits_{x \to 0} \dfrac{\cos 2x + a\cos x}{x^2} = $ finite, $a = -6$, limit $= \infty$ Ans. (b) $\lim\limits_{x \to 0} \dfrac{a\tan x + b\sin x - 2}{x^3} = $ finite, $a = 6, b = -6$ and limit $= \infty$ Ans.

(c) $\lim\limits_{x \to 0} \dfrac{ae^x + b\log(1+x) - ce^{-x}}{x\tan x} = $ finite, $a = 1, b = -\dfrac{6}{5}$ and $c = \dfrac{1}{5}$ Ans.

(d) $\lim\limits_{x \to 0} \dfrac{ax\cos x + b\log(1+x) - cxe^x}{x^2 \sin x} = 2$, $a = -\dfrac{34}{15}, b = -\dfrac{8}{5}$ and $c = -\dfrac{2}{3}$ Ans.

(13) (a) $\lim\limits_{x \to 0}\left\{\sin\left(\dfrac{\pi}{2}+x\right)\right\}^{\frac{1}{x}} = \lim\limits_{x \to 0} (\cos x)^{\frac{1}{x}} = y$ (say)

$\Rightarrow \log y = \lim\limits_{x \to 0} \dfrac{1}{x}.\log(\cos x)$, use L'Hospital rule. or $\log y = \lim\limits_{x \to 0} \dfrac{-\sin x}{\cos x} = 0 = \log 1$ $\therefore\ y = 1$ Ans.

(b) $\lim\limits_{x \to \infty}\left(\dfrac{x+3}{x+2}\right)^{x+4} = \lim\limits_{x \to \infty}\left(1 + \dfrac{1}{x+2}\right)^{(x+2)+2}$, $\left(\text{use formula,}\quad \lim\limits_{x \to \infty}\left(1 + \dfrac{p}{x}\right)^x = e^p\right)$

$\Rightarrow \lim\limits_{x \to \infty}\left(1 + \dfrac{1}{x+2}\right)^{(x+2)} . \lim\limits_{x \to \infty}\left(1 + \dfrac{1}{x+2}\right)^2 = e^1.1 = e$ proved.

(c) $\lim\limits_{x \to \infty}\left(\dfrac{x+1}{x-2}\right)^{x+3} = e^3$ $\left(\text{Hint:}- \lim\limits_{x \to \infty}\left(1 + \dfrac{3}{x-2}\right)^{(x-2)+5} = \lim\limits_{x \to \infty}\left(1 + \dfrac{3}{x-2}\right)^{(x-2)} . \lim\limits_{x \to \infty}\left(1 + \dfrac{3}{x-2}\right)^5 = e^3\right)$

(d) $\lim\limits_{x \to 0}\left(\dfrac{1+2x^2}{1+3x^2}\right)^{\frac{1}{x^2}} = \lim\limits_{x \to 0} \dfrac{(1+2x^2)^{\frac{1}{x^2}}}{(1+3x^2)^{\frac{1}{x^2}}} = \dfrac{e^2}{e^3} = e^{2-3} = e^{-1} = \dfrac{1}{e}$ proved. $\left(\text{use formula,}\ \lim\limits_{x \to 0}(1+px)^{\frac{1}{x}} = e^p\right)$

IInd method: − Let $y = \lim\limits_{x\to 0}\left(\dfrac{1+2x^2}{1+3x^2}\right)^{\frac{1}{x^2}}$ $\Rightarrow \log y = \lim\limits_{x\to 0}\dfrac{1}{x^2}\cdot\left(\dfrac{1+2x^2}{1+3x^2}-1\right) = \lim\limits_{x\to 0}\dfrac{1}{x^2}\cdot\left(\dfrac{-x^2}{1+3x^2}\right) = -1 = \log e^{-1}$

$$\text{or}\quad \log y = \log e^{-1}\quad \therefore\ y = e^{-1} = \frac{1}{e}\ \text{ proved.}$$

(e) $\lim\limits_{x\to 0}\left\{\dfrac{2}{3}\left(\dfrac{3+x}{2+x}\right)\right\}^{\frac{1}{x}} = \lim\limits_{x\to 0}\left\{\dfrac{2}{3}\cdot\dfrac{3}{2}\left(\dfrac{1+\frac{x}{3}}{1+\frac{x}{2}}\right)\right\}^{\frac{1}{x}} = \lim\limits_{x\to 0}\left(\dfrac{1+\frac{x}{3}}{1+\frac{x}{2}}\right)^{\frac{1}{x}} = \lim\limits_{x\to 0}\dfrac{\left(1+\frac{x}{3}\right)^{\frac{1}{x}}}{\left(1+\frac{x}{2}\right)^{\frac{1}{x}}}$ $\left(\text{formula}\ \lim\limits_{x\to 0}\left(1+\dfrac{x}{p}\right)^{\frac{1}{x}}= e^{\frac{1}{p}}\right)$

$$\therefore\ \lim\limits_{x\to 0}\dfrac{\left(1+\frac{x}{3}\right)^{\frac{1}{x}}}{\left(1+\frac{x}{2}\right)^{\frac{1}{x}}} = \dfrac{e^{\frac{1}{3}}}{e^{\frac{1}{2}}} = e^{\frac{1}{3}-\frac{1}{2}} = e^{\frac{2-3}{6}} = e^{-\frac{1}{6}}\ \text{or}\ \dfrac{1}{e^{\frac{1}{6}}}\quad \text{proved.}$$

(f) $\lim\limits_{x\to 0}\dfrac{\log(1+x)}{x} - 2 = 1 - 2 = -1$ Ans. (g) $\lim\limits_{x\to\infty}\left(\dfrac{x-5}{x+3}\right)^x = e^{-8}$ or $\dfrac{1}{e^8}$ Ans. $\left(\text{use formula},\ \lim\limits_{x\to\infty}\left(1+\dfrac{p}{x}\right)^x = e^p\right)$

(h) $\lim\limits_{x\to 0}\left(\dfrac{\sin 2x}{2x}\right)^{\frac{1}{x^2}} = e^{-\frac{2}{3}}$ Ans. (i) e^3 (j) 1 (k) 1 (l) ∞ (m) ∞

(14) Ans: − (a) $\dfrac{2}{3}$ (b) 3 (c) 1 (d) $\sqrt{\dfrac{2}{3}}$ (e) 1 (f) $\dfrac{\sqrt{3}}{2}$ (15) Ans: − (a) 1 (b) 2 (c) $\dfrac{5}{6}$ (d) $\log\left(\dfrac{3}{5}\right)$ (e) $\dfrac{3}{2}$ (f) 1

(16) Ans: − (a) $\dfrac{1}{2}$ (b) 0 (c) $\log\left(\dfrac{1}{3}\right)$ (d) $\dfrac{1}{2\sqrt{2}}$ (e) 1 (f) 2 (g) 0 (h) $-\dfrac{3}{16}$

(17) Ans: − (a) $-4\sqrt{2}$ (b) $-\dfrac{9}{2}$ (c) $\dfrac{1-\sqrt{3}}{2}$ (d) $-\dfrac{1}{2}$ (e) 0 (f) 2 (g) 0

(18) Ans: − (a) limit does not exist. (b) 1 (c) limit does not exist. (d) limit does not exist. (e) 1 (f) 2 (g) ∞ (h) 0

(19) Ans: − (a) $\dfrac{1}{2}$ (b) $\dfrac{1}{3}$ (c) ∞ (d) 1 (e) 1 (without use L'Hospital rule) (f) e^{-1} or $\dfrac{1}{e}$ (g) e^2 (h) $e^{\frac{1}{3}}$

(20) Ans: − (a) e^{-3} (b) -1 (c) $2e^{-\pi}$ (d) -10 (e) $\dfrac{1}{\log 4}$ (f) 1 (g) 0 (h) $\dfrac{\log(e+1)-1}{\log(2e+3)}$

(21) Ans: − (a) $\dfrac{\log\left(\frac{21}{5}\right)}{\log 6}$ or $\log_6\left(\dfrac{21}{5}\right)$ (b) 2a (c) 2 (d)

(22) Ans: − (a) 1 (b) 3a (c) $\dfrac{1}{2}$ (d) 0

(24) (a)   (b) $\theta = \pm\dfrac{\pi}{2}$ (c) $a = 2$, $L = \dfrac{1}{64}$

Exercise − A4

(1) (a) A function f is defined as $f(x) = \begin{cases} \dfrac{x^2-5x+6}{x^2-4} & \text{for } x\neq 2 \\[2mm] \dfrac{1}{4} & \text{for } x = 2 \end{cases}$ function $f(x)$ is continuous at $x = 2$.

(b) A function f is defined as $f(x) = \begin{cases} \dfrac{2x^2+x-3}{3x^2+5x-2}, & x\neq 1 \\[2mm] 0, & x = 1 \end{cases}$ show that $f(x)$ is differentiable at $x = 1$ and find its value.

(c) $f(x) = \begin{cases} \dfrac{1 - \cos 2x}{x^2}, & x < 0 \\ a, & x = 0 \\ \dfrac{\sqrt{x}}{\sqrt{25 + \sqrt{x}} - 5}, & x > 0 \end{cases}$ find the value of a so that the function may be continuous at $x = 0$.

(2) (a) Discuss the continuity and differentiability of the function $f(x) = \begin{cases} \dfrac{x}{2 + |x|}, & |x| \geq 2 \\ \dfrac{x}{2 - |x|}, & |x| < 2 \end{cases}$

(b) Discuss the continuity and differentiability of the function $f(x) = \begin{cases} 3 + \sqrt{4 - x^2}, & |x| \leq 2 \\ 3e^{(2-x)^2}, & |x| > 2 \end{cases}$

(c) Discuss the differentiability of $\tan\{\pi(x - [x])\}$ in $(0, \pi)$.

(3) (a) Show that the value of the derivative of $|x - 2| + |x - 4|$ at $x = 3$ is 0.

(b) The function $f(x) = (x^2 - 1)|x^2 + 5x + 6| + \sin|x|$ is not differentiable at (i) 0 (ii) -2 (iii) -3 (iv) 1 .

(c) If $f(x) = \begin{cases} 2x - 3, & 0 \leq x \leq 1 \\ 3x - a, & 1 < x \leq 2 \end{cases}$ and $f(x)$ be continuous at $x = 1$ and find the value of a .

(4) (a) Prove that $f(x) = |\log(x + 1)|$ is continuous at $x = 0$ but is not differentiable at $x = 0$.

(b) If $f(x) = \dfrac{2}{x + 1}$, then determine the points of discontinuity of $f\big(f(f(x))\big)$. (c) Find the points of discontinuity of $f(x) = \dfrac{1}{x + |x|}$.

(5) (a) $f(x) = \begin{cases} -\cos x, & -\dfrac{\pi}{2} \leq x < 0 \\ a\cos x + b, & 0 < x < \dfrac{\pi}{2} \\ 2\sin x, & \dfrac{\pi}{2} \leq x \leq \pi \end{cases}$ If $f(x)$ is continuous on $\left[0, \dfrac{\pi}{2}\right]$ then show that $a = -3$, $b = 2$.

(b) $f(x) = \begin{cases} \dfrac{9^x - 3^x - 5^x + 1}{x}, & x > 0 \\ e^x \cos x - \pi x + a\log 3, & x \leq 0 \end{cases}$ If $f(x)$ is continuous at $x = 0$. then show that $a = \dfrac{\log 3 - \log 5 - 1}{\log 3}$.

(c) Let $f(x) = \begin{cases} x^2 \left|\cos\left(\dfrac{\pi}{x}\right)\right|, & x \neq 0 \\ 0, & x = 0 \end{cases}$ $x \in R$, then f is ... ...?

(6) (a) If $f(x) = x\big[\sqrt{x + 1} - \sqrt{x + 3}\big]$, then $f(x)$ is continuous and differentiable at $x = 0$.

(b) Discuss the continuity and differentiability of $f(x) = \begin{cases} |x - 3|, & x \geq 2 \\ \dfrac{x^2}{3} - \dfrac{4x}{2} + \dfrac{9}{3}, & x < 2 \end{cases}$ at $x = 2, 3$.

(c) Discuss the continuity and differentiability of $f(x) = \begin{cases} 2x, & x < 2 \\ 3 - 2x, & 2 \leq x \leq 4 \\ -3 + 2x - x^2, & x > 4 \end{cases}$ at $x = 2$ and $x = 4$.

(7) (a) Let $f(x) = \begin{cases} 2xe^{\left(\frac{1}{|x|} - \frac{1}{x}\right)}, & x \neq 0 \\ 0, & x = 0 \end{cases}$ The function $f(x)$ is continuous and differentiable at $x = 0$.

(b) If $f(x) = \begin{cases} \left(x - \dfrac{\pi}{4}\right) \cdot \dfrac{1 - \tan x}{1 + \tan x}, & x \neq \dfrac{\pi}{4} \\ 0, & x = \dfrac{\pi}{4} \end{cases}$ then $f(x)$ is continuous and differentiable at $x = \dfrac{\pi}{4}$.

(c) If $f(x) = \begin{cases} ax(x + 1) - b, & x < -1 \\ 2x - 1, & -1 \leq x \leq 0 \\ 2cx^2 + 3dx - 2, & x > 0 \end{cases}$ then $f(x)$ is continuous at $x = -1$ and $x = 0$. Determine the constant a, b, c and d.

(8) (a) If $f(x) = \cos x$ and $g(x) = \sqrt{|x|} - 1$ then calculate $(fog)(x)$ and $(gof)(x)$. discuss the differentiability of $(gof)(x)$ at $x = 0$

and discuss the continuous of $(fog)(x)$ at $x = 0$.

(b) If $f(x) = -2 + |x + 2|, -2 \le x \le 0$ and $g(x) = 3 - |x - 2|, -3 \le x \le 3$. then calculate $(fog)(x)$ and $(gof)(x)$.

Discuss the continuity of $(fog)(x)$ at $x = -2$ and differentiability of $(gof)(x)$ at $x = 2$.

(c) Discuss the continuity and differentiability of the function $f(x) = x + |x - 3|$ at $x = 3$.

(9) (a) $f(x) = \dfrac{1 + \cos x}{\left(\dfrac{\pi}{2} - \dfrac{x}{2}\right)^2} \cdot \dfrac{\log(\cos x)}{\log(1 + 2\pi^2 - 3\pi x + x^2)}$, $x \ne \pi$. Determine $f(\pi)$, if $f(x)$ is continuous at $x = \pi$.

(b) $f(x) = \dfrac{1 + \cos x}{(\pi - x)^4} \cdot \sin x \, (4x^2 - \pi^2)$, $x \ne \pi$. Determine $f(\pi)$, if $f(x)$ is continuous at $x = \pi$.

(c) If $f(x) = 3x - \dfrac{x - 1}{|x + 3|}$ then determine the point of discontinuity of the function $f(x)$.

(10) (a) If $f(x) = \begin{cases} \dfrac{\log(1 + |x|)}{x}, & x \ne 0 \\ 0, & x = 0 \end{cases}$ then $f(x)$ is discontinuous at $x = 0$.

(b) Let $f(x) = \begin{cases} x^2|x|, & x \ne 0 \\ 0, & x = 0 \end{cases}$ Discuss the function $f(x)$ is continuity and differentiability at $x = 0$.

(c) Let $f(x) = \begin{cases} ae^{\cos x} - b, & x < \dfrac{\pi}{2} \\ ax + b, & x > \dfrac{\pi}{2} \\ 1, & x = \dfrac{\pi}{2} \end{cases}$ then the function $f(x)$ is continuous at $x = \dfrac{\pi}{2}$ find the value of a and b.

(11) (a) Discuss the continuity and differentiability of $f(x) = \begin{cases} \sec x - \tan x, & 0 \le x \le \dfrac{\pi}{4} \\ \sqrt{2} \cot x - 1, & \dfrac{\pi}{4} < x \le \dfrac{\pi}{2} \end{cases}$ at $x = \dfrac{\pi}{4}$.

(b) If $f(x) = \begin{cases} e^{|1+x|} + 2, & x \ne -1 \\ 2, & x = -1 \end{cases}$ then the function $f(x)$ is discontinuous at $x = -1$.

(c) If $f(x) = \begin{cases} \log(ax + b) + 1, & -1 \le x < 2 \\ \log x, & 2 < x \le 3 \\ -1, & x = 2 \end{cases}$ then the function $f(x)$ is continuous at $x = 2$.

and the function $f(x)$ is not differentiable at $x = 2$.

(12) (a) If $f(x) = \begin{cases} \dfrac{1 + e^{-x}}{1 - e^x}, & x \ne 0 \\ 0, & x = 0 \end{cases}$ prove that $f(x)$ is not differentiable at $x = 0$.

(b) If $f(x) = \begin{cases} \dfrac{|x|}{x}, & x \ne 0 \\ 0, & x = 0 \end{cases}$ Discuss its continuity at $x = 0$.

(c) Discuss the continuity and differentiable of the following function $f(x) = \begin{cases} x^3, & x < -3 \\ 9, & -3 \le x \le 3 \\ x^3, & x > 3 \end{cases}$

(13) (a) which of the following function is not differentiable at $x = 0$, $f(x) = 2 \sin|x| - |x|$.

(b) If $f(x) = \cos|x| - |x|$, Discuss its continuity at $x = 0$.

(c) Discuss the continuity and differentiability at $x = 2$, if $f(x) = \begin{cases} 2^x, & -2 \le x \le 2 \\ 6 - x, & 2 < x < 5 \end{cases}$.

(14) (a) Discuss the continuity and differentiability at $x = \dfrac{\pi}{4}$ if $f(x) = \begin{cases} e^{a\tan x} - 1, & 0 < x \le \dfrac{\pi}{4} \\ \log_2(\sin x), & \dfrac{\pi}{4} < x < \dfrac{\pi}{2} \end{cases}$ find the value of a.

(b) Let $f(x) = \begin{cases} x^2 + ax + b, & 0 < x < 1 \\ 2ax - 3, & 1 \le x \le 2 \\ ax + b, & 2 < x < 3 \end{cases}$ then f(x) is continuous at x = 1,2 and differentiable at x = 1,2 also find the value of a and b.

(c) Discuss the continuity and differentiable at x = 0, if $f(x) = \begin{cases} \dfrac{2}{x^2 - 2x + 1}, & x \ne 0 \\ 2, & x = 0 \end{cases}$

(15) (a) If $f(x) = \dfrac{(5^x - 1)^2}{\cos x \cdot [\log(1 + x)]^2}$, $x \ne 0$ is continuous at x = 0 then find f(0).

(b) If the derivative of the function $f(x) = \begin{cases} 2ax^2 + 4b, & x < -2 \\ bx^2 + 2ax + 3, & x \ge -2 \end{cases}$ is everywhere continuous. then show that $a = \dfrac{1}{4}$ and $b = \dfrac{5}{8}$.

(c) Examine the continuity and differentiability in $-\infty < x < \infty$ of the following function $f(x) = \begin{cases} 2, & -\infty < x < 0 \\ 1 + \cos x, & 0 \le x \le \dfrac{\pi}{2} \\ 1 + \left(x - \dfrac{\pi}{2}\right), & \dfrac{\pi}{2} \le x < \infty \end{cases}$

(16) Test the following functions for continuity: − (a) $f(x) = x \tan\left(\dfrac{1}{x}\right)$, $x \ne 0$. $f(0) = 0$ at x = 0

(b) $f(x) = \dfrac{e^{-\frac{1}{x}}}{1 + e^{\frac{1}{x}}}$, $x \ne 0$. $f(0) = 0$ at x = 0   (c) $f(x) = \dfrac{|x|}{x}$, $x \ne 0$. $f(0) = 0$ at x = 0

(17) (a) If $f(x) = \begin{cases} \dfrac{\cot x + \tan x}{\sec x}, & -\dfrac{\pi}{4} < x \le \dfrac{\pi}{4} \\ \csc x, & \dfrac{\pi}{4} < x \le \dfrac{\pi}{2} \\ \sqrt{2}\sin x + \cos x, & \dfrac{\pi}{2} < x < \pi \end{cases}$ then the function f(x) is continuous at $x = \dfrac{\pi}{4}$ and $x = \dfrac{\pi}{2}$.

(b) If $f(x) = \begin{cases} e^{|x|}, & x \ne 0 \\ 1, & x = 0 \end{cases}$ then the function f(x) is continuous at x = 0 and not differentiable at x = 0.

(c) If $f(x) = \begin{cases} e^{-\frac{1}{|x|}}, & x \ne 0 \\ 0, & x = 0 \end{cases}$ then the function f(x) is discontinuity at x = 0 and not differentiable at x = 0

(18) (a) If $f(x) = \begin{cases} \dfrac{2^x}{1 + 2^{-x}}, & -1 < x \le 1 \\ \dfrac{x + 3}{x + 2}, & x > 1 \end{cases}$ then the function f(x) is continuous at x = 1.

(b) If $f(x) = \begin{cases} \log(\sin x + \cos x), & 0 < x \le \dfrac{\pi}{4} \\ \dfrac{a}{2}, & x > \dfrac{\pi}{4} \end{cases}$ then the function f(x) is continuous at $x = \dfrac{\pi}{4}$ and prove that $a = 2 + \log 2$.

(c) If $f(x) = \begin{cases} \dfrac{e^{|x+1|}}{1 + e^{x+1}}, & x \ne -1 \\ \dfrac{1}{2}, & x = -1 \end{cases}$ then the function f(x) is discontinuous at x = 0 and not differentiable at x = 0.

(19) (a) If $f(x) = \begin{cases} \dfrac{\sin x}{|x|}, & x \ne 0 \\ 1, & x = 0 \end{cases}$ then the function f(x) is discontinuous at x = 0 and not differentiable at x = 0.

(c) Discuss the continuity and differentiability at $x = \pi$ if $f(x) = \begin{cases} \dfrac{\sqrt{1 + x - \pi} - 1}{|x - \pi|}, & x \ne \pi \\ 0, & x = \pi \end{cases}$

(20) (a) Discuss the continuity and differentiability of the following function $f(x) = \begin{cases} \sqrt{1 + \sin x}, & -\dfrac{\pi}{2} < x \le 0 \\ \sqrt{1 + \cos 2x}, & 0 < x \le \dfrac{\pi}{4} \\ \dfrac{\sec x + \csc x}{2}, & \dfrac{\pi}{4} < x < \dfrac{\pi}{2} \end{cases}$

(b) If $f(x) = \begin{cases} \dfrac{x^2 + 3x + 2}{x + 1}, & -2 < x \le -1 \\ ae^{-x}, & -1 < x < 0 \end{cases}$ then the function $f(x)$ is continuous at $x = -1$ and its value of a.

(c) If $f(x) = \begin{cases} \dfrac{e^{\sin x} - 1}{\sin x}, & -1 < x \le 0 \\ x + 1, & 0 < x < 1 \end{cases}$ Discuss the function $f(x)$ is continuity and differentiability at $x = 0$.

(21) (a) If $f(x) = \begin{cases} \sin x, & x < \dfrac{\pi}{2} \\ ax - b, & x \ge \dfrac{\pi}{2} \end{cases}$ then the function $f(x)$ is continuous and differentiable at $x = \dfrac{\pi}{2}$. find the value of a and b.

(b) Discuss the continuity and differentiability at $x = 0$, if $f(x) = \begin{cases} 2\cos x - 1, & x < 0 \\ \dfrac{x}{\sin x \cos x}, & x \ge 0 \end{cases}$

(c) Discuss the continuity and differentiability at $x = -2$ and $x = -3$, if $f(x) = |x^2 + 5x + 6|$

(22) (a) If $f(x) = \begin{cases} \dfrac{(3^{x-1} - 5^{x-1}).\log x. \cos x}{(x - 1)^2}, & x \ne 1 \\ ax + \log 3, & x = 1 \end{cases}$ then the function $f(x)$ is continuous at $x = 1$ and prove that $a = \log\left(\dfrac{1}{5}\right)$.

(b) Discuss the continuity and differentiability at $x = \dfrac{\pi}{8}$, if $f(x) = \begin{cases} \dfrac{e^{\tan 2x} - 1}{\tan 2x + \cot 2x}, & x \ne \dfrac{\pi}{8} \\ \dfrac{1}{2}, & x = \dfrac{\pi}{8} \end{cases}$

(c) Discuss the continuity and differentiability at $x = \dfrac{\pi}{2}$, if $f(x) = \begin{cases} \dfrac{(\cos x)^{\sin x} - 1}{x}, & x \le \dfrac{\pi}{2} \\ -\dfrac{1}{x}, & x > \dfrac{\pi}{2} \end{cases}$

(23) Discuss the continuity and differentiability at $x = \dfrac{\pi}{2}$

(a) $f(x) = \begin{cases} \dfrac{\sqrt[3]{1 + x - \dfrac{\pi}{2}} - 1}{x - \dfrac{\pi}{2}}, & x \ne \dfrac{\pi}{2} \\ \dfrac{1}{3}, & x = \dfrac{\pi}{2} \end{cases}$ (b) $f(x) = \begin{cases} \dfrac{\cos^2 x}{1 - \sin x}, & x \ne \dfrac{\pi}{2} \\ 2, & x = \dfrac{\pi}{2} \end{cases}$ (c) $f(x) = \begin{cases} \dfrac{\sin\left(x - \dfrac{\pi}{2}\right)}{\left|x - \dfrac{\pi}{2}\right|}, & x \ne \dfrac{\pi}{2} \\ 1, & x = \dfrac{\pi}{2} \end{cases}$

Maxima and Minima

Rule: – It is clear that at P the function $y = f(x)$ is maximum and at Q it is minimum. At these points tangent is parallel to X – axis so that its slope is zero. (from the given below figure).

Draw graph

Criteria for maxima and minima: – Let $x = a, b, c$ be the values of x given by $\frac{dy}{dx} = 0$. consider the point $x = a$ i.e P where y is maximum. it is clear that tangent at any point

$x = a - h$ will make an acute angle with X – axis as at L , and tangent at $x = a + h$ will make

an obtuse angle as at M. (from the above figure)

Thus , for $x = a$ $\therefore$ $\frac{dy}{dx} = 0$ for x slightly $< a$, $\frac{dy}{dx} = +$ve for x slightly $> a$, $\frac{dy}{dx} = -$ve

Hence if y is maximum at $x = a$ then $\frac{dy}{dx}$ changes sign from $+$ ve to $-$ ve for values of $x < a$ and $x > a$ in that

order, Now consider the point $x = b$ i.e Q where y is minimum. it is clear that tangent at any point $x = b - h$

will make an obtuse angle with X – axis as at M and tangent at $x = b + h$ will make an acute angle as at N.

Thus for $x = b$ $\therefore$ $\frac{dy}{dx} = 0$, for x slightly $< b$ $\frac{dy}{dx} = -$ve (obtuse) for x slightly $> b$, $\frac{dy}{dx} = +$ve (acute)

Hence if y is minimum at $x = b$ then $\frac{dy}{dx}$ changes sign from $-$ ve to $+$ ve for value of $x < b$ and $x > b$ in that order .

Working Rule: – Calculate $\frac{dy}{dx} = 0$ and solve for x and say $x = a, b, c$ etc.

put values of $x < a$ in $\frac{dy}{dx}$ and values of $x > a$ in $\frac{dy}{dx}$.

If $\frac{dy}{dx}$ changes sign from $+$ ve to $-$ ve , then maximum at $x = a$.

If $\frac{dy}{dx}$ changes sign from $-$ ve to $+$ ve , then minimum at $x = a$.

In case there is no change of sign , then neither a maximuma nor a minimum.

	$-\infty$	$+$ ve	a	$-$ ve	b	$+$ ve	∞
			(Maximum)		(Minimum)		

- ■ $-$ ve sign to $+$ ve sign at $x = b$ that is minimum.

- ■ $+$ ve sign to $-$ ve sign at $x = a$ that is maximum.

Example: – Let $y = x^3 - 3x^2$ $\therefore$ $\frac{dy}{dx} = 3x^2 - 6x = 3x(x - 2)$

The critical points are $0, 2$ $\therefore$ $\frac{dy}{dx} = 0$ or $3x(x - 2) = 0$ then $x = 0, 2$

$-\infty$	$+$ ve	0	$-$ ve	2	$+$ ve	∞

At x = 0 there will be a change of sign from + ve to − ve and hence there will be Maximum at x = 0 .

At x = 2 there will be a change of sign from − ve to + ve and hence there will be Minimum at x = 2 .

Second Rule: − Calculate $\frac{dy}{dx} = 0$ and solve for x. suppose one root of $\frac{dy}{dx} = 0$ is at x = a.

If $\frac{d^2y}{dx^2} = -ve$ for x = a then Maximum at x = a.

If $\frac{d^2y}{dx^2} = +ve$ for x = a then Minimum at x = a.

If $\frac{d^2y}{dx^2} = 0$ at x = a then find $\frac{d^3y}{dx^3}$. if $\frac{d^3y}{dx^3} \neq 0$ at x = a

then neither Max. nor Min. at x = a if $\frac{d^3y}{dx^3} = 0$ at x = a then find $\frac{d^4y}{dx^4}$. if $\frac{d^4y}{dx^4} > 0$ $i.e + ve$ at x = a then y is Min. at x = a .

and if $\frac{d^4y}{dx^4} < 0$ $i.e - ve$ at x = a then y is Max. at x = a and so on.

Parametric Form of a Function: − Let a function y = f(x) be represented in parametric form by the equations x = θ(t) , y = φ(t)

where θ(t) and φ(t) have derivatives both of first and second orders within a certain interval of t.

Let at t = t_0 , $φ'(t) = 0$, then

(a) if $φ''(t_0) < 0$, $f(x)$ has a Max. at x = x_0 .

(b) if $φ''(t_0) > 0$, $f(x)$ has a Min. at x = x_0 .

(c) if $φ''(t_0) = 0$, f(x) has an extreme value at x = x_0 .

Important Points: − (1) According to the given conditions of the problem determine the function whose Maximum and Minimum

value are to be found.

(2) The above function is not of single variable but contains move than a function of the form: −

$$kf(x), k + f(x), [f(x)]^k \text{ or } [f(x)]^{\frac{1}{k}}$$

where k is a + ve constant if f(x) > 0 then the function f (x) is Maximum or Minimum.

If y is Maximum and Minimum then $\log y = z$ is also Maximum and Minimum at y > 0 .

Also y = f(x) is Max. or Min. then $z = \frac{1}{f(x)}$ is also Min. or Max.

(3) Maximum and Minimum occur alternately it may be noted that Maximum may be less than the Minimum.

Remember Geometrical Formula: −

Area of Squar = x^2 and Perimeter = 4x

Area of rectangle = xy , Perimeter = 2(x + y)

Area of equilateral = $\frac{\sqrt{3}}{4}x^2$, Perimeter = 3x

Area of trapezium = $\frac{1}{2}$ (sum of parallel side) × distance between them.

Area of circle = $πr^2$, Perimeter = 2πr

Volume of a sphere $= \frac{4}{3}\pi r^3$ and surface $= 4\pi r^2$

Volume of right cone $= \frac{1}{3}\pi r^2 h$, Total surface $= \pi r(r + l)$, curved surface $= \pi r l$

Volume of a cylinder $= \pi r^2 h$, Total surface $= 2\pi r(r + h)$, curved surface $= 2\pi r h$

Volume of a cuboid $= xyz$ and surface $= 2(xy + yz + zx)$

Finding the Greatest and the Least value of a function: —

The greatest (least) value of a continuous function $f(x)$ on an interval $[a, b]$ is attained either at the critical points or at the end points of the interval. To find the greatest (least) value of the function we have comput its values at all the critical points on the interval $[a, b]$ the values of $f(a)$, $f(b)$ of the function at the end points of the interval If a function $f(x)$ is defined and continuous in same interval. if the interval is not closed then the function may have neither the greatest nor the least value.

Example: — (1) Find the greatest and least values of the following function on the given intervals: —

(a) $f(x) = 3x^3 - 6x^2 - 12x + 6$ on $[0,1]$ (b) $f(x) = x\log x$ on $[1, e^2]$

Solution: — (a) $f(x) = 3x^3 - 6x^2 - 12x + 6$ on the points $[0,1]$

$f'(x) = 9x^2 - 12x - 12$ ∴ $f'(x) = 0$ ⇒ $9x^2 - 12x - 12 = 0$ ⇒ $3x^2 - 4x - 4 = 0$

⇒ $3x^2 - 6x + 2x - 4 = 0$ ∴ $3x(x - 2) + 2(x - 2) = 0$ ∴ $(3x + 2)(x - 2) = 0$ ∴ $x = -\frac{2}{3}, 2$

critical points are $-\frac{2}{3}, 0, 1, 2$

$-\infty$	$-$ ve	$-\frac{2}{3}$	$+$ ve	0	$-$ ve	1	$+$ ve	2	$-$ ve	∞

At $x = -\frac{2}{3}$ then $f(x)$ is Minimum. ∴ $f(x) = 3x^3 - 6x^2 - 12x + 6$

∴ $f\left(-\frac{2}{3}\right) = 3 \times \left(-\frac{2}{3}\right)^3 - 6 \times \left(-\frac{2}{3}\right)^2 - 12 \times \left(-\frac{2}{3}\right) + 6 = \frac{94}{9}$

At $x = 0$, then $f(x)$ is Maximum.   ∴ $f(x) = 3x^3 - 6x^2 - 12x + 6$ ∴ $f(0) = 6$

At $x = 1$, then $f(x)$ is Minimum.   ∴ $f(x) = 3x^3 - 6x^2 - 12x + 6$ ∴ $f(1) = 3 - 6 - 12 + 6 = -9$

At $x = 2$, then $f(x)$ is Maximum.

∴ $f(x) = 3x^3 - 6x^2 - 12x + 6$ ∴ $f(2) = 3(2)^3 - 6(2)^2 - 12(2) + 6 = 24 - 24 - 24 + 6 = -18$

Hence the greatest value is $f\left(-\frac{2}{3}\right) = \frac{94}{9}$ and the least value is $f(2) = -18$ Ans.

(b) $f(x) = x\log x$ on the points $[1, e^2]$ ∴ $f'(x) = x.\frac{1}{x} + \log x . 1 = 1 + \log x$ ∴ $f'(x) = 0$ or $1 + \log x = 0$

or $\log x = -1 = \log e^{-1}$ ⇒ $\log x = \log e^{-1}$ ∴ $x = e^{-1} = \frac{1}{e}$ critical points are $\frac{1}{e}, 1, e^2$

$-\infty$	$-$ ve	$\frac{1}{e}$	$+$ ve	1	$-$ ve	e^2	$+$ ve	∞

At $x = \dfrac{1}{e}$ then f(x) is Minimum. $\quad \therefore \ f(x) = x \log x \ \Rightarrow \ f\left(\dfrac{1}{e}\right) = \dfrac{1}{e} \log_e e^{-1} = -\dfrac{1}{e}$

At $x = 1$ then f(x) is Maximum. $\quad \therefore \ f(x) = x \log x \ \Rightarrow \ f(1) = 1. \log 1 = 0$

At $x = e^2$ then f(x) is Minimum. $\quad \therefore \ f(x) = x \log x \ \Rightarrow \ f(e^2) = e^2 \log e^2 = 2e^2$

Hence, the greatest value is $f(e^2) = 2e^2$ and the least value is $f(1) = 0.$ Ans.

Solved Example

(1) Find the Max. and Min. value of the following: $-$ (a) Find the Maximum and Minimum value of $2x^3 - 24x + 54$ in the interval $(-3,3)$.

(b) Find the Maximum and Minimum value of the function $f(x) = \cos x \,(1 + \sin x)$ in the interval $\left[-\dfrac{\pi}{2}, \dfrac{\pi}{2}\right]$.

Solution: $-$ (a) Let $f(x) = 2x^3 - 24x + 54$, $f'(x) = 6x^2 - 24$

For Maximum or Minimum , $f'(x) = 0$ i.e $3x^2 - 24 = 0$ or $x^2 = 4$ or $x = \pm 2$

Now , $f''(x) = 12x \ \Rightarrow f''(2) = 24 > 0$ *it is minimum.* $\quad \Rightarrow \ f''(-2) = -24 < 0$ *it is maximum.*

The function f(x) has a maximum at $x = -2$,

Required maximum value $= 2(-2)^3 - 24 \times (-2) + 54 = 86$ Ans.

The function f(x) has a minimum at $x = 2$, Required minimum value $= 2(2)^3 - 24 \times 2 + 54 = 22$ Ans.

(b) We have $f(x) = \cos x \,(1 + \sin x)$

$f'(x) = \cos x \,(0 + \cos x) + (1 + \sin x)(-\sin x) = \cos^2 x - \sin x - \sin^2 x = (1 - \sin^2 x) - \sin x - \sin^2 x = -2\sin^2 x - \sin x + 1$

For stationary points , $f'(x) = 0 \ \Rightarrow \ -2\sin^2 x - \sin x + 1 = 0 \ \Rightarrow \ 2\sin^2 x + \sin x - 1 = 0$

or $\sin x = \dfrac{-1 \pm \sqrt{1+8}}{2 \times 2} = \dfrac{-1 \pm 3}{4} = -1 \ \text{or} \ \dfrac{1}{2} \quad \left[\text{formula}, \ ax^2 + bx + c = 0 \ \text{then} \ x = \dfrac{-b \pm \sqrt{b^2 - 4ac}}{2a}\right]$

$$\text{or} \ \sin x = -1, \dfrac{1}{2} \quad \text{or} \ x = -\dfrac{\pi}{2}, \dfrac{\pi}{6}$$

Now , $f\left(-\dfrac{\pi}{2}\right) = 0$ and $f\left(\dfrac{\pi}{6}\right) = \dfrac{3\sqrt{3}}{4} > 0$

f(x) has maximum value $= \dfrac{3\sqrt{3}}{4}$ at $x = \dfrac{\pi}{6}$ and minimum value $= 0$ at $x = -\dfrac{\pi}{2}$ and also minimum value $= 0$ at $x = \dfrac{\pi}{2}$ Ans.

(2) Find the Maxima or Minima of the following function.

(a) $x^3 - 6x^2 + 9x + 7$ (b) $2x^3 - 54x + 108$ (c) $x^3 + 4x^2 - 3x + 2$ (d) $x^4 - 62x^2 + 120x + 9$

(e) $x^4 - 2x^2$ (f) $x^3 - 12x$ (g) $3x^2 - 4x + 5$ (h) $x^3 + 2x^2 - 4x + 1$

Solution: $-$ (a) Let $f(x) = x^3 - 6x^2 + 9x + 7$, $f'(x) = 3x^2 - 12x + 9$

For stationary points, $f'(x) = 0 \ \therefore \ 3x^2 - 12x + 9 = 0 \ \Rightarrow x^2 - 4x + 3 = 0 \ \Rightarrow \ x^2 - 3x - x + 4 = 0$

$$\Rightarrow \ x(x - 3) - 1(x - 3) = 0 \ \Rightarrow \ (x - 1)(x - 3) = 0 \quad \therefore \ x = 1,3$$

$-\infty$	$+\,ve$	1	$-\,ve$	3	$+\,ve$	∞

At $x = 1$ then $f(x) = x^3 - 6x^2 + 9x + 7$, $f(1) = 1 - 6 + 9 + 7 = 11 > 0$

and $f'(1) = 3x^2 - 12x + 9 = 3 - 12 + 9 = 0$ and $f''(1) = 6x - 12 = 6 - 12 = -6 < 0$

At $x = 3$ then $f(x) = x^3 - 6x^2 + 9x + 7$, $f(3) = 27 - 54 + 27 + 7 = 7 > 0$, $f'(3) = 27 - 36 + 9 = 0$

and $f''(3) = 6x - 12 = 18 - 12 = 6 > 0$ $f(x)$ is maximum at $x = 1$ and minimum at $x = 3$. Ans.

(b) Ans: $-$ Hint $-$ Let $f(x) = 2x^3 - 54x + 108$, $f'(x) = 6x^2 - 54$ (solve same as above question)

stationary points are ± 3 , $f(x)$ is minimum at $x = 3$ and maximum at $x = -3$.

(c) Let $f(x) = x^3 + 4x^2 - 3x + 2$, $f'(x) = 3x^2 + 8x - 3$ $\therefore$ $f'(x) = 0$

For stationary points, $f'(x) = 0$ or $3x^2 + 8x - 3 = 0$ or $3x^2 + 9x - x - 3 = 0$ or $3x(x + 3) - 1(x + 3) = 0$

or $(3x - 1)(x + 3) = 0$ $\therefore x = -3, \dfrac{1}{3}$ or $f(x)$ is maximum at $x = -3$ and minimum at $x = -\dfrac{1}{3}$. Ans.

(d) Let $f(x) = x^4 - 62x^2 + 120x + 9$, $f'(x) = 4x^3 - 124x + 120$ $\therefore f'(x) = 0$

For stationary points, $f'(x) = 0$ or $4x^3 - 124x + 120 = 0$ or $x^3 - 31x + 30 = 0$

or $x^2(x - 1) + x(x - 1) - 30(x - 1) = 0$ or $(x - 1)(x^2 + x - 30) = 0$ or $(x - 1)(x^2 + 6x - 5x - 30) = 0$

or $(x - 1)[x(x + 6) - 5(x + 6)] = 0$ or $(x - 1)(x - 5)(x + 6) = 0$ $\therefore$ $x = -6, 1, 5$

At $x = -6$ then $f''(x) = 12x^2 - 124$ or $f''(-6) = 12(-6)^2 - 124 = 432 - 124 = 108 > 0$ (+ve) Min.

At $x = 1$ then $f''(x) = 12x^2 - 124$ or $f''(1) = 12 \times 1 - 124 = 12 - 124 = -112 < 0$ (−ve) Max.

At $x = 5$ then $f''(x) = 12x^2 - 124$ or $f''(5) = 12(5)^2 - 124 = 300 - 124 = 176 > 0$ (+ve) Min.

$\qquad\qquad\qquad$ $f(x)$ is maximum at $x = 1$ and minimum at $x = -6, 5$.   Ans.

(e) $f(x)$ is maximum at $x = 0$ and minimum at $x = \pm 1$.   Ans.   (Do yourself, same as above question)

(f) $f(x)$ is maximum at $x = -2$ and minimum at $x = 2$.   Ans.   (Do yourself, same as above question)

(g) $f(x)$ is minimum at $x = \dfrac{2}{3}$. Ans.   (h) $f(x)$ is maximum at $x = -2$ and minimum at $x = \dfrac{2}{3}$. Ans.

(3) Prove that the function $f(x) = 3x + 5$ is Monotonic for all value of $x \in R$.

Solution: $-$ Consider two values of x (say) $x_1, x_2 \in R$ such that $x_2 > x_1$ … … … … … … … … … .. (i)

Multiplying both sides of (i) by 3 , we have $3x_2 > 3x_1$ … … … … … … … … … .. (ii)

Adding 5 to both sides of (ii) , we get $3x_2 + 5 > 3x_1 + 5$ we have $f(x_2) > f(x_1)$

Thus , we find $f(x_2) > f(x_1)$ whenever $x_2 > x_1$

Hence the given function $f(x) = 3x + 5$ is monotonic function.   Proved.

(4) Show that $f(x) = x^2$ is a strictly decreasing function for all $x < 0$.

Solution: $-$ Consider x_1, x_2 two value of x such that $x_2 > x_1$, $x_1 x_2 < 0$ … … … … .. (i)

it is multiplied by a negative number in equation (i) by x_2 , we have $x_2 . x_2 < x_1 . x_2$ or $x_2{}^2 < x_1 . x_2$ … … … … … … … … … (ii)

Now , Multiplying (i) by x_1 , we have $x_1 . x_2 < x_1 . x_1$ or $x_1 . x_2 < x_1{}^2$ … … … … … .. … .. … … … . (iii)

from (ii) and (iii) , we have $x_2{}^2 < x_1 . x_2 < x_1{}^2$ or $x_2{}^2 < x_1{}^2$ or $f(x_2) < f(x_1)$ … … … .. … . (iv)

Thus , from (i) and (iv) , we have $x_2 > x_1$, $f(x_2) < f(x_1)$

Hence , the given function is strictly decreasing for all $x < 0$. *Proved.*

(5) Find the values of x , the function $f(x) = x^2 - 2x + 3$ is increasing and decreasing.

Solution: $-$ $f(x) = x^2 - 2x + 3$, $\quad$ $f'(x) = 2x - 2$

For f(x) to be increasing $f'(x) > 0$ $i.e$ $2x - 2 > 0$ or $2(x - 1) > 0$ or $x - 1 > 0$ or $x > 1$, $\ The\ function\ increases\ for\ x > 1.$

For f(x) to be decreasing $f'(x) < 0$ $i.e$ $2x - 2 < 0$ or $2(x - 1) < 0$ or $x - 1 < 0$ or $x < 1$, $\ Thus\ , the\ function\ decreases\ for\ x < 1.$

(6) Find the interval in which $f(x) = x^3 + 3x^2 - 9x + 5$ is increasing or decreasing.

Solution: $-$ $f(x) = x^3 + 3x^2 - 9x + 5$, $\ f'(x) = 3x^2 + 6x - 9 = 3(x^2 + 2x - 3) = 3(x - 1)(x + 3)$

The critical points are $f'(x) = 0$ or $3(x - 1)(x + 3) = 0$ $\ \therefore\ $ $x = -3, 1$

	increasing		decreasing		increasing	
$-\infty$	$+\ ve$	-3	$-\ ve$	1	$+\ ve$	∞

For f(x) to be increasing $f'(x) > 0$ and $f(x)$ to be decreasing $f'(x) < 0$

The function f(x) is increasing for $x > 1\ or\ x < -3$ and $f(x)$ is decreasing for $-3 < x < 1$. $\ $ Ans.

(7) Determine the intervals for which the function $f(x) = \dfrac{x}{x^2 + 3}$ is increasing or decreasing.

Solution: $-$ $f(x) = \dfrac{x}{x^2 + 3}$, $\ f'(x) = \dfrac{(x^2 + 3).1 - x.2x}{(x^2 + 3)^2} = \dfrac{x^2 + 3 - 2x^2}{(x^2 + 3)^2} = \dfrac{3 - x^2}{(x^2 + 3)^2}$

$\left[formula, \quad y = \dfrac{f(x)}{g(x)} , \quad y'\ or\ \dfrac{dy}{dx} = \dfrac{g(x).f'(x) - f(x).g'(x)}{[g(x)]^2} \right]$

As $(x^2 + 3)^2$ is positive for all real value of x.

The stationary point are $f'(x) = 0$ or $\dfrac{3 - x^2}{(x^2 + 3)^2} = 0$ or $3 - x^2 = 0$ or $x^2 = 3$ $\ \therefore\ $ $x = \pm\sqrt{3}$

	decreasing		increasing		decreasing	
$-\infty$	$-\ ve$	$-\sqrt{3}$	$+\ ve$	$\sqrt{3}$	$-\ ve$	∞

The function f(x) is increasing $f'(x) > 0$ for $-\sqrt{3} < x < \sqrt{3}$ and f(x) is decreasing $f'(x) < 0$ for $x > \sqrt{3}$ or $x < -\sqrt{3}$ $\quad$ Ans.

(8) Find the Maximum (Local Maximum) and Minimum (Local Minimum) Points of the function $f(x) = 2x^3 - 6x^2 - 18x + 3$

Solution: $-$ Here, $f(x) = 2x^3 - 6x^2 - 18x + 3$, $\ f'(x) = 6x^2 - 12x - 18 = 6(x^2 - 2x - 3)$

or $f'(x) = 0$, gives us $6(x^2 - 2x - 3) = 0$ $\ \therefore\ x^2 - 2x - 3 = 0$ or $(x + 1)(x - 3) = 0$ $\ \therefore\ x = 3, -1$

Critical points are $x = 3$ and $x = -1$

		Maximum		Minimum		
$-\infty$	$+\ ve$	-1	$-\ ve$	3	$+\ ve$	∞

At $x > 3$ $then\ f'(x) = +ve > 0$ $Value + ve\ to - ve\ then\ the\ function\ is\ maximum\ at\ x = -1$

At $x < 3$ $then\ f'(x) = -ve < 0$ $\ Value - ve\ to + ve\ then\ the\ function\ is\ minimum\ at\ x = 3$

At $x < -1$ $then\ f'(x) = +ve > 0$

At $x > -1$ $then\ f'(x) = -ve < 0$

The function f(x) has minimum value at $x = 3$, $\ $ minimum value $= f(3) = 2(3)^3 - 6(3)^2 - 18(3) + 3 = -51$

The function f(x) has maximum value at $x = -1$, $\ $ maximum value $= f(-1) = 2(-1)^3 - 6(-1)^2 - 18(-1) + 3 = 13$

or $(-1, 13)$ and $(3, -51)$ are points of local maxima and local minima respectively. $\ $ Ans.

(9) Find the Local Maximum and the Local Minimum of the function $f(x) = x^2 - 6x$.

Solution: $-\ f(x) = x^2 - 6x$, $f'(x) = 2x - 6$ putting $f'(x) = 0$ or $2x - 6 = 0$ $\therefore$ $x = 3$

we have to examine whether $x = 3$ is the point of local maximum or local minimum or neither maximum nor minimum.

Let us take $x = 2.9$ which is to the left of 3 and $x = 3.1$ which is to the right of 3 and find $f'(x)$ at these points.

$$f'(2.9) = 2 \times 2.9 - 6 < 0 \ \ and \ f'(3.1) = 2 \times 3.1 - 6 > 0$$

Since $f'(x) < 0$ $\ as \ we \ approach \ 3 \ from \ the \ left \ and \ f'(x) > 0 \ \ as \ we \ approach \ 3 \ from \ the \ right.$

$\therefore$ there is a local minimum at $x = 3$

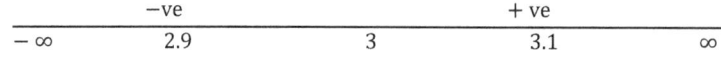

Sign of $f'(x)$ and Point $x = 3$

Left of 3	Right of 3
$f'(x) < 0$	$f'(x) > 0$

Local Minimum at $x = 3$ Ans.

(10) Find all Local Maxima and Local Minima of the function $f(x) = x^3 + 3x^2 - 9x + 7$.

Solution: $-\ f(x) = x^3 + 3x^2 - 9x + 7$, $f'(x) = 3x^2 + 6x - 9$

Putting $f'(x) = 0$, $3x^2 + 6x - 9 = 0$ or $x^2 + 2x - 3 = 0$ or $(x - 1)(x + 3) = 0$ $\therefore x = 1, -3$

The critical points are $x = 1$ and $x = -3$

we examine whether these points are points of local maximum or local minimum or neither of them.

solve, same as above question.

Ans: $-\ f'(0.9) < 0$, $f'(1.1) > 0$ $\ and \ f'(-2.9) < 0$, $f'(-3.1) > 0$

Then the function is a local minimum at $x = 1$ and local maximum at $x = -3$.

	+ve	$x = -3$	$-$ve	$-$ve	$x = 1$	$+$ve	
$-\infty$	-3.1	-3	-2.9	0.9	1	1.1	∞

Local Maximum   Local Minimum

(11) Find the Local Maximum and Local Minimum, if any of the following function $f(x) = \dfrac{x + 1}{x^2 + 3}$.

Solution: $-\ f(x) = \dfrac{x + 1}{x^2 + 3}$, $f'(x) = \dfrac{(x^2 + 3).1 - (x + 1).2x}{(x^2 + 3)^2} = \dfrac{x^2 + 3 - 2x^2 - 2x}{(x^2 + 3)^2} = \dfrac{3 - 2x - x^2}{(x^2 + 3)^2}$

As $(x^2 + 3)^2$ is Positive for all real value of x.

For stationary Points are $f'(x) = 0$ or $\dfrac{3 - 2x - x^2}{(x^2 + 3)^2} = 0$ or $3 - 2x - x^2 = 0$

or $3(1 - x) + x(1 - x) = 0$ or $(1 - x)(x + 3) = 0$ $\therefore$ critical points are $x = 1$, $x = -3$

At $x > 1$ $\ then \ f'(x) = -$ve , $f'(x) < 0$

		Local Minimum		Local Maximum	
$-\infty$	$-$ve	-3	$+$ve	1	$-$ve	∞

At $x < -3$ $\ then \ f'(x) = -$ve , $f'(x) < 0$

At $x < 1$ $\ then \ f'(x) = +$ve , $f'(x) > 0$

At $x > -3$ *then* $f'(x) = +ve$, $f'(x) > 0$

The function f(x) is Local Maximum at $x = 1$, Local Maximum value $= \dfrac{1+1}{1+3} = \dfrac{2}{4} = \dfrac{1}{2}$ Ans.

and function f(x) is Local Minimum at $x = -3$, Local Minimum value $= \dfrac{-3+1}{9+3} = \dfrac{-2}{12} = -\dfrac{1}{6}$ Ans.

(12) Find the Local Maximum and Local Minimum , if any for the function $f(x) = \dfrac{x}{2} - \sin x$, $0 \le x \le \dfrac{\pi}{2}$.

Solution: $-$ We have $f(x) = \dfrac{x}{2} - \sin x$, $f'(x) = \dfrac{1}{2} - \cos x$

For Local Maxima or Minima $f'(x) = 0$ or $\dfrac{1}{2} - \cos x = 0$ or $\cos x = \dfrac{1}{2}$ $\therefore x = \dfrac{\pi}{3}$ in $0 \le x \le \dfrac{\pi}{2}$

At $x = \dfrac{\pi}{3}$, for $x > \dfrac{\pi}{3}$ then $f'(x) = \dfrac{1}{2} > 0$, $\dfrac{1}{2} - \cos x > 0$

For $x < \dfrac{\pi}{3}$, $f'(x) < 0$ or $\dfrac{1}{2} - \cos x < 0$

$-\infty$	$-ve$	$\dfrac{\pi}{3}$	$+ve$	∞

$f'(x)$ changes sign from negative to positive in the neithbourhood of $\dfrac{\pi}{3}$.

At $x = \dfrac{\pi}{3}$ is a point of Local Minima, then the Minimum value $= f\left(\dfrac{\pi}{3}\right) = \dfrac{\pi}{6} - \sin\left(\dfrac{\pi}{3}\right) = \dfrac{\pi}{6} - \dfrac{\sqrt{3}}{2} = \dfrac{\pi - 3\sqrt{3}}{6}$

$$\therefore \text{ Point of Local Minima is } \left(\dfrac{\pi}{3}, \dfrac{\pi - 3\sqrt{3}}{6}\right) \text{ Ans.}$$

(13) Find the Local Minimum or Maximum of the following function $f(x) = 2x^3 - 9x^2 + 12x - 8$.

Solution: $-$ We have $f(x) = 2x^3 - 9x^2 + 12x - 8$

then $f'(x) = 6x^2 - 18x + 12 = 6(x^2 - 3x + 2) = 6(x - 1)(x - 2)$

For Local Maximum or Minimum $f'(x) = 0$ $\therefore 6(x - 1)(x - 2) = 0$ or $x = 1, 2$

Now, $f''(x) = \dfrac{d[f'(x)]}{dx} = \dfrac{d[6(x^2 - 3x + 2)]}{dx} = 12x - 18 = 6(2x - 3)$

For $x = 1$, $f''(1) = 6(2 - 3) = -6 < 0$ *then* $f(x)$ is a Local Maximum at $x = 1$.

and $f(1) = 2(1)^3 - 9(1)^2 + 12(1) - 8 = 2 - 9 + 12 - 8 = -3$ is a Local Maximum value.

For $x = 2$, $f''(2) = 6(4 - 3) = 6 > 0$ *then* $f(x)$ is a Local Minimum at $x = 2$.

and $f(2) = 2(2)^3 - 9(2)^2 + 12(2) - 8 = 16 - 36 + 24 - 8 = -4$ is a Local Minimum value.

$\therefore$ $(1, -3)$ and $(2, -4)$ are Points of Local Maxima and Local Minima respectively.

		Local Maximum		Local Minimum		
$-\infty$	$+ve$	1	$-ve$	2	$+ve$	∞

(14) Find Local Maxima and Minima (if any) for the function $f(x) = \sin 3x$, $0 < x < \dfrac{\pi}{2}$.

Solution: $-$ $f(x) = \sin 3x$ then $f'(x) = 3\cos 3x$ $\therefore$ putting $f'(x) = 0$ $\therefore 3\cos 3x = 0$ or $\cos 3x = 0$

or $3x = \dfrac{\pi}{2}, \dfrac{3\pi}{2}, \dfrac{5\pi}{2}$ or $x = \dfrac{\pi}{6}, \dfrac{\pi}{2}, \dfrac{5\pi}{6}$ $\therefore$ only $x = \dfrac{\pi}{6}$ lie in the interval $0 < x < \dfrac{\pi}{2}$

Now, $f''(x) = -9 \sin 3x$

At $x = \dfrac{\pi}{6}$, $f''(x) = -9\sin 3x$ then $f''\left(\dfrac{\pi}{6}\right) = -9\sin\left(\dfrac{3\pi}{6}\right) = -9\sin\dfrac{\pi}{2} = -9 < 0$

The function $f(x)$ is Maximum at $x = \dfrac{\pi}{6}$ and Maximum value $= f\left(\dfrac{\pi}{6}\right) = \sin\left(\dfrac{3\pi}{6}\right) = \sin\left(\dfrac{\pi}{2}\right) = 1$ Ans.

(15) Find the Minimum and Maximum value of the function $f(x) = 4x^3 - 48x + 105$ in the interval $[1,3]$ and $[-3,0]$ respectively.

Solution: $-$ $f(x) = 4x^3 - 48x + 105$ then $f'(x) = 12x^2 - 48$

For Local Maximum or Minimum $f'(x) = 0$ i.e $12x^2 - 48 = 0$ or $x^2 = 4$ $\therefore$ $x = \pm 2$

$f(x)$ is Minimum in the interval $[1,3]$ only $x = 2$ belong to the interval $[1,3]$. find Minimum if any at $x = 2$ only.

Now, $f''(x) = 24x$ $\therefore$ $f''(2) = 24 \times 2 = 48 > 0$ (+ve) which implies the function

$f(x)$ has a Minimum at $x = 2$, $\therefore$ Required Minimum value $= 4(2)^3 - 48(2) + 105 = 32 - 96 + 105 = 41$

Thus the Point of Minimum belonging to the given interval $[1,3]$ is 2 and the Minimum value of the function is 41.

Now, $f''(x) = 24x$, $f''(-2) = 24(-2) = -48 < 0$ (−ve) $\left[-2 \text{ lies in } [-3,0]\right]$

which implies the function $f(x)$ shall have a Maximum at $x = -2$.

Required Maximum value $= 4(-2)^3 - 48(-2) + 105 = -32 + 96 + 105 = 169$ Ans.

(16) Find the Maximum and Minimum value of the function $f(x) = \cos x\,(1 + \sin x)$ in $(0, \pi)$.

Solution: $-$ We have $f(x) = \cos x\,(1 + \sin x)$ then $f'(x) = \cos x.(0 + \cos x) + (1 + \sin x).(-\sin x)$

$f'(x) = \cos^2 x - \sin x - \sin^2 x = -2\sin^2 x - \sin x + 1$

For stationary points $f'(x) = 0$ $\therefore$ $-2\sin^2 x - \sin x + 1 = 0$ $\therefore$ $\sin x = -1, \dfrac{1}{2}$ $\therefore x = -\dfrac{\pi}{2}, \dfrac{\pi}{6}, \dfrac{5\pi}{6}$

Now, $f\left(\dfrac{\pi}{6}\right) = \cos\dfrac{\pi}{6}\left(1 + \sin\dfrac{\pi}{6}\right) = \dfrac{\sqrt{3}}{2}\left(1 + \dfrac{1}{2}\right) = \dfrac{3\sqrt{3}}{4} > 0$

$f\left(\dfrac{5\pi}{6}\right) = \cos\dfrac{5\pi}{6}\left(1 + \sin\dfrac{5\pi}{6}\right) = -\dfrac{\sqrt{3}}{2}\left(1 + \dfrac{1}{2}\right) = -\dfrac{3\sqrt{3}}{4} < 0$

$f(x)$ has Maximum value $\dfrac{3\sqrt{3}}{4}$ at $x = \dfrac{\pi}{6}$ and Minimum value $-\dfrac{3\sqrt{3}}{4}$ at $x = \dfrac{5\pi}{6}$ and also Minimum value 0 at $x = -\dfrac{\pi}{2}$ Ans.

(17) Find the two positive real numbers whose sum is 50 and their product is Maximum.

Solution: $-$ Let one number be x and other number is $(50 - x)$ as two numbers are positive.

we have $x > 0, 50 - x > 0$ $\Rightarrow 50 - x > 0$ or $-x > -50$ or $x < 50$ $\therefore$ $0 < x < 50$

Let their product be $f(x) = x(50 - x) = 50x - x^2$ we have maximize the product $f(x)$.

find $f'(x)$ and put that equal to zero. $f'(x) = 50 - 2x$ then $f'(x) = 0$ or $50 - 2x = 0$ or $-2x = -50$ or $2x = 50$ $\therefore$ $x = 25$

Now, $f'(x) = -2$ which is negative. Hence $f(x)$ is Maximum at $x = 35$

then the other number is $70 - x = 35$ or $-x = 35 - 70 = -35$ $\therefore x = 35$

Hence the required numbers are $35, 35$. Ans.

(18) Show that among rectangles of given area, the square has the least perimeter.

Solution: $-$ Let x, y be the length and breadth of the rectangle respectively.

its area = length × breadth = x. y   ∴ Area (A) = xy   ∴ A = xy or $y = \dfrac{A}{x}$

Now, perimeter (P) of the rectangle = 2(x + y)   ∴ $P = 2(x + y) = 2\left(x + \dfrac{A}{x}\right)$

Differentiate with respect to x then $\dfrac{dP}{dx} = 2\left(1 - \dfrac{A}{x^2}\right)$

For Minimum $\dfrac{dP}{dx} = 0$ then $2\left(1 - \dfrac{A}{x^2}\right) = 0$ or $\dfrac{A}{x^2} = 1$ or $A = x^2$ ∴ $x = \sqrt{A}$

Now, $\dfrac{d^2P}{dx^2} = 2\{0 - (-2)Ax^{-3}\} = \dfrac{4A}{x^3}$ which is positive.

Hence perimeter (P) is Minimum when $x = \sqrt{A}$ ∴ $y = \dfrac{A}{x} = \dfrac{x^2}{x} = x$ ∴ $y = x$ $(A = x^2)$

Thus the perimeter is Minimum when rectangle is a square. proved.

(19) An open box with a square base is to be mode out of a given quantity of area a. show that the Maximum volume of the box is $\dfrac{a\sqrt{a}}{6\sqrt{3}}$.

Solution: − Let x be the side of the square base of the box and y its height.

Total surface area of the box = $x^2 + 4xy$ ∴ $a = x^2 + 4xy$ ∴ $y = \dfrac{a - x^2}{4x}$

Volume of the box (V) = base area × height = $x^2 × y = x^2\left(\dfrac{a - x^2}{4x}\right)$ ∴ $V = \dfrac{ax - x^3}{4}$ ……………… (i)

Differentiating equation (i) with respect to x , we have   ∴ $\dfrac{dV}{dx} = \dfrac{a - 3x^2}{4}$

For Maxima or Minima $\dfrac{dV}{dx} = 0$ then $\dfrac{a - 3x^2}{4} = 0$ or $a = 3x^2$ or $x^2 = \dfrac{a}{3}$ ∴ $x = \pm\sqrt{\dfrac{a}{3}}$ ……………… (ii)

From (i) and (ii) , we have $V = \dfrac{1}{4}(ax - x^3) = \dfrac{1}{4}\left[a.\sqrt{\dfrac{a}{3}} - \left(\sqrt{\dfrac{a}{3}}\right)^3\right] = \dfrac{1}{4}\left[\dfrac{a\sqrt{a}}{\sqrt{3}} - \dfrac{a\sqrt{a}}{3\sqrt{3}}\right] = \dfrac{1}{4}\left[\dfrac{3a\sqrt{a} - a\sqrt{a}}{3\sqrt{3}}\right]$

$$\therefore \ V = \dfrac{1}{4}\left[\dfrac{2a\sqrt{a}}{3\sqrt{3}}\right] = \dfrac{a\sqrt{a}}{6\sqrt{3}}$$

Again , $\dfrac{d^2V}{dx^2} = \dfrac{d\left[\dfrac{a - 3x^2}{4}\right]}{dx} = \dfrac{1}{4}(0 - 6x) = -\dfrac{3x}{2}$, x being the length of the side is positive $\dfrac{d^2V}{dx^2} < 0$

∴   The Volume is Maximum. hence maximum volume of the box = $\dfrac{a\sqrt{a}}{6\sqrt{3}}$ Ans.

Putting area (a) = Any real numbers Let Area (a) = 4 then maximum value of the box = $\dfrac{a\sqrt{a}}{6\sqrt{3}} = \dfrac{4\sqrt{4}}{6\sqrt{3}} = \dfrac{4}{3\sqrt{3}}$ Ans.

Now, put a = 3 then maximum value of the box = $\dfrac{a\sqrt{a}}{6\sqrt{3}} = \dfrac{3\sqrt{3}}{6\sqrt{3}} = \dfrac{1}{2}$ Ans.

(20) Show that of all rectangle inscribed in a given circle , the square has the maximum area.

Solution: − Let ABCD be a rectangle inscribed in a circle of radius r then diameter AC = 2r.

Let AB = x and BC = y then $AB^2 + BC^2 = AC^2$ or $x^2 + y^2 = (2r)^2$ or $x^2 + y^2 = 4r^2$ …………….. (i)

Now , area A of the rectangle = xy

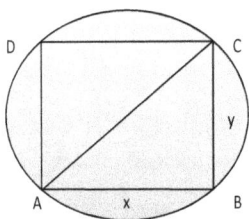

or $A = x\sqrt{4r^2 - x^2}$, $\dfrac{dA}{dx} = x.\dfrac{1}{2\sqrt{4r^2 - x^2}}.(-2x) + \sqrt{4r^2 - x^2}.1 = \dfrac{-x^2 + 4r^2 - x^2}{\sqrt{4r^2 - x^2}} = \dfrac{4r^2 - 2x^2}{\sqrt{4r^2 - x^2}}$

For Maxima or Minima , $\dfrac{dA}{dx} = 0$ or $\dfrac{4r^2 - 2x^2}{\sqrt{4r^2 - x^2}} = 0$ or $4r^2 - 2x^2 = 0$ or $x^2 = 2r^2$ $\therefore$ $x = \sqrt{2}\,r$

Now, $\dfrac{d^2A}{dx^2} = \dfrac{\sqrt{4r^2 - x^2}.(-4x) - (4r^2 - 2x^2).\dfrac{-2x}{2\sqrt{4r^2 - x^2}}}{\left(\sqrt{4r^2 - x^2}\right)^2} = \dfrac{(4r^2 - x^2)(-4x) + x(4r^2 - 2x^2)}{(4r^2 - x^2)^{\frac{3}{2}}} = \dfrac{-16xr^2 + 4x^3 + 4xr^2 - 2x^3}{(4r^2 - x^2)^{\frac{3}{2}}}$

$\quad\quad\quad = \dfrac{2x^3 - 12xr^2}{(4r^2 - x^2)^{\frac{3}{2}}}$

Putting $x = \sqrt{2}\,r$ then $\dfrac{d^2A}{dx^2} = \dfrac{4\sqrt{2}r^3 - 12\sqrt{2}r^3}{(4r^2 - 2r^2)^{\frac{3}{2}}} = \dfrac{-8\sqrt{2}\,r^3}{(2r^2)^{\frac{3}{2}}} = \dfrac{-8\sqrt{2}\,r^3}{2\sqrt{2}\,r^3} = -4 < 0$

Thus , A is Maximum when $x = \sqrt{2}\,r$.

Now from (i) , $x^2 + y^2 = 4r^2$ or $y^2 = 4r^2 - x^2$ $\left(\text{putting } x = \sqrt{2}\,r\right)$

or $y^2 = 4r^2 - 2r^2 = 2r^2$ $\therefore$ $y = \sqrt{2}\,r = x$ or $x = y$ hence rectangle ABCD is a square. $\therefore$ $x = y = \sqrt{2}\,r$ Ans.

(21) Show that the height of a closed right circular cylinder of a given volume and least surface is equal to its diameter.

Solution: $-$ Let v = volume , r = radius and h = height of the cylinder then $v = \pi r^2 h$ or $h = \dfrac{v}{\pi r^2}$ ... ... (i)

Now , surface area (s) $= 2\pi rh + 2\pi r^2$ ... ... (ii)

Putting $h = \dfrac{v}{\pi r^2}$ in equation (ii) , we have $s = 2\pi r.\dfrac{v}{\pi r^2} + 2\pi r^2 = \dfrac{2v}{r} + 2\pi r^2$

Now, $\dfrac{ds}{dr} = \dfrac{-2v}{r^2} + 4\pi r$

For Minimum surface area , $\dfrac{ds}{dr} = 0$ or $\dfrac{-2v}{r^2} + 4\pi r = 0$ or $-2v = -4\pi r^3$ $\therefore$ $v = 2\pi r^3$ ... ... (iii)

Put value of v in equation (i) , we get $h = \dfrac{v}{\pi r^2}$ or $h = \dfrac{2\pi r^3}{\pi r^2} = 2r$ or $h = 2r$ ... ... (iv)

Aganin , $\dfrac{d^2s}{dr^2} = \dfrac{d}{dr}\left[\dfrac{-2v}{r^2} + 4\pi r\right] = \dfrac{4v}{r^3} + 4\pi$, put $v = 2\pi r^3$ then $\dfrac{d^2s}{dr^2} = \dfrac{4v}{r^3} + 4\pi = \dfrac{4.2\pi r^3}{r^3} + 4\pi = 12\pi > 0$

s is least at $h = 2r$ thus height of the cylinder = diameter of the cylinder.   Ans.

(22) A square metal sheet of side 96 cm has four equal squares removed from the corners and the sides are then

turned up so as to form an open box. Determine the size of the square cut so that volume of the box is maximum.

Solution: $-$ Let the side of each of the small squares cut be x cm, so that each side of the box to be made is $(96 - 2x)$ cm and height x cm.

Now, $x > 0$, $96 - 2x > 0$ $i.e$ $x < 48$ $\therefore x$ lies between 0 and 48 or $0 < x < 48$.

Now , volume v of the box = $(96 − 2x)(96 − 2x)x$.

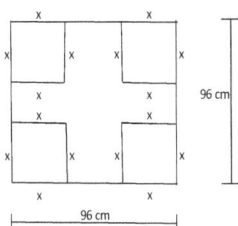

or $v = (96 − 2x)^2.x$ then $\dfrac{dv}{dx} = (96 − 2x)^2. 1 + x. 2(96 − 2x)(−2) = (96 − 2x)(96 − 2x − 4x)$ ∴ $\dfrac{dv}{dx} = (96 − 2x)(96 − 6x)$

For Minimum or Maximum is $\dfrac{dv}{dx} = 0$ i.e $(96 − 2x)(96 − 6x) = 0$ ∴ $96 − 2x = 0$ or $96 − 6x = 0$

or $x = 48$ and $x = 16$ ∴ $0 < x < 48$ ∴ *Rejecting* $x = 48$, *we have* $x = 16\,cm$

Now, $\dfrac{d^2v}{dx^2} = \dfrac{d}{dx}[(96 − 2x)(96 − 6x)] = (96 − 2x).(−6) + (96 − 6x)(−2) = 24x − 768$

$\left(\dfrac{d^2v}{dx^2}\right)_{x=16} = 24 \times 16 − 768 = 444 − 768 = −324 < 0,$ *Hence for* $x = 16$, *the volume is maximum.*

Hence the square of the side 16 cm should be cut from each corners.   Ans.

(23) The profit function $P(x)$ of a firm, selling x items per day is given by $P(x) = (100 − x)x − 1200$.

find the number of items the firm should manufacture to get maximum profit also find the maximum profit.

Solution: − It is given that x is the number of items produced and sold out by the firm every day.

In order to maximize profit, $P'(x) = 0$ i.e $\dfrac{dP}{dx} = 0$

or $P(x) = (100 − x)x − 1200$ then $P'(x) = 100 − 2x$, $P'(x) = 0$ i.e $100 − 2x = 0$ ∴ $2x = 100$ ∴ $x = 50$

Now , $P''(x) = −2$ is a negative quantity hence $P(x)$ is maximum at $x = 50$.

Thus , the firm should manufacture only 50 item a day to make maximum profit.

Now, Maximum Profit $= P(x) = (100 − x)x − 1200$ or Maximum Profit at $x = 50$

then Maximum Profit $= P(50) = (100 − 50)50 − 1200 = 2500 − 1200 = 1300$ or Maximum Profit is Rs. 1300 Ans.

Exercise − A6

Find the intervals for which the following function are increasing or decreasing: −

(1) (a) $f(x) = x^2 − 7x + 12$ (b) $f(x) = 3x^2 − 7x − 6$ (c) $f(x) = x^2 − 5x + 6$

(2) (a) $f(x) = x^3 + 5x^2 + 8x − 12$ (b) $f(x) = 2x^3 − 9x^2 − 24x + 7$ (c) $f(x) = x^3 + 3x − 5$

(3) (a) $f(x) = −3x^2 − 18x + 10$ (b) $f(x) = 2 − 36x − 15x^2 − 2x^3$ (c) $f(x) = (x − 1)^2(x + 1)$

(4) (a) $f(x) = \dfrac{x − 3}{x + 2}$, $x \neq −2$ (b) $y = \dfrac{x}{x − 2}$, $x \neq 2$ (c) $y = \dfrac{x}{3} − \dfrac{3}{x}$, $x \neq 0$

(5) (a) Prove that the function $\sin x$ is increasing in the interval $\left[\dfrac{\pi}{2}, 2\pi\right]$.

(b) Prove that the function $\log(\cos x)$ is decreasing in the interval $[0, \pi]$.

(c) Find the interval in which the function $\cos\left(2x + \dfrac{\pi}{4}\right), 0 < x < \pi$ is decreasing or increasing. find also the points on the graph

of the function at which the tangents are parallel to $x -$ axis. find all points of local maxima and local minima of the following function

also, find the maxima and minima of such points.

(6) (a) $x^2 - 6x + 7$ (b) $x^3 - 9x^2 + 12x + 18$ (c) $2x^3 - 24x^2 + 42x - 20$

(7) (a) $x^4 - 62x^2 + 120x + 18$ (b) $(x - 2)(x + 1)^2$ (c) $\dfrac{x + 2}{x^2 + x + 2}$

(8) (a) $x^3 - 7x^2 + 11x - 6$ (b) $2x^3 - 4x^2 - 8x - 3$ (c) $6x^2 - x - 2$

(9) (a) $(x - 1)(x + 2)^2$ (b) $\dfrac{x^2 - 3x + 2}{x^2 + 2}$ (c) $x^2 - 2x + \dfrac{3}{4}$

Find local maximum and local minimum for each of the following function using second order derivatives.

(10) (a) $2x^3 + 6x^2 - 18x + 12$ (b) $-x^3 + 6x^2 - 7$ (c) $(x + 1)(x - 2)^2$

(11) (a) $\cos 2x - x, \quad -\dfrac{\pi}{2} \le x \le \dfrac{\pi}{2}$ (b) $\cos x\,(1 - \sin x), \; 0 < x < \dfrac{\pi}{2}$ (c) $\sin x - \cos x, \; 0 < x < \dfrac{\pi}{2}$

(12) (a) $x^5 - 10x^4 + 20x^3 - 5$ (b) $x \log x$ (c) $(x + 1)e^x$ (d) $(1 + \log x)x$

(13) (a) $x^4 - 2x^3 - 3x^2 + 3$ (b) $(x - 3)(x + 2)^2$ (c) $\dfrac{2x - 1}{x^2 + 1}$ (d) $\dfrac{x^3 + 1}{x - 1}$

Find the local maxima or minima of the following function: −

(14) (a) $x^3 - 3x^2 - 9x + 1$ (b) $x^4 + 4x^2 - 12x + 9$ (c) $4x^2 - 12x + 9$

(15) (a) $\dfrac{1}{x^2 - 2}$ (b) $\dfrac{x}{(x - 1)(x - 3)}$ (c) $x\sqrt{2 + x}, \; x > -2$

(16) (a) $\sin x + \dfrac{1}{2}\cos 2x, \; 0 \le x \le \dfrac{\pi}{2}$ (b) $\cos 2x, \; 0 \le x \le 2\pi$ (c) $2 \sin x + x, \; 0 \le x \le 2\pi$

(17) For the what value of x lying in the close interval $[0,6]$, the slope of the tangent to $x^3 - 12x^2 + 36x + 18$

is maximum. Also find the point.

(18) Find the value of the greatest slope of a tangent to $-x^3 + 6x^2 + 3x - 15$ at a point of the other curve. find also the point.

(19) Find the two numbers whose sum is 15 and the square of one multiplied by the cube of the other is maximum.

(20) Prove that the perimeter of a right angled triangle of given hypotenuse is maximum when the triangle is isosceles.

(21) A movie theatre'smanagement is considering reducing the price of tickets from Rs. 55 in order to get more customers. After checking

out various facts they decide that the average number of customers per day "P" is given by the function where x is the amount of ticket

price reduced. Find the ticket price 0 that result in maximum revenue. $P = 500 + 100x$

Where x is the amount of ticket price reduced. Find the ticket price that result is maximum revenue.

(22) (a) Find the maximum and minimum value of $\dfrac{1 - x + x^2}{1 + x + x^2}$ for all real value of x.

(b) On the interval $[0,1]$ the function $x^{25}(1 - x)^{75}$ takes its maximum value at the point.

(c) Let f: IR $\to$ IR be defined as $f(x) = |x| + |x^2 - 1|$. The total number of points at which f attains either a local maximum

or a local minimum.

(d) Let p(x) be a real polynomial of least degree which has a local maximum at x = 1 and a local Minimum at x = 3.

If $p(1) = 6$ and $p(3) = 2$, then $p'(0)$ is

Answer – A6

(1) (a) Increasing for $x > 4$, $x < 3$ *and Decreasing for* $3 < x < 4$ (b) Increasing for $x > 3$, $x < -\dfrac{2}{3}$ and Decreasing for $-\dfrac{2}{3} < x < 3$

(c) Increasing for $x > 3$, $x < 2$ *and Decreasing for* $2 < x < 3$

(2) (a) Increasing for $x > -\dfrac{4}{3}$, $x < -2$ *and Decreasing for* $-2 < x < -\dfrac{4}{3}$ (b) Increasing for $x > 4$, $x < -1$ *and Decreasing for* $-1 < x < 4$

(3) (a) Increasing for $x < -3$ *and Decreasing* $x > -3$ (b) Increasing for $-3 < x < -2$ *and Decreasing for* $x > -2$, $x < -3$

(c) Increasing for $x > \dfrac{2}{3}$, $x < -\dfrac{1}{2}$ and Decreasing for $-\dfrac{1}{2} < x < \dfrac{2}{3}$

(4) (a) Increasing for $x < -2$ *and Decreasing for* $x > -2$ (b) Increasing for $x > 2$ *and Decreasing for* $x < 2$

(c) Increasing for $x > 3$, $x < 0$ *and Decreasing for* $0 < x < 3$

(6) (a) Local minimum at point x = 3, Point is (3, −2)

(b) Local minimum at $x = 3 + \sqrt{5}$ and maximum at $x = 3 - \sqrt{5}$, Points $\left(3 + \sqrt{5}, -22.36\right), \left(3 - \sqrt{5}, 22.36\right)$.

(c) Local minimum at x = 7 and maximum at x = 1, Points (1,0) & (7, −216).

(7) (a) Local minimum at x = −6, 5 and local maximum at x = 1, Points are (−6, −1638), (1,77) and (5, −307).

(b) Local minimum at x = 1 and local maximum at x = −1, Points are (−1,0) & (1, −4).

(c) Local minimum at x = −4 and local maximum at x = 0, Points are $\left(-4, -\dfrac{1}{7}\right)$ & (0,1).

(8) (a) Local minimum at $x = \dfrac{11}{3}$ and local maximum at x = 1, Points are (1, −1) & $\left(\dfrac{11}{3}, \dfrac{20611}{27}\right)$.

(b) Local minimum at x = 2 and local maximum at $x = -\dfrac{2}{3}$, Points are $\left(-\dfrac{2}{3}, -\dfrac{1}{27}\right)$ and (2, −19).

(c) Local minimum at $x = \dfrac{1}{12}$, Point is $\left(\dfrac{1}{12}, -\dfrac{49}{24}\right)$.

(9) (a) Local minimum at x = 0 and local maximum at x = −2, Points are (0, −4) & (−2,0).

(b) Local minimum at $x = \sqrt{2}$ and local maximum at $x = -\sqrt{2}$, Points are $\left(-\sqrt{2}, \dfrac{4 + 3\sqrt{2}}{4}\right)$ and $\left(\sqrt{2}, \dfrac{4 - 3\sqrt{2}}{4}\right)$.

(c) Local minimum at x = 1, Point is $\left(1, -\dfrac{1}{4}\right)$.

(10) (a) Local minimum at x = −1   (b) Local maximum at x = 2   (c) Local minimum at x = 1

(12) (a) Local minimum at $x = 0, 3 + \sqrt{3}$ and local maximum at $x = 3 - \sqrt{3}$

(13) (a) Local minimum at $x = \dfrac{1 + \sqrt{3}}{2}$ and local maximum at $x = \dfrac{1 - \sqrt{3}}{2}$ (b) Local minimum at $x = -\dfrac{1}{3}$ (c)

(14) (a) Local minima at x = 3 and local maxima at x = −1 (b) Local minima at x = 1 (c) Local minima at $x = \dfrac{3}{2}$

(17) Maximum at x = 2, Point is (2,50).   (22) (a) Maximum value = 3 and minimum value = $\dfrac{1}{3}$ (b) $\dfrac{1}{4}$ (c) 5

Tangent and Normal

(A) Tangent at (x, y): – Let $y = f(x)$ be a given curve and $P(x, y)$ and $Q(x + \delta x, y + \delta y)$ be two neighbouring point on it.

Equation of the line PQ is $Y - y = \dfrac{y + \delta y - y}{x + \delta x - x}(X - x)$ or $Y - y = \dfrac{\delta y}{\delta x}(X - x) \dots \dots \dots \dots \dots \dots (A)$

The line (A) will be a tangent to the given curve at P. If $Q \to P$, $\delta x \to 0$

we know that $\displaystyle\lim_{\delta x \to 0} \dfrac{\delta y}{\delta x} = \dfrac{dy}{dx}$

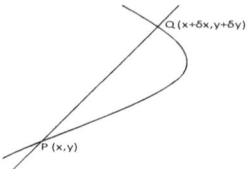

The equation of the tangent is $Y - y = \dfrac{dy}{dx}(X - x) \dots \dots \dots \dots \dots (B)$ (Formula)

Normal at (x, y): – The normal at (x, y) being perpendicular to tangent will have its slope is $-1 \Big/ \dfrac{dy}{dx}$ and

hence its equation is $Y - y = -\dfrac{1}{\dfrac{dy}{dx}}(X - x) \dots \dots \dots \dots \dots \dots (C)$ (Formula)

Geometrical meaning of $\dfrac{dy}{dx}$: – $\dfrac{dy}{dx}$ represent the slope of the tangent to the given curve

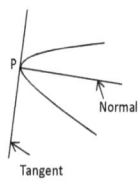

$y = f(x)$ at any point (x, y)

$$\therefore \quad \dfrac{dy}{dx} = \tan \Psi$$

where Ψ is the angle which the tangent to the curve makes with $+$ ve direction of x – axis.

find the tangent at any point (x_1, y_1) then $\left(\dfrac{dy}{dx}\right)_{(x_1, y_1)}$.

The value of $\dfrac{dy}{dx}$ at (x_1, y_1) will represent the slope of the tangent and hence its equation in this case will be

$$\boxed{y - y_1 = \left(\dfrac{dy}{dx}\right)_{(x_1, y_1)}(x - x_1)}$$

Normal equation is $\boxed{y - y_1 = \dfrac{-1}{\left(\dfrac{dy}{dx}\right)_{(x_1, y_1)}}(x - x_1)}$ slope of the tangent $= m_1 = \dfrac{dy}{dx} = \left(\dfrac{dy}{dx}\right)_{(x_1, y_1)}$

slope of the normal $= \dfrac{-1}{m_1} = -\dfrac{1}{\dfrac{dy}{dx}} = \dfrac{-1}{\left(\dfrac{dy}{dx}\right)_{(x_1, y_1)}}$

■ If a tangent is parallel to $x -$ axis or normal is perpendicular to $x -$ axis then $m = 0$ so that $\dfrac{dy}{dx} = 0$.

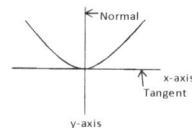

■ If a tangent is perpendicular to $x -$ axis or normal is parallel to $x -$ axis then $m = \infty$, $\dfrac{dy}{dx} = \infty$ or its reciprocal $\dfrac{dx}{dy} = 0$

Parametric form: $- \quad \dfrac{dy}{dx} = \dfrac{dy}{dt} \div \dfrac{dx}{dt}$

If the equation of the curve be $x = a \sin t$, $y = b \cos t$ then $x = a \sin t \dots \dots \dots (i)$

and $y = b \cos t \dots \dots \dots (ii)$

Differentiate equation (i) and (ii) with respect to t, we get $\quad \therefore \quad \dfrac{dx}{dt} = a \cos t$ and $\dfrac{dy}{dt} = -b \sin t$

then $\dfrac{dy}{dx} = \dfrac{dy}{dt} \div \dfrac{dx}{dt} = -b \sin t \div a \cos t = \dfrac{-b \sin t}{a \cos t} = -\dfrac{b}{a} \tan t \dots \dots \dots \dots \dots \dots (iii)$

Divide equation $^{(i)}/_{(ii)}$ we get, $\dfrac{x}{y} = \dfrac{a \sin t}{b \cos t} = \dfrac{a}{b} \tan t \quad \therefore \quad \tan t = \dfrac{bx}{ay}$

put value $\tan t = \dfrac{bx}{ay}$ in equation (iii), we have $\quad \dfrac{dy}{dx} = -\dfrac{b}{a} \tan t = -\dfrac{b}{a} \cdot \dfrac{bx}{ay} = -\dfrac{b^2}{a^2}\left(\dfrac{x}{y}\right) \quad \therefore \quad \dfrac{dy}{dx} = -\dfrac{b^2}{a^2}\left(\dfrac{x}{y}\right)$

Tangent is $y - b \cos t = -\dfrac{b^2 x}{a^2 y}(x - a \sin t)$ and Normal is $y - b \cos t = \dfrac{a^2 y}{b^2 x}(x - a \sin t)$

Tangent is parallel to $x -$ axis if $y = 0$ and Tangent is perpendicular to $x -$ axis if $x = 0$.

Partial Differentiation: $-$ If the equation of the curve is of the form $f(x, y) = c$ or 0 then a convenient method of finding $\dfrac{dy}{dx}$ is by the

help of partial derivatives $\dfrac{dy}{dx} = -\dfrac{f_x}{f_y}$, where f_x is differentiation of $f(x, y)$ with respect to x but y constant and f_y is differentiation

of $f(x, y)$ w. r. t y but x constant.

e.g. if $x^2 + y^2 - 2axy = 0 = f(x, y)$ then $f_x = 2x - 2ay$, $f_y = 2y - 2ax$ or $\dfrac{dy}{dx} = -\dfrac{f_x}{f_y} = -\dfrac{2x - 2ay}{2y - 2ax} = -\dfrac{2(x - ay)}{2(y - ax)} = -\dfrac{x - ay}{y - ax}$

Equation of tangent can be written as $Y - y = -\dfrac{f_x}{f_y}(X - x)$ or $(X - x)f_x + (Y - y)f_y = 0 \dots \dots \dots \dots \dots (i)$

or $(X - x)f_{x(a,b)} + (Y - b)f_{y(a,b)} = 0$ at a point (a, b)

Normal will be $\dfrac{X - x}{f_x} = \dfrac{Y - y}{f_y} \dots \dots \dots \dots (ii)$ or $\dfrac{X - a}{f_{x(a,b)}} = \dfrac{Y - b}{f_{y(a,b)}}$ at the point (a, b).

(B) Angle of intersection of two curves: $-$ By angle of intersection of two curves we mean the angle between the tangent to the two

curves at their common point of intersection hence if θ be the acute angle between the tangents then $\tan \theta = \left|\dfrac{m_1 - m_2}{1 + m_1 m_2}\right|$

where $m_1 = \dfrac{dy}{dx}$ at the common point for 1st curve, $m_2 = \dfrac{dy}{dx}$ at the common point for 2nd curve.

■ Condition for orthogonal intersection: − Two curves are said to cut orthogonally if the angle between them is a right angle.

i.e $\theta = 90^0$ ∴ $\tan 90^0 = \infty$ or $1 + m_1 m_2 = 0$ or $m_1 m_2 = -1$ or $\left(\dfrac{dy}{dx}\right)_I \left(\dfrac{dy}{dx}\right)_{II} = -1$

■ Condition for the two curves to touch: − If the two curves touch then $\theta = 0^0$, $\tan\theta = \left|\dfrac{m_1 - m_2}{1 + m_1 m_2}\right|$

$$\text{or } \tan 0^0 = \left|\frac{m_1 - m_2}{1 + m_1 m_2}\right| \quad \therefore \ m_1 - m_2 = 0 \quad \text{or } m_1 = m_2 \quad \text{or } \left(\frac{dy}{dx}\right)_I = \left(\frac{dy}{dx}\right)_{II}$$

If the two curves be $f(x, y) = 0$ and $\phi(x, y) = 0$ then $\left(\dfrac{dy}{dx}\right)_I = -\dfrac{f_x}{f_y}$ and $\left(\dfrac{dy}{dx}\right)_{II} = -\dfrac{\phi_x}{\phi_y}$

$$\tan\theta = \left|\frac{\left(-\frac{f_x}{f_y}\right)\left(-\frac{\phi_x}{\phi_y}\right)}{1 + \left(-\frac{f_x}{f_y}\right)\left(-\frac{\phi_x}{\phi_y}\right)}\right| \quad \therefore \ \tan\theta = \left|\frac{f_x \phi_y - f_y \phi_x}{f_x \phi_x + f_y \phi_y}\right|$$

■ Condition for touching: − If $\theta = 0^0$ then $\tan\theta = 0$ ∴ $\left|\dfrac{f_x \phi_y - f_y \phi_x}{f_x \phi_x + f_y \phi_y}\right| = 0$ or $f_x \phi_y - f_y \phi_x = 0$

$$\text{or } f_x \phi_y = f_y \phi_x \quad \text{or } \frac{f_x}{f_y} = \frac{\phi_x}{\phi_y}$$

■ Condition to cut orthogonally: − If $\theta = 90^0$ then $\tan\theta = \infty$ ∴ $\left|\dfrac{f_x \phi_y - f_y \phi_x}{f_x \phi_x + f_y \phi_y}\right| = \infty$ or $f_x \phi_x + f_y \phi_y = 0$.

(C) Intercepts of tangent on the axes: − Find the equation of the tangent put $y = 0$ and find the value of x which will be intercept on axis of x. Then put $x = 0$ and find the value of y which will be intercept on y − axis.

Length of tangent and normal: − Length of tangent = PT and length of normal = PN

where P is the point of contact, T and N is the point where T (Tangent) and N (Normal) meets the axis of x.

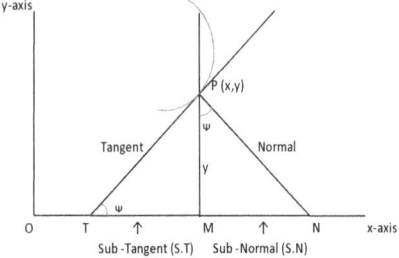

From the figure , $\dfrac{y}{PT} = \sin\Psi$ ∴ $PT = y \csc\Psi$ ……………. (i) and $\dfrac{y}{PN} = \cos\Psi$ ∴ $PN = y \sec\Psi$ ………… (ii)

Now, $\tan\Psi = \dfrac{dy}{dx} = y'$ (say) ∴ $\sec\Psi = \sqrt{1 + \tan^2\Psi} = \sqrt{1 + y'^2}$ ………………… (iii)

$\csc\Psi = \sqrt{1 + \cot^2\Psi} = \sqrt{1 + \dfrac{1}{\tan^2\Psi}} = \sqrt{\dfrac{1 + \tan^2\Psi}{\tan^2\Psi}} = \dfrac{\sqrt{1 + \tan^2\Psi}}{\tan\Psi} = \dfrac{\sqrt{1 + y'^2}}{y'}$ ……………… (iv)

put value $\csc\Psi$ in equation (i) , we have ∴ $PT = y \csc\Psi = y \dfrac{\sqrt{1 + y'^2}}{y'}$ or $\boxed{\therefore PT = \dfrac{y}{y'}\sqrt{1 + y'^2}}$

put value $\sec\Psi$ in equation (ii) , we have ∴ $PN = y \sec\Psi = y\sqrt{1 + y'^2}$ or $\boxed{\therefore PN = y\sqrt{1 + y'^2}}$

- Condition for a given line to touch a given curve: $-\ Y - y = \dfrac{dy}{dx}(X - x)$

compare this with the given line $aX + bY + C = 0$ and eliminate x and y.

- Length of sub $-$ tangent (TM) $= y \cot \Psi = \dfrac{y}{\tan \Psi}$ $\quad \therefore\ TM = \dfrac{y}{\dfrac{dy}{dx}} = \dfrac{y}{y'}$ $\quad \left\{\tan \Psi = \dfrac{dy}{dx} = y'\ (\text{say})\right\}$

- Length of sub $-$ normal (MN) $= y \tan \Psi = y\dfrac{dy}{dx} = yy'$ $\quad \therefore\ MN = \left(\dfrac{dy}{dx}\right)y = yy'$

- Tangent to the curve $y = f(x)$, which is parallel to the line $ax + by + c = 0$ then slope of the line $= -\dfrac{a}{b} = $ slope of the tangent

If perpendicular to the line $ax + by + c = 0$ then slope of the line is $-\dfrac{a}{b}$ but slope of the tangent is $\dfrac{-1}{-\dfrac{a}{b}} = \dfrac{b}{a}$.

- Normal to the curve $y = f(x)$, which is parallel to the line $ax + by + c = 0$ then slope of the line is $-\dfrac{a}{b} = $ slope of the normal

If perpendicular to the $ax + by + c = 0$ then slope of the line is $\left(-\dfrac{a}{b}\right)$ but slope of the normal is $\left(\dfrac{b}{a}\right)$.

Solved Example

(1) Find the equation of the tangent to the curve at any point (x, y), $\quad \dfrac{x^p}{a^p} + \dfrac{y^p}{b^p} = 1 \ldots\ldots\ldots\ldots$ (i)

Solution: $-\quad \dfrac{x^p}{a^p} + \dfrac{y^p}{b^p} = 1\quad$ Differentiating w. r. t x, we have

$$p.\dfrac{x^{p-1}}{a^p} + p.\dfrac{y^{p-1}}{b^p}.\dfrac{dy}{dx} = 0 \quad \therefore\ \dfrac{dy}{dx} = -\dfrac{p.\dfrac{x^{p-1}}{a^p}}{p.\dfrac{y^{p-1}}{b^p}} = -\left(\dfrac{b}{a}\right)^p.\left(\dfrac{x}{y}\right)^{p-1} = -\left(\dfrac{x}{a}\right)^{p-1}.\dfrac{1}{a}.b\left(\dfrac{b}{y}\right)^{p-1}$$

Equation of tangent is $Y - y = \dfrac{dy}{dx}(X - x)\ $ or $\ Y - y = -\left(\dfrac{b}{a}\right)^p.\left(\dfrac{x}{y}\right)^{p-1}(X - x)$

or $Y - y = -\left(\dfrac{x}{a}\right)^{p-1}.\dfrac{1}{a}.b\left(\dfrac{b}{y}\right)^{p-1}(X - x)\ $ or $\ \dfrac{Y - y}{-b\left(\dfrac{b}{y}\right)^{p-1}} = \left(\dfrac{x}{a}\right)^{p-1}.\dfrac{1}{a}(X - x)$

or $-\dfrac{Y}{b}\left(\dfrac{y}{b}\right)^{p-1} + \dfrac{y}{b}\left(\dfrac{y}{b}\right)^{p-1} = \dfrac{X}{a}\left(\dfrac{x}{a}\right)^{p-1} - \dfrac{x}{a}\left(\dfrac{x}{a}\right)^{p-1}\ $ or $\ \dfrac{X}{a}\left(\dfrac{x}{a}\right)^{p-1} - \left(\dfrac{x}{a}\right)^p = -\dfrac{Y}{b}\left(\dfrac{y}{b}\right)^{p-1} + \left(\dfrac{y}{b}\right)^p$

or $\dfrac{X}{a}\left(\dfrac{x}{a}\right)^{p-1} + \dfrac{Y}{b}\left(\dfrac{y}{b}\right)^{p-1} = \left(\dfrac{y}{b}\right)^p + \left(\dfrac{x}{a}\right)^p\ \therefore\ \dfrac{X}{a}\left(\dfrac{x}{a}\right)^{p-1} + \dfrac{Y}{b}\left(\dfrac{y}{b}\right)^{p-1} = 1$ by equation (i)

Another method: $-\quad f_x = p.\dfrac{x^{p-1}}{a^p}, \quad f_y = p.\dfrac{y^{p-1}}{b^p}$

Equation of tangent is $(X - x)f_x + (Y - y)f_y = 0\quad$ (formula) $\quad$ or $(X - x)p.\dfrac{x^{p-1}}{a^p} + (Y - y)p.\dfrac{y^{p-1}}{b^p} = 0$

$$\text{or } \dfrac{X}{a}\left(\dfrac{x}{a}\right)^{p-1} + \dfrac{Y}{b}\left(\dfrac{y}{b}\right)^{p-1} = \left(\dfrac{y}{b}\right)^p + \left(\dfrac{x}{a}\right)^p = 1 \text{ by equation (i) Ans.}$$

(2) Find the equation of the tangent to the curve at any point (x, y)

(a) $\left(\dfrac{x}{a}\right)^{\frac{1}{3}} + \left(\dfrac{y}{b}\right)^{\frac{1}{3}} = 1$ (b) $\left(\dfrac{x}{a}\right)^{m/m-2} + \left(\dfrac{y}{b}\right)^{m/m-2} = 1$ (c) $x^{\frac{1}{3}} + y^{\frac{1}{3}} = a^{\frac{1}{3}}$ (d) $\dfrac{x^3}{a^3} + \dfrac{y^3}{b^3} = 1$

Solution: $-$ (a) $\left(\dfrac{x}{a}\right)^{\frac{1}{3}} + \left(\dfrac{y}{b}\right)^{\frac{1}{3}} = 1$, $\dfrac{x^p}{a^p} + \dfrac{y^p}{b^p} = 1$ (above question) and equation of tangent is $\dfrac{X}{a}\left(\dfrac{x}{a}\right)^{p-1} + \dfrac{Y}{b}\left(\dfrac{y}{b}\right)^{p-1} = 1$

put $p = \dfrac{1}{3}$ equation of tangent is $\dfrac{X}{a}\left(\dfrac{x}{a}\right)^{p-1} + \dfrac{Y}{b}\left(\dfrac{y}{b}\right)^{p-1} = 1$ or $\dfrac{X}{a}\left(\dfrac{x}{a}\right)^{\frac{1}{3}-1} + \dfrac{Y}{b}\left(\dfrac{y}{b}\right)^{\frac{1}{3}-1} = 1$

or $\dfrac{X}{a}\left(\dfrac{x}{a}\right)^{-\frac{2}{3}} + \dfrac{Y}{b}\left(\dfrac{y}{b}\right)^{-\frac{2}{3}} = 1$ or $\dfrac{X}{a}\left(\dfrac{a}{x}\right)^{\frac{2}{3}} + \dfrac{Y}{b}\left(\dfrac{b}{y}\right)^{\frac{2}{3}} = 1$ or $\dfrac{X}{x^{\frac{2}{3}}.a^{\frac{1}{3}}} + \dfrac{Y}{y^{\frac{2}{3}}.a^{\frac{1}{3}}} = 1$ Ans.

(b) put $p = \dfrac{m}{m-2}$ $\therefore$ equation of tangent is $\dfrac{X}{a}\left(\dfrac{x}{a}\right)^{m/(m-2)-1} + \dfrac{Y}{b}\left(\dfrac{y}{b}\right)^{m/(m-2)-1} = 1$ or $\dfrac{X}{a}\left(\dfrac{x}{a}\right)^{2/(m-2)} + \dfrac{Y}{b}\left(\dfrac{y}{b}\right)^{2/(m-2)} = 1$ Ans.

(c) put $p = \dfrac{1}{3}$ $\therefore$ equation of tangent is $\dfrac{X}{a}\left(\dfrac{x}{a}\right)^{p-1} + \dfrac{Y}{b}\left(\dfrac{y}{b}\right)^{p-1} = 1$ $\left[x^{\frac{1}{3}} + y^{\frac{1}{3}} = a^{\frac{1}{3}} \text{ or } \left(\dfrac{x}{a}\right)^{\frac{1}{3}} + \left(\dfrac{y}{b}\right)^{\frac{1}{3}} = 1 \right]$

here $a = b$, then $\dfrac{X}{a}\left(\dfrac{x}{a}\right)^{\frac{1}{3}-1} + \dfrac{Y}{a}\left(\dfrac{y}{a}\right)^{\frac{1}{3}-1} = 1$ [put b = a] or $\dfrac{X}{a}\left(\dfrac{x}{a}\right)^{-\frac{2}{3}} + \dfrac{Y}{a}\left(\dfrac{y}{a}\right)^{-\frac{2}{3}} = 1$ or $\dfrac{X}{x^{\frac{2}{3}}.a^{\frac{1}{3}}} + \dfrac{Y}{y^{\frac{2}{3}}.a^{\frac{1}{3}}} = 1$ or $\dfrac{X}{x^{\frac{2}{3}}} + \dfrac{Y}{y^{\frac{2}{3}}} = a^{\frac{1}{3}}$ Ans.

(d) put $p = 3$ $\therefore$ equation of tangent is $\dfrac{X}{a}\left(\dfrac{x}{a}\right)^{p-1} + \dfrac{Y}{b}\left(\dfrac{y}{b}\right)^{p-1} = 1$ or $\dfrac{X}{a}\left(\dfrac{x}{a}\right)^{3-1} + \dfrac{Y}{b}\left(\dfrac{y}{b}\right)^{3-1} = 1$

$$\dfrac{X}{a}\left(\dfrac{x}{a}\right)^{2} + \dfrac{Y}{b}\left(\dfrac{y}{b}\right)^{2} = 1 \quad \text{Ans.}$$

(3) Does the straight line $\dfrac{x}{a} + \dfrac{y}{b} = 1$ touch the curve $\left(\dfrac{y}{b}\right)^2 + \left(\dfrac{x}{a}\right)^2 = 1$?

If it touches then determine the co − ordinates of the point of contact.

Solution: − $\left(\dfrac{y}{b}\right)^2 + \left(\dfrac{x}{a}\right)^2 = 1 \ldots\ldots\ldots\ldots\ldots$ (i) Differentiating equation (i) with respect to x, we have

$$\therefore \dfrac{1}{a^2}.2x + \dfrac{1}{b^2}.2y.\dfrac{dy}{dx} = 0 \text{ or } \dfrac{1}{b^2}.2y.\dfrac{dy}{dx} = -\dfrac{1}{a^2}.2x \text{ or } \dfrac{dy}{dx} = \dfrac{\frac{2x}{a^2}}{\frac{2y}{b^2}} = -\dfrac{b^2 x}{a^2 y}$$

Equation of tangent is $Y - y = \dfrac{dy}{dx}(X - x)$ or $Y - y = -\dfrac{b^2 x}{a^2 y}(X - x)$

or $Y - y = -\dfrac{b^2}{y}.\dfrac{x}{a^2}(X - x)$ or $\dfrac{Yy}{b^2} - \dfrac{y^2}{b^2} = \dfrac{x^2}{a^2} - \dfrac{Xx}{a^2}$ or $\dfrac{Yy}{b^2} + \dfrac{Xx}{a^2} = \left(\dfrac{x}{a}\right)^2 + \left(\dfrac{y}{b}\right)^2 = 1$ by equation (i)

or $\dfrac{Y}{b}\dfrac{y}{b} + \dfrac{X}{a}\dfrac{x}{a} = 1$,   comparing with straight line $\dfrac{x}{a} + \dfrac{y}{b} = 1$ and $\dfrac{Y}{b}\dfrac{y}{b} + \dfrac{X}{a}\dfrac{x}{a} = 1$ then $\dfrac{x}{a} = 1, \dfrac{y}{b} = 1$ or $x = a, y = b$

and equation of tangent becomes $\dfrac{X}{a} + \dfrac{Y}{b} = 1$ hence the given line touches and the point of contact is (a, b).

(4) (a) $x = a\sin^2\theta$, $y = a\cos^2\theta$ or $x + y = a$ then find equation of tangent and normal.

Solution: − Parametic form , $x = a\sin^2\theta$, $y = a\cos^2\theta$ then $\dfrac{dx}{d\theta} = 2a\cos\theta$, $\dfrac{dy}{d\theta} = -2a\sin\theta$ $\therefore$ $\dfrac{dy}{dx} = -\dfrac{2a\sin\theta}{2a\cos\theta} = -\tan\theta$

Eliminating θ form , $x = a\sin^2\theta$, $\left(\dfrac{x}{a}\right)^{\frac{1}{2}} = \sin\theta$ and $y = a\cos^2\theta$, $\left(\dfrac{y}{a}\right)^{\frac{1}{2}} = \cos\theta$ or $\sin^2\theta + \cos^2\theta = 1$ or $\dfrac{x}{a} + \dfrac{y}{a} = 1$ or $x + y = a$

Equation of the tangent is $Y - y = \dfrac{dy}{dx}(X - x)$ or $y - a\cos^2\theta = -\tan\theta\,(x - a\sin^2\theta)$

$$\text{or } y - a\cos^2\theta = -\dfrac{\sin\theta}{\cos\theta}(x - a\sin^2\theta) \text{ or } y\cos\theta - a\cos^3\theta = a\sin^3\theta - x\sin\theta$$

or $x\sin\theta + y\cos\theta = a\sin^3\theta + a\cos^3\theta = a(\sin^3\theta + \cos^3\theta)$ Ans.

or $x\sin\theta + y\cos\theta = a[(\sin\theta + \cos\theta)^3 - 3\sin\theta\cos\theta\,(\sin\theta + \cos\theta)]$ $[\because a^3 + b^3 = (a + b)^3 - 3ab(a + b)]$

or $x \sin\theta + y \cos\theta = a(\sin\theta + \cos\theta)[(\sin\theta + \cos\theta)^2 - 3\sin\theta\cos\theta] = a(\sin\theta + \cos\theta)[\sin^2\theta + \cos^2\theta + 2\sin\theta\cos\theta - 3\sin\theta\cos\theta]$

or $x \sin\theta + y \cos\theta = a(\sin\theta + \cos\theta)(1 - \sin\theta\cos\theta) = a(\sin^3\theta + \cos^3\theta)$

Equation of the normal is $Y - y = -\dfrac{1}{\frac{dy}{dx}}(X - x)$ or $y - a\cos^2\theta = \dfrac{1}{\tan\theta}(x - a\sin^2\theta)$

or $y - a\cos^2\theta = \dfrac{\cos\theta}{\sin\theta}(x - a\sin^2\theta)$ or $y \sin\theta - a\cos^2\theta\sin\theta = x\cos\theta - a\sin^2\theta\cos\theta$

or $x\cos\theta - y\sin\theta = a\sin^2\theta\cos\theta - a\cos^2\theta\sin\theta = a\sin\theta\cos\theta(\sin\theta - \cos\theta)$ or $x\cos\theta - y\sin\theta = \dfrac{a}{2}\sin 2\theta(\sin\theta - \cos\theta)$ Ans.

(b) Normal to $x + y = a$ in the form $x\cos\theta - y\sin\theta = \dfrac{a}{2}\sin 2\theta(\sin\theta - \cos\theta)$ where θ is the angle which the normal

makes with the axis of x.

Solution: $- x + y = a$ $\therefore$ Differentiate, $1 + \dfrac{dy}{dx} = 0$ $\therefore$ $\dfrac{dy}{dx} = -1$

slope of the normal is $1 = \tan\theta$ or $\tan\theta = 1$ or $\dfrac{\sin\theta}{\cos\theta} = 1$ or $\sin\theta = \cos\theta$

squaring $\sin^2\theta = \cos^2\theta$ $\therefore$ $\sin^2\theta + \cos^2\theta = 1$, $x = a\sin^2\theta$, $y = a\cos^2\theta$

Hence the normal whose slope is $\tan\theta$ is given by $y - a\cos^2\theta = \dfrac{\sin\theta}{\cos\theta}(x - a\sin^2\theta)$

or $x\cos\theta - y\sin\theta = \dfrac{a}{2}\sin 2\theta(\sin\theta - \cos\theta)$ as in Q. No. $-(4)(a)$

(5) (a) The equation of the tangent to the curve $x^3 + y^2 = 3xy$ at the point $(2,2)$.

Solution: $-$ $x^3 + y^2 = 3xy$ $\therefore$ $3x^2 + 2y\dfrac{dy}{dx} = 3x\dfrac{dy}{dx} + 3y$ or $2y\dfrac{dy}{dx} - 3x\dfrac{dy}{dx} = 3y - 3x^2$

or $\dfrac{dy}{dx}(2y - 3x) = 3(y - x^2)$ $\therefore$ $\dfrac{dy}{dx} = \dfrac{3(y - x^2)}{2y - 3x}$ at a point $(2,2)$ $\therefore$ $\left(\dfrac{dy}{dx}\right)_{(2,2)} = \dfrac{3(2 - 4)}{4 - 6} = \dfrac{-6}{-2} = 3$ $\therefore$ $\dfrac{dy}{dx} = 3$

Equation of the tangent is $y - 2 = 3(x - 2)$ or $y - 2 = 3x - 6$ $\boxed{\therefore\ 3x - y = 4\ \text{Ans.}}$

(b) The equation of the normal to the curve $y = 4 - x^2$ at the point $(1,3)$.

Solution: $-$ $y = 4 - x^2$ $\therefore$ $\dfrac{dy}{dx} = 0 - 2x$ or $\dfrac{dy}{dx} = -2x$ at a point $(1,3)$

$\therefore$ $\left(\dfrac{dy}{dx}\right)_{(1,3)} = -2 \times 1 = -2$ slope of tangent $(m_1) = -2$ and slope of normal $(m_2) = -\dfrac{-1}{2} = \dfrac{1}{2}$

Equation of normal is $y - 3 = \dfrac{1}{2}(x - 1)$ or $2y - 6 = x - 1$ $\boxed{\therefore\ x - 2y = -5\ \text{Ans.}}$

(6) (a) Tangent to the parabola $x^2 = 4ay$ in the form $y = mx - 2am^2 + am^2$ or $y = m(x - am)$.

Solution: $-$ $x^2 = 4ay$ $\therefore$ $2x = 4a\dfrac{dy}{dx}$ or $\dfrac{dy}{dx} = \dfrac{x}{2a} = m$ (say)

slope of tangent $m = \dfrac{x}{2a}$ $\therefore$ $x = 2am$, $y = \dfrac{x^2}{4a} = \dfrac{4a^2m^2}{4a} = am^2$ $\therefore$ $x = 2am$, $y = am^2$

Hence, the tangent whose slope is m will be the point $(2am, am^2)$

so that its equation is $y - am^2 = m(x - 2am)$ $\therefore$ $y - am^2 = mx - 2am^2$ or $y = mx - 2am^2 + am^2$

$$\boxed{\therefore\ y = mx - am^2\ \text{or}\ y = m(x - am)\ \text{Ans.}}$$

(b) Tangent to the curve $y^2 = 4x$ which passes through the point (1,2).

Solution: $-\quad y^2 = 4x \quad \therefore \quad 2y\dfrac{dy}{dx} = 4 \quad$ or $\quad \dfrac{dy}{dx} = \dfrac{2}{y} = m \quad$ or $y = \dfrac{2}{m}$, $x = \dfrac{y^2}{4} = \dfrac{4}{4m^2} = \dfrac{1}{m^2}$

Tangent is $\quad y - \dfrac{2}{m} = m\left(x - \dfrac{1}{m^2}\right)$, if it passes through the point (1,2) then $\quad 2 - \dfrac{2}{m} = m\left(1 - \dfrac{1}{m^2}\right)$

or $\quad \dfrac{2m - 2}{m} = m\left(\dfrac{m^2 - 1}{m^2}\right) \quad$ or $\quad 2m - 2 = m^2 - 1 \quad$ or $\quad m^2 - 2m + 1 = 0 \quad$ or $\quad (m-1)^2 = 0 \quad$ or $\quad m = 1$

$$\therefore \quad \text{Required tangent is} \quad y - 2 = 1(x-1) \quad \text{or} \quad y - 2 = x - 1 \quad \boxed{\therefore \; x - y = -1 \quad \text{Ans.}}$$

(7) (a) The equation of the tangent at the point P(t), where t is any parameter to the parabola $x^2 = 2ay$.

Solution: $-\quad x^2 = 2ay \quad \therefore \quad 2x = 2a\dfrac{dy}{dx} \quad$ or $\quad \dfrac{dy}{dx} = \dfrac{x}{a}$

Equation of the tangent is $\quad Y - y = \dfrac{dy}{dx}(X - x) \quad$ or $\quad Y - y = \dfrac{x}{a}(X - x) \quad$ or $\quad aY - ay = Xx - x^2$

or $\quad Xx = aY - ay + x^2 \quad$ or $\quad Xx = aY - ay + 2ay \quad$ or $\quad Xx = aY + ay \quad$ at a point $P(t)$, $(at^2, 2at)$

where t is a parameter. $\quad \therefore \quad Xat^2 = aY + a.2at \quad$ or $\quad aXt^2 = aY + 2a^2t \quad$ [Put X = x and Y = y]

$$\text{or} \quad axt^2 = ay + 2a^2t \quad \boxed{\therefore \; xt^2 = y + 2at \quad \text{Ans.}}$$

(b) The parametric equation of a curve are $x = \sqrt{2}\,a\cos 2t$, $y = 2a\sin t$. find the equation of tangent in the form

$mx - y = \dfrac{a}{2\sqrt{2}\,m}(1 + 4m^2)$ where m is the slope of the tangent. hence prove that two perpendicular tangent meet on the line $x = \dfrac{5a}{2\sqrt{2}}$.

Solution: $-\quad x = \sqrt{2}\,a\cos 2t \quad \therefore \quad \dfrac{dx}{dt} = -2\sqrt{2}\,a\sin 2t \quad$ and $\quad y = 2a\sin t \quad \therefore \quad \dfrac{dy}{dt} = 2a\cos t$

or $\quad \dfrac{dy}{dx} = \dfrac{\dfrac{dy}{dt}}{\dfrac{dx}{dt}} = \dfrac{2a\cos t}{-2\sqrt{2}\,a\sin 2t} = \dfrac{2a\cos t}{-4\sqrt{2}\,a\sin t\cos t} = -\dfrac{1}{2\sqrt{2}\sin t} \quad \therefore \quad \dfrac{dy}{dx} = -\dfrac{1}{2\sqrt{2}\sin t} = m \quad$ or $\quad \sin t = -\dfrac{1}{2\sqrt{2}\,m}$

$\therefore \quad x = \sqrt{2}\,a\cos 2t = \sqrt{2}\,a(1 - 2\sin^2 t) = \sqrt{2}\,a\left[1 - 2\left(-\dfrac{1}{2\sqrt{2}\,m}\right)^2\right] = \sqrt{2}\,a\left(1 - \dfrac{2}{8m^2}\right) = \sqrt{2}\,a\left(1 - \dfrac{1}{4m^2}\right)$

$$\therefore \quad x = \sqrt{2}\,a\left(1 - \dfrac{1}{4m^2}\right), \quad y = 2a\sin t = 2a\left(-\dfrac{1}{2\sqrt{2}\,m}\right) = -\dfrac{a}{\sqrt{2}\,m}$$

Hence, equation of tangent is $\quad Y - y = m(X - x) \quad$ or $\quad y + \dfrac{a}{\sqrt{2}m} = m\left[x - \sqrt{2}\,a\left(1 - \dfrac{1}{4m^2}\right)\right]$

or $\quad y + \dfrac{a}{\sqrt{2}\,m} = mx - \sqrt{2}am + \dfrac{\sqrt{2}am}{4m^2} \quad$ or $\quad y + \dfrac{a}{\sqrt{2}\,m} = mx - \sqrt{2}am + \dfrac{a}{2\sqrt{2}m}$

or $\quad mx - y = \dfrac{a}{\sqrt{2}\,m} + \sqrt{2}am - \dfrac{a}{2\sqrt{2}m} \quad$ or $\quad mx - y = \dfrac{a}{\sqrt{2}\,m}\left(1 - \dfrac{1}{2}\right) + \sqrt{2}am = \dfrac{a}{2\sqrt{2}\,m} + \sqrt{2}am = \dfrac{a + 2\sqrt{2}\,m.\sqrt{2}am}{2\sqrt{2}\,m} = \dfrac{a + 4am^2}{2\sqrt{2}\,m}$

or $\quad mx - y = \dfrac{a}{2\sqrt{2}\,m}(1 + 4m^2) \quad$ or $\quad mx - y = \dfrac{a}{2\sqrt{2}\,m}(1 + 4m^2)$

Equation of tangent in terms of its slope m, if it passes through the point (x_1, y_1) then

or $\quad mx_1 - y_1 = \dfrac{a}{2\sqrt{2}\,m}(1 + 4m^2) \quad$ or $\quad 2\sqrt{2}\,x_1 m^2 - 2\sqrt{2}\,my_1 = a + 4am^2$

or $\quad 2\sqrt{2}\,m^2 x_1 - 4am^2 - 2\sqrt{2}\,y_1 m - a = 0 \quad$ or $\quad m^2(2\sqrt{2}\,x_1 - 4a) - 2\sqrt{2}\,y_1 m - a = 0$

Above is a quadratic equation in m which shows that there will be two tangent which pass through (x_1, y_1)

if the two tangent be perpendicular, then $m_1 m_2 = -1$ or $\dfrac{-a}{(2\sqrt{2}\, x_1 - 4a)} = -1$

or $-a = -(2\sqrt{2}\, x_1 - 4a)$ or $-a = -2\sqrt{2}\, x_1 + 4a$ or $2\sqrt{2}\, x_1 = 4a + a = 5a$ $\therefore$ $x_1 = \dfrac{5a}{2\sqrt{2}}$

Thus, the two perpendicular tangent meet at $x = \dfrac{5a}{2\sqrt{2}}$ Ans.

(8) Find the equation of tangent and normal to the curve: −

(a) $y^2(2a - y) = x^2(3a + y)$ at the points where $y = a$. (b) $x^2(2a + x) = y^2(3a - x)$ at the point $x = \dfrac{a}{2}$.

Solution: −   (a) $y^2(2a - y) = x^2(3a + y)$ at the points where $y = a$ then

or $a^2(2a - a) = x^2(3a + a)$ or $a^3 = 4ax^2$ or $x^2 = \dfrac{a^3}{4a} = \dfrac{a^2}{4}$ $\therefore$ $x = \pm\dfrac{a}{2}$

Hence, the point are $P\left(\dfrac{a}{2}, a\right)$ and $Q\left(-\dfrac{a}{2}, a\right)$

or $y^2(2a - y) = x^2(3a + y)$ or $2ay^2 - y^3 = 3ax^2 + x^2 y$ $\therefore$ $4ay\dfrac{dy}{dx} - 3y^2\dfrac{dy}{dx} = 6ax + x^2\dfrac{dy}{dx} + y.2x$

or $\dfrac{dy}{dx}(4ay - 3y^2 - x^2) = 6ax + 2xy$ or $\dfrac{dy}{dx} = \dfrac{6ax + 2xy}{4ay - 3y^2 - x^2}$ at a point $P\left(\dfrac{a}{2}, a\right)$

or $\left(\dfrac{dy}{dx}\right)_{\left(\frac{a}{2},a\right)} = \dfrac{6.a.\frac{a}{2} + 2.\frac{a}{2}.a}{4aa - 3a^2 - \left(\frac{a}{2}\right)^2} = \dfrac{4a^2}{4a^2 - 3a^2 - \frac{a^2}{4}} = \dfrac{4a^2}{\frac{16a^2 - 12a^2 - a^2}{4}} = \dfrac{16a^2}{3a^2} = \dfrac{16}{3} = m$

slope of the tangent is m then, equation of the tangent is $y - a = \dfrac{16}{3}\left(x - \dfrac{a}{2}\right)$

or $3y - 3a = 16x - 8a$ $\therefore$ $16x - 3y = -3a + 8a$ $\boxed{\text{or } 16x - 3y = 5a \quad \text{Ans.}}$

slope of the normal is $-\dfrac{1}{m}$ then equation of normal is $y - a = -\dfrac{1}{\frac{16}{3}}\left(x - \dfrac{a}{2}\right)$

or $y - a = -\dfrac{3}{16}\left(x - \dfrac{a}{2}\right)$ or $16y - 16a = -3a + \dfrac{3a}{2}$ or $32y - 32a = -6x + 3a$

$\boxed{\therefore\ 6x + 32y = 35a \quad \text{Ans.}}$

Similarly, point $Q\left(-\dfrac{a}{2}, a\right)$ slope of the tangent is $\left(\dfrac{dy}{dx}\right)_{\left(-\frac{a}{2},a\right)} = \dfrac{6a\left(-\frac{a}{2}\right) + 2\left(-\frac{a}{2}\right)a}{4aa - 3a^2 - \left(-\frac{a}{2}\right)^2} = \dfrac{-4a^2}{\frac{16a^2 - 12a^2 - a^2}{4}}$

or $\left(\dfrac{dy}{dx}\right)_{\left(-\frac{a}{2},a\right)} = -\dfrac{16a^2}{3a^2} = -\dfrac{16}{3} = m_1$ then the equation of tangent is $y - a = -\dfrac{16}{3}\left(x + \dfrac{a}{2}\right)$

or $3y - 3a = -16x - 8a$ $\boxed{\therefore\ 16x + 3y = -5a \quad \text{Ans.}}$

slope of the normal $-\dfrac{1}{m_1}$ or $-\dfrac{1}{\frac{-16}{3}}$ or $\dfrac{3}{16}$ then the equation of the normal is $y - a = \dfrac{3}{16}\left(x + \dfrac{a}{2}\right)$

or $16y - 16a = 3x + \dfrac{a}{2}$ or $32y - 32a = 6x + a$ $\boxed{\therefore\ 6x - 32y = -33a \quad \text{Ans.}}$

(b) $x^2(2a + x) = y^2(3a - x)$ at the point $x = \dfrac{a}{2}$ then or $\frac{a^2}{4}\left(2a + \frac{a}{2}\right) = y^2\left(3a - \frac{a}{2}\right)$ or $\frac{a^2}{4}\left(\frac{5a}{2}\right) = y^2\left(\frac{5a}{2}\right)$ or $y^2 = \frac{a^2}{4}$ $\therefore$ $y = \pm\frac{a}{2}$

Hence, the point are $P\left(\frac{a}{2}, \frac{a}{2}\right)$ and $Q\left(\frac{a}{2}, -\frac{a}{2}\right)$, $x^2(2a + x) = y^2(3a - x)$ or $2ax^2 + x^3 = 3ay^2 - xy^2$

or $4ax + 3x^2 = 6ay\frac{dy}{dx} - x.2y\frac{dy}{dx} - y^2.1$ $\therefore$ $\frac{dy}{dx}(6ay - 2xy) = 4ax + 3x^2 + y^2$ or $\frac{dy}{dx} = \frac{4ax + 3x^2 + y^2}{6ay - 2xy}$ at a point $P\left(\frac{a}{2}, \frac{a}{2}\right)$

then slope of the tangent (m) $= \left(\frac{dy}{dx}\right)_{\left(\frac{a}{2}, \frac{a}{2}\right)} = \frac{4a\frac{a}{2} + 3\frac{a^2}{4} + \frac{a^2}{4}}{6a\frac{a}{2} - 2\frac{a}{2}.\frac{a}{2}}$ or $m = \frac{8a^2 + 3a^2 + a^2}{6a^2 - a^2} \times \frac{2}{4} = \frac{12a^2}{10a^2} = \frac{6}{5}$

Equation of the tangent is $y - \frac{a}{2} = \frac{6}{5}\left(x - \frac{a}{2}\right)$ or $\frac{2y - a}{2} = \frac{6}{5}\left(\frac{2x - a}{2}\right)$ or $10y - 5a = 12x - 6a$

or $12x - 10y = a$ $\boxed{\therefore \ 6x - 5y = \frac{a}{2} \ \text{Ans.}}$

slope of the normal $-\frac{1}{m}$ or $\frac{-1}{\frac{6}{5}}$ or $-\frac{5}{6}$ then, equation of the normal is $y - \frac{a}{2} = -\frac{5}{6}\left(x - \frac{a}{2}\right)$

or $2y - a = -\frac{10}{6}\left(\frac{2x - a}{2}\right)$ or $12y - 6a = -10x + 5a$ $\boxed{\therefore \ 10x + 12y = 11a \ \text{Ans.}}$

Similarly, At a point $Q\left(\frac{a}{2}, -\frac{a}{2}\right)$, slope of the tangent $(m_1) = -\frac{6}{5}$

then equation of the tangent is $y + \frac{a}{2} = -\frac{6}{5}\left(x - \frac{a}{2}\right)$ or $2y + 2a = -\frac{12}{5}\left(\frac{2x - a}{2}\right)$

or $10y + 10a = -12x + 6a$ or $12x + 10y = -4a$ $\boxed{\therefore \ 6x + 5y = -2a \ \text{Ans.}}$

slope of the normal is $-\frac{1}{m_1}$ or $-\frac{1}{\frac{-6}{5}} = \frac{5}{6}$ then equation of the normal is $y + \frac{a}{2} = \frac{5}{6}\left(x - \frac{a}{2}\right)$

or $\frac{2y + a}{2} = \frac{5}{6}\left(\frac{2x - a}{2}\right)$ or $12y + 6a = 10x - 5a$ or $10x - 12y = 11a$ $\boxed{\therefore \ 5x - 6y = \frac{11a}{2} \ \text{Ans.}}$

(9) (a) Normal to $x = y^2 - 2y$ which is parallel to $3x + 4y = 6$.

Solution:- $x = y^2 - 2y$ $\therefore$ $1 = 2y\frac{dy}{dx} - 2\frac{dy}{dx}$ or $\frac{dy}{dx}(2y - 2) = 1$ $\therefore$ $\frac{dy}{dx} = \frac{1}{2y - 2}$

slope of the tangent $m = \frac{dy}{dx} = \frac{1}{2y - 2}$ and slope of the normal $\left(-\frac{1}{m}\right) = -\frac{1}{\frac{dy}{dx}} = -\frac{1}{\frac{1}{2y - 2}} = -(2y - 2)$

Since, it is parallel to $3x + 4y = 6$ whose slope is $-\frac{3}{4}$ then $-(2y - 2) = -\frac{3}{4}$

or $2y = \frac{3}{4} + 2$ or $y = \frac{11}{8}$ $\therefore$ $x = y^2 - 2y$

put $y = \frac{11}{8}$ then $x = \frac{121}{64} - \frac{22}{8} = \frac{121 - 176}{64} = \frac{-55}{64}$ $\therefore$ $x = -\frac{55}{64}$, $y = \frac{11}{8}$

Hence, the point is $P\left(\frac{-55}{64}, \frac{11}{8}\right)$ then slope of the normal at which is $-\frac{3}{4}$

hence equation of the normal is

$$y - \frac{11}{8} = -\frac{3}{4}\left(x + \frac{55}{64}\right) \text{ or } \frac{8y - 11}{8} = -\frac{3}{4}\left(\frac{64x + 55}{64}\right) \text{ or } 32(8y - 11) = -3(64x + 55)$$

or $256y - 352 = -192x - 165$ or $192x + 256y = 352 - 165$ or $192x + 256y = 187$

$$\therefore \quad 3x + 4y = \frac{187}{64} \quad \text{Ans.}$$

(b) Normal to $2y^2 + 3y = x - 2$ which is perpendicular to $x - 2y = 3$.

Solution: $- \quad 2y^2 + 3y = x - 2 \quad \therefore \quad 4y\dfrac{dy}{dx} + 3\dfrac{dy}{dx} = 1 \quad \text{or} \quad \dfrac{dy}{dx}(4y + 3) = 1 \quad \text{or} \quad \dfrac{dy}{dx} = \dfrac{1}{4y + 3}$

$$\text{slope of normal is } -\frac{1}{\dfrac{dy}{dx}} = -\frac{1}{\left(\dfrac{1}{4y+3}\right)} = -(4y + 3)$$

Since, it is perpendicular to $x - 2y = 3$ whose slope of the line is $\dfrac{1}{2}$.

slope of the normal is $-\dfrac{1}{\dfrac{1}{2}} = -2.$ $\quad \therefore \quad -(4y + 3) = -2 \quad \text{or} \quad 4y + 3 = 2 \quad \text{or} \quad 4y = -1 \quad \therefore \quad y = -\dfrac{1}{4}$

Put $y = -\dfrac{1}{4}$ in $2y^2 + 3y = x - 2$ then find x, $\quad 2y^2 + 3y = x - 2 \quad \text{or} \quad 2.\dfrac{1}{16} - \dfrac{3}{4} = x - 2$

$$\text{or} \quad \frac{1}{8} - \frac{3}{4} = x - 2 \quad \therefore \quad x = \frac{1}{8} - \frac{3}{4} + 2 = \frac{1 - 6 + 16}{8} = \frac{11}{8}$$

Hence the point is $P\left(\dfrac{11}{8}, -\dfrac{1}{4}\right)$ slope of the normal at which is -2

Hence equation of the normal is $\quad y + \dfrac{1}{4} = -2\left(x - \dfrac{11}{8}\right) \quad \text{or} \quad 4y + 1 = -8\left(\dfrac{8x - 11}{8}\right)$

or $4y + 1 = -8x + 11 \quad \therefore \quad 8x + 4y = 11 - 1 \quad \therefore \quad 8x + 4y = 10 \quad \text{or} \quad \boxed{\therefore \quad 4x + 2y = 5 \quad \text{Ans.}}$

(10) (a) Tangent to $2y^2 + 3x^2 + 2x + 3y = 0$ which are parallel to the line $2x + 3y = 4$.

Solution: $- \quad 2y^2 + 3x^2 + 2x + 3y = 0 \quad \therefore \quad 4y\dfrac{dy}{dx} + 6x + 2 + 3\dfrac{dy}{dx} = 0 \quad \therefore \quad \dfrac{dy}{dx}(4y + 3) = -(6x + 2)$

$$\therefore \quad \frac{dy}{dx} = \frac{-(6y + 2)}{(4y + 3)} \quad , \quad \text{slope of the tangent is } \frac{-(6y + 2)}{(4y + 3)}.$$

Since, it is parallel to the line $2x + 3y = 4$ whose slope of the tangent is $-\dfrac{2}{3}$.

$$\therefore \quad \frac{-(6y + 2)}{(4y + 3)} = -\frac{2}{3} \quad \text{or} \quad 18x + 6 = 8y + 6 \quad \text{or} \quad 18x = 8y \quad \therefore \quad 9x = 4y \quad \text{or} \quad y = \frac{9}{4}x$$

put $y = \dfrac{9}{4}x$ in the given curve $2y^2 + 3x^2 + 2x + 3y = 0$ then find x,

or $2y^2 + 3x^2 + 2x + 3y = 0 \quad \text{or} \quad 2.\dfrac{81}{16}x^2 + 3x^2 + 2x + 3.\dfrac{9}{4}x = 0 \quad \text{or} \quad \dfrac{81}{8}x^2 + 3x^2 + 2x + \dfrac{27}{4}x = 0$

or $81x^2 + 24x^2 + 16x + 54x = 0 \quad \text{or} \quad 105x^2 + 70x = 0 \quad \text{or} \quad x(105x + 70) = 0 \quad \therefore \quad x = 0 \text{ or } x = -\dfrac{70}{105} = -\dfrac{2}{3}$

$$\therefore \quad x = 0, \quad y = \frac{9}{4}x = 0 \text{ and } x = -\frac{2}{3}, \quad y = \frac{9}{4}x = \frac{9}{4}\left(-\frac{2}{3}\right) = -\frac{3}{2}$$

Hence the two points are $(0,0), \left(-\dfrac{2}{3}, -\dfrac{3}{2}\right)$ then equation of the tangent their points

$$y - 0 = -\frac{2}{3}(x - 0) \quad \text{or} \quad y + \frac{3}{2} = -\frac{2}{3}\left(x + \frac{2}{3}\right) \quad \therefore \quad 3y = -2x \text{ or } 2y + 3 = -\frac{4}{3}\left(\frac{3x + 2}{3}\right) = -\frac{4}{9}(3x + 2)$$

$$\boxed{\therefore \quad 2x + 3y = 0 \text{ or } 12x + 18y = -35 \quad \text{Ans.}}$$

(b) Tangent to $x^2 + y^2 + x + y = 0$ which are perpendicular to the line $-x + y = 3$.

Solution: — Given, $x^2 + y^2 + x + y = 0$ $\therefore$ $2x + 2y\dfrac{dy}{dx} + 1 + \dfrac{dy}{dx} = 0$ or $\dfrac{dy}{dx}(2y + 1) = -(2x + 1)$ $\therefore$ $\dfrac{dy}{dx} = -\dfrac{2x + 1}{2y + 1}$

slope of the tangent is $-\dfrac{2x + 1}{2y + 1}$ and slope of the line is 1 whose slope of the tangent is -1. because tangent is perpendicular to the line.

$$\text{or } -\dfrac{2x + 1}{2y + 1} = -1 \text{ or } 2x + 1 = 2y + 1 \text{ or } 2x = 2y \text{ or } x = y$$

sloving with the given curve, we have $x^2 + y^2 + x + y = 0$ or $y^2 + y^2 + y + y = 0$

or $2y^2 + 2y = 0$ or $2y(y + 1) = 0$ $\therefore$ $2y = 0$ or $y + 1 = 0$ $\therefore$ $y = 0, -1$

At $y = 0, x = 0$ and $y = -1, x = -1$ hence the two points are $(0,0), (-1, -1)$.

slope of the tangent is -1. hence the two points are $(0,0), (-1, -1)$

then their equation is $y - 0 = -1(x - 0)$ or $y + 1 = -1(x + 1)$

$$\therefore \quad x + y = 0 \quad \text{or } y + 1 = -x - 1 \quad \boxed{\therefore \ x + y = 0 \quad \text{or } x + y = -2 \quad \text{Ans.}}$$

(11) (a) Tangent and normal to parabola $x^2 = -4y + 3$ which is parallel to $4y = 2x - 5$.

Solution: — Given, $x^2 = -4y + 3$ $\therefore$ $2x = -4\dfrac{dy}{dx}$ or $\dfrac{dy}{dx} = -\dfrac{2x}{4} = -\dfrac{x}{2}$

$$\text{slope of the tangent } \dfrac{dy}{dx} \text{ or } -\dfrac{x}{2}, \text{ slope of the normal } -\dfrac{1}{\frac{dy}{dx}} \text{ or } \dfrac{2}{x}$$

Since it is parallel to the line $4y = 2x - 5$ whose slope of tangent and normal is same $\dfrac{1}{2}$.

Tangent: $-\dfrac{x}{2} = \dfrac{1}{2}$ or $x = -1$ $\therefore$ $1 = -4y + 3$ or $y = \dfrac{1}{2}$

then the equation of the tangent is $y - \dfrac{1}{2} = \dfrac{1}{2}(x + 1)$ $\therefore$ $2y - 1 = x + 1$ $\therefore$ $x - 2y = -2$ Ans.

Normal: $-\dfrac{2}{x} = \dfrac{1}{2}$ or $x = 4$ $\therefore$ $16 = -4y + 3$ or $y = -\dfrac{13}{4}$

then the equation of the normal is $y + \dfrac{13}{4} = \dfrac{1}{2}(x - 4)$ $\therefore$ $4y + 13 = 2x - 8$ or $2x - 4y = 21$ Ans.

(b) Tangent and normal to parabola $2y^2 + 4 = -8x$ which is perpendicular to the line $6x + 3y = 4$.

Solution: — Tangent: — $2y^2 + 4 = -8x$ $\therefore$ $4y\dfrac{dy}{dx} = -8$ or $\dfrac{dy}{dx} = \dfrac{-8}{4y}$ or $\dfrac{dy}{dx} = -\dfrac{2}{y} = $ slope of the tangent

Since it is perpendicular to the line $6x = -3y + 4$ whose slope of the given line is -2

then the slope of the tangent is $\dfrac{1}{2}$. $\therefore$ $-\dfrac{2}{y} = \dfrac{1}{2}$ or $y = -4$ $\therefore$ $2 \times 16 + 4 = -8x$ or $x = -\dfrac{36}{8} = -\dfrac{9}{2}$

Equation of the tangent is $y + 4 = \dfrac{1}{2}\left(x + \dfrac{9}{2}\right)$ or $2y + 8 = \dfrac{2x + 9}{2}$ or $4y + 16 = 2x + 9$ or $2x - 4y = 7$ Ans.

Normal: — slope of the normal is $-\dfrac{1}{\frac{dy}{dx}} = -\dfrac{1}{\frac{-2}{y}} = \dfrac{y}{2}$

Since it is perpendicular to the line $6x = -3y + 4$ whose slope of the given line is -2,

---

slope of the normal is $\dfrac{1}{2}$ or $\dfrac{y}{2} = \dfrac{1}{2}$ or $y = 1$ then $2 + 4 = -8x$ or $x = -\dfrac{6}{8} = -\dfrac{3}{4}$

Equation of the normal is $y - 1 = \dfrac{1}{2}\left(x + \dfrac{3}{4}\right)$ or $2y - 2 = \dfrac{4x + 3}{4}$ or $8y - 8 = 4x + 3$

$$\text{or } 4x - 8y = -11 \quad \text{Ans.}$$

(12) (a) On the ellipse $16x^2 + 25y^2 = 1$ the points at which the tangents are perpendicular to the line $5x = 4y$.

Solution: − Given $16x^2 + 25y^2 = 1$ $\therefore 32x + 50y\dfrac{dy}{dx} = 0$ or $\dfrac{dy}{dx} = -\dfrac{32x}{50y}$ or $\dfrac{dy}{dx} = -\dfrac{16x}{25y}$ = slope of the tangent

Since it is perpendicular to the line $5x = 4y$ whose slope of the tangent is $-\dfrac{4}{5}$.

$$\text{or } -\dfrac{16x}{25y} = -\dfrac{4}{5} \text{ or } 80x = 100y \text{ or } 4x = 5y \quad \therefore x = \dfrac{5}{4}y$$

put $x = \dfrac{5}{4}y$ then find y, $16x^2 + 25y^2 = 1$ or $16 \times \dfrac{25}{16}y^2 + 25y^2 = 1$ or $50y^2 = 1$ or $y^2 = \dfrac{1}{50}$ $\therefore y = \pm\dfrac{1}{5\sqrt{2}}$

At $y = \dfrac{1}{5\sqrt{2}}$ then $x = \dfrac{5}{4}y = \dfrac{5}{4} \times \dfrac{1}{5\sqrt{2}} = \dfrac{1}{4\sqrt{2}}$ and $y = -\dfrac{1}{5\sqrt{2}}$ then $x = \dfrac{5}{4}y = \dfrac{5}{4}\left(-\dfrac{1}{5\sqrt{2}}\right) = -\dfrac{1}{4\sqrt{2}}$

$$\text{Points are } \left(\dfrac{1}{4\sqrt{2}}, \dfrac{1}{5\sqrt{2}}\right) \text{ and } \left(-\dfrac{1}{4\sqrt{2}}, -\dfrac{1}{5\sqrt{2}}\right) \quad \text{Ans.}$$

(b) Determine the equation of normal at vertex of the parabola $(y - 3)^2 = 4(x + 5)$.

Solution: − vertex is $(-5,3)$ $\therefore (y - 3)^2 = 4(x + 5)$ or $2(y - 3)\dfrac{dy}{dx} = 4$

or $\dfrac{dy}{dx} = \dfrac{4}{2(y - 3)} = \dfrac{2}{y - 3}$ at a point $(-5,3)$ $\therefore \left(\dfrac{dy}{dx}\right)_{(-5,3)} = \dfrac{2}{3 - 3} = \dfrac{2}{0} = \infty$

slope of normal is $-\dfrac{1}{\frac{dy}{dx}} = -\dfrac{1}{\infty} = 0$. equation of the normal at a point $(-5,3)$

$$\text{or } y - 3 = 0(x + 5) \quad \text{or } y - 3 = 0 \quad \therefore y = 3 \quad \text{Ans.}$$

(c) If the tangent to the curve $y = f(x + a)$ at the point $(1,3)$ makes an angle $\dfrac{\pi}{3}$ with the positive x − axis then find $f'(1 + a)$.

Solution: − $y = f(x + a)$ $\therefore \dfrac{dy}{dx} = f'(x + a)$ at a point $(1,3)$

or $\left(\dfrac{dy}{dx}\right)_{(1,3)} = f'(1 + a)$, $\theta = \dfrac{\pi}{3}$ $\therefore$ slope $(m) = \tan\theta = \tan\dfrac{\pi}{3} = \sqrt{3}$ $\boxed{\therefore f'(1 + a) = \sqrt{3} \quad \text{Ans.}}$

Normal: − $\left(\dfrac{dy}{dx}\right)_{(1,3)} = f'(1 + a)$, slope of the normal is $-\dfrac{1}{\frac{dy}{dx}} = -\dfrac{1}{f'(1 + a)}$

or $\theta = \dfrac{\pi}{3}$ then $\tan\theta = -\dfrac{1}{f'(1 + a)}$ or $\tan\dfrac{\pi}{3} = -\dfrac{1}{f'(1 + a)}$ or $\sqrt{3} = -\dfrac{1}{f'(1 + a)}$ $\boxed{\text{or } f'(1 + a) = -\dfrac{1}{\sqrt{3}} \quad \text{Ans.}}$

(13) (a) Tangent and normal to the curve $x = \dfrac{2at}{1 + t}, y = \dfrac{2at^2}{1 + t}$ at the point for which $t = \dfrac{3}{2}$.

Solution: − Given , $x = \dfrac{2at}{1 + t}, y = \dfrac{2at^2}{1 + t}$ at the point $t = \dfrac{3}{2}$ $\therefore x = \dfrac{6a}{5}, y = \dfrac{9a}{5}$

or $\dfrac{dx}{dt} = \dfrac{(1 + t)2a - 2at}{(1 + t)^2} = \dfrac{2a + 2at - 2at}{(1 + t)^2} = \dfrac{2a}{(1 + t)^2}$

---

or $\dfrac{dy}{dt} = \dfrac{(1+t)4at - 2at^2}{(1+t)^2} = \dfrac{4at + 4at^2 - 2at^2}{(1+t)^2} = \dfrac{4at + 2at^2}{(1+t)^2}$

or $\dfrac{dy}{dx} = \dfrac{\dfrac{dy}{dt}}{\dfrac{dx}{dt}} = \dfrac{\dfrac{4at + 2at^2}{(1+t)^2}}{\dfrac{2a}{(1+t)^2}} = \dfrac{4at + 2at^2}{2a} = \dfrac{2a(2t + t^2)}{2a} = 2t + t^2$

slope of the tangent $\left(\dfrac{dy}{dx}\right)_{t=\frac{3}{2}} = 2 \times \dfrac{3}{2} + \dfrac{9}{4} = \dfrac{21}{4}$ and slope of the normal $-\dfrac{1}{\dfrac{dy}{dx}} = -\dfrac{1}{\dfrac{21}{4}} = -\dfrac{4}{21}$

Equation of their equation is $y - \dfrac{9a}{5} = \dfrac{21}{4}\left(x - \dfrac{6a}{5}\right)$ and $y - \dfrac{9a}{5} = -\dfrac{4}{21}\left(x - \dfrac{6a}{5}\right)$

or $20y - 36a = 105x - 126a$ and $105y - 189a = -20x + 24a$ $\quad \therefore \ 21x - 4y = 18a$ and $20x + 105y = 213a$ Ans.

(b) Normal at any point θ to the curve $x = \sin\theta - \theta\cos\theta$, $y = \cos\theta + \theta\sin\theta$ also show that it is at a constant distance from the origin.

Solution: $-$ Given, $x = \sin\theta - \theta\cos\theta$, $y = \cos\theta + \theta\sin\theta$ $\quad \therefore \ \dfrac{dy}{dx} = \dfrac{\dfrac{dy}{d\theta}}{\dfrac{dx}{d\theta}} = \dfrac{\theta\cos\theta}{\theta\sin\theta} = \cot\theta$

slope of the normal is $-\dfrac{1}{\dfrac{dy}{dx}} = -\dfrac{1}{\cot\theta} = -\tan\theta$ then the equation of the normal is $y - (\cos\theta + \theta\sin\theta) = -\dfrac{\sin\theta}{\cos\theta}[x - (\sin\theta - \theta\cos\theta)]$

solve, $x\sin\theta + y\cos\theta = \sin^2\theta + \cos^2\theta = 1$ its distance from the origin is clearly $\dfrac{1}{\sqrt{\sin^2\theta + \cos^2\theta}} = 1$ i.e constant Proved.

(14) (a) The rectangular co$-$ordinates of a point on the curve are given by $x = 2\tan\theta - \tan^2\theta$ and $y = 2\cot\theta - \cot^2\theta$ find the

equation of the tangent at any point on the curve and show that at the point P where $\theta = \dfrac{\pi}{4}$, the tangent passes through the origin.

Solution: $-$ Given, $x = 2\tan\theta - \tan^2\theta$, $y = 2\cot\theta - \cot^2\theta$

or $\dfrac{dx}{d\theta} = 2\sec^2\theta - 2\tan\theta\sec^2\theta$ and $\dfrac{dy}{d\theta} = -2\csc^2\theta + 2\cot\theta\csc^2\theta$ then $\dfrac{dy}{dx} = \dfrac{\dfrac{dy}{d\theta}}{\dfrac{dx}{d\theta}}$

or $\dfrac{dy}{dx} = \dfrac{-2\csc^2\theta + 2\cot\theta\csc^2\theta}{2\sec^2\theta - 2\tan\theta\sec^2\theta} = \dfrac{2\csc^2\theta\,(\cot\theta - 1)}{2\sec^2\theta\,(1 - \tan\theta)} = \dfrac{\cos^2\theta}{\sin^2\theta}\cdot\dfrac{(\cos\theta - \sin\theta)}{(\cos\theta - \sin\theta)}\cdot\dfrac{\cos\theta}{\sin\theta} = \dfrac{\cos^3\theta}{\sin^3\theta}$

slope of the tangent is $\dfrac{dy}{dx} = \dfrac{\cos^3\theta}{\sin^3\theta}$ then equation of the tangent is $y - (2\cot\theta - \cot^2\theta) = \dfrac{\cos^3\theta}{\sin^3\theta}[x - (2\tan\theta - \tan^2\theta)]$

or $y - 2\dfrac{\cos\theta}{\sin\theta} + \dfrac{\cos^2\theta}{\sin^2\theta} = \dfrac{\cos^3\theta}{\sin^3\theta}\left(x - 2\dfrac{\sin\theta}{\cos\theta} + \dfrac{\sin^2\theta}{\cos^2\theta}\right)$

or $y\sin^2\theta - 2\sin\theta\cos\theta + \cos^2\theta = \dfrac{\cos\theta}{\sin\theta}(x\cos^2\theta - 2\sin\theta\cos\theta + \sin^2\theta)$

or $y\sin^3\theta - 2\sin^2\theta\cos\theta + \cos^2\theta\sin\theta = x\cos^3\theta - 2\sin\theta\cos^2\theta + \sin^2\theta\cos\theta$

Divide $\sin^3\theta\cos^3\theta$ both of side , we get

or $y\sec^3\theta - 2\csc\theta\sec^2\theta + \csc^2\theta\sec\theta = x\csc^3\theta - 2\csc^2\theta\sec\theta + \csc\theta\sec^2\theta$

or $x\csc^3\theta - y\sec^3\theta = 3\csc^2\theta\sec\theta - 3\csc\theta\sec^2\theta = 3\sec\theta\csc\theta\,(\csc\theta - \sec\theta)$

$\therefore \ x\csc^3\theta - y\sec^3\theta = 3\sec\theta\csc\theta\,(\csc\theta - \sec\theta)$

If $\theta = \dfrac{\pi}{4}$ then $\csc\theta = \csc\dfrac{\pi}{4} = \sqrt{2} = \sec\dfrac{\pi}{4}$, Tangent is $2\sqrt{2}x - 2\sqrt{2}y = 0$

or $x - y = 0$ it clearly passes through the origin. Proved.

(b) The parametric equation of a curve are $x = a \sin t$, $y = a \cos 2t$. prove that the equation of tangent and normal at any point t are

$x \sin 2t + y \cos 2t = a$ and $x \cos 2t = y \sin 2t$ respectively. also show that S. T (sub − tangent) $= -a \cos 2t . \cot 2t$ and

sub − normal (S. N) $= -a \sin 2t$.

Solution: − Given, $x = a \sin 2t$, $y = a \cos 2t$ then $\dfrac{dx}{dt} = 2a \cos 2t$ and $\dfrac{dy}{dt} = -2a \sin 2t$

$$\therefore \frac{dy}{dx} = \frac{dy}{dt} \div \frac{dx}{dt} = \frac{-2a \sin 2t}{2a \cos 2t} = -\frac{\sin 2t}{\cos 2t} = -\tan 2t$$

slope of the tangent is $-\tan 2t$ and slope of the normal is $\cot 2t$ then, equation of the tangent is

$$\therefore \quad y - a \cos 2t = -\frac{\sin 2t}{\cos 2t}(x - a \sin 2t) \quad \text{or } y \cos 2t - a \cos^2 2t = -x \sin 2t + a \sin^2 2t$$

or $x \sin 2t + y \cos 2t = a \sin^2 2t + a \cos^2 2t$ or $x \sin 2t + y \cos 2t = a(\sin^2 2t + \cos^2 2t)$ $\therefore$ $x \sin 2t + y \cos 2t = a$ Proved.

Equation of the normal is $y - a \cos 2t = \dfrac{\cos 2t}{\sin 2t}(x - a \sin 2t)$

or $y \sin 2t - a \sin 2t \cos 2t = x \cos 2t - a \sin 2t \cos 2t$ $\therefore$ $x \cos 2t = y \sin 2t$ Proved.

Also, Sub − Tangent (S. T) $= \dfrac{y}{y'} = \dfrac{a \cos 2t}{-\dfrac{\sin 2t}{\cos 2t}} = -\dfrac{a \cos^2 2t}{\sin 2t} = -a \cos 2t . \cot 2t$ Proved.

$$\text{Sub − Normal (S. N)} = yy' = a \cos 2t . \left(-\frac{\sin 2t}{\cos 2t}\right) = -a \sin 2t \quad \text{Proved.}$$

(15) (a) Tangent and normal to the curve $y = (x^2 - 4)(x - 1)$ at the points where the curve cuts the x − axis.

Solution: − Given, $y = (x^2 - 4)(x - 1)$ the curve cuts the x − axis $y = 0$

or $0 = (x^2 - 4)(x - 1)$ $\therefore$ $(x^2 - 4) = 0$ or $(x - 1) = 0$ $\therefore$ $x = \pm 2$ or $x = 1$, Points are $(1,0), (2,0), (-2,0)$

At a point $(1,0)$, $\therefore$ $y = (x^2 - 4)(x - 1)$ $\therefore$ $\dfrac{dy}{dx} = (x^2 - 4).1 + (x - 1).2x$ $\therefore$ $\left(\dfrac{dy}{dx}\right)_{(1,0)} = (1 - 4) + 0 = -3$

slope of tangent is -3 then the equation of tangent and normal at a point $(1,0)$

$y - 0 = -3(x - 1)$ and $y - 0 = \dfrac{1}{3}(x - 1)$ or $y = -3x + 3$ and $3y = x - 1$ $\therefore$ $3x + y = 3$ and $x - 3y = 1$ Ans.

Similiarly, point $(2,0)$ then slope of the tangent $\left(\dfrac{dy}{dx}\right)_{(2,0)} = 3x^2 - 2x - 4 = 4$ and slope of the normal is $-\dfrac{1}{4}$.

Equation of the tangent and normal is $y - 0 = 4(x - 2)$ and $y - 0 = -\dfrac{1}{4}(x - 2)$

$$\boxed{\therefore \ \ 4x - y = 2 \ \ \text{and } x + 4y = 2 \ \ \text{Ans.}}$$

At a point $(-2,0)$, slope of the tangent $\left(\dfrac{dy}{dx}\right)_{(-2,0)} = 3x^2 - 2x - 4 = 12$ and slope of the normal is $-\dfrac{1}{12}$.

Equation of the tangent and normal is $y - 0 = 12(x + 2)$ and $y - 0 = -\dfrac{1}{12}(x + 2)$

$$\boxed{\therefore \ \ 12x - y = -24 \ \ \text{and } x + 12y = -2 \ \ \text{Ans.}}$$

(b) Find the slopes of the tangent of the curve $x = \left(y - \dfrac{3}{2}\right)(y + 1)$ at the points where it cuts the axis of y.

Solution: − Given, $x = \left(y - \dfrac{3}{2}\right)(y + 1)$ it cuts the axis of y i.e $x = 0$

or $x = 0$, $y = \dfrac{3}{2}, -1$ then the points are $(0, -1), \left(0, \dfrac{3}{2}\right)$ ∴ $x = \left(y - \dfrac{3}{2}\right)(y + 1)$ or $x = y^2 + y - \dfrac{3}{2}y - \dfrac{3}{2}$

Differentiating, $1 = 2y\dfrac{dy}{dx} + \dfrac{dy}{dx} - \dfrac{3}{2}\dfrac{dy}{dx} - 0$ or $\dfrac{dy}{dx} = \dfrac{1}{2y + 1 - \dfrac{3}{2}} = \dfrac{1}{2y - \dfrac{1}{2}} = \dfrac{2}{4y - 1}$

slope of the tangent $\left(\dfrac{dy}{dx}\right)_{(0,-1)} = \dfrac{2}{-4 - 1} = -\dfrac{2}{5}$ and $\left(\dfrac{dy}{dx}\right)_{\left(0,\frac{3}{2}\right)} = \dfrac{2}{4 \times \dfrac{3}{2} - 1} = \dfrac{2}{6 - 1} = \dfrac{2}{5}$

Then the two slope are $-\dfrac{2}{5}, \dfrac{2}{5}$ Ans.

(c) Tangent to the curve $y^2 = 2x^2 + 4x + 1$ at the point whose abscissa x is 4.

Solution: − Given, $x = 4$ then $y^2 = 2x^2 + 4x + 1$ or $y^2 = 32 + 16 + 1 = 49$ ∴ $y = \pm 7$

Points are $(4, 7), (4, -7)$

curve is $y^2 = 2x^2 + 4x + 1$ ∴ Differentiating, $2y\dfrac{dy}{dx} = 4x + 4$ or $\dfrac{dy}{dx} = \dfrac{2x + 2}{y}$

slopf of the tangent $= \left(\dfrac{dy}{dx}\right)_{(4,7)} = \dfrac{10}{7}$ and $\left(\dfrac{dy}{dx}\right)_{(4,-7)} = -\dfrac{10}{7}$, slope of the normal $= -\dfrac{7}{10}$ and $\dfrac{7}{10}$

Equaiton of the tangent is $y - 7 = \dfrac{10}{7}(x - 4)$ and $y - 7 = -\dfrac{10}{7}(x - 4)$

or $7y - 49 = 10x - 40$ and $7y - 49 = -10x + 40$ ∴ $10x - 7y = -9$ and $10x + 7y = 89$ Ans.

Equation of the normal is $y - 7 = -\dfrac{7}{10}(x - 4)$ and $y - 7 = \dfrac{7}{10}(x - 4)$

or $10y - 70 = -7x + 28$ and $10y - 70 = 7x - 28$ ∴ $7x + 10y = 98$ and $7x - 10y = -42$ Ans.

(16) (a) The points on the curve $x^4 + 3y^2 = 16x^2$ where the tangent is horizantal is (are) … … …

Solution: − $x^4 + 3y^2 = 16x^2$

Differentiating $4x^3 + 6y\dfrac{dy}{dx} = 32x$ or $\dfrac{dy}{dx} = \dfrac{32x - 4x^3}{6y} = \dfrac{16x - 2x^3}{3y}$

If tangent is horizantal $\dfrac{dy}{dx} = 0$, $\dfrac{16x - 2x^3}{3y} = 0$ or $16x - 2x^3 = 0$ or $2x^3 = 16x$ or $2x^2 = 16$

or $x^2 = 8$ ∴ $x = \pm 2\sqrt{2}$, now $x = -2\sqrt{2}$ is a negative value so x^2 is already positive then, $x = -2\sqrt{2}$ is ruled out

Hence, $x = 2\sqrt{2}$, $x^4 + 3y^2 = 16x^2$ ∴ $\left(2\sqrt{2}\right)^4 + 3y^2 = 16\left(2\sqrt{2}\right)^2$ or $3y^2 = 128 - 64$ or $y^2 = \dfrac{64}{3}$

or $y = \pm\dfrac{8}{\sqrt{3}}$, point is $\left(2\sqrt{2}, \pm\dfrac{8}{\sqrt{3}}\right)$ Ans.

(b) The curve $2y^2 = \log y^2 - 2x$ has a vertical tangent at the point.

Solution: − $2y^2 = \log y^2 - 2x$ ∴ $2y^2 + 2x = 2\log y$ or $4y\dfrac{dy}{dx} + 2 = 2.\dfrac{1}{y}\dfrac{dy}{dx}$ or $2y\dfrac{dy}{dx} + 1 = \dfrac{1}{y}\dfrac{dy}{dx}$

or $\dfrac{dy}{dx} = \dfrac{1}{\dfrac{1}{y} - 2y} = \dfrac{y}{1 - 2y^2}$

If tangent is vertical $\dfrac{dy}{dx} = \infty$ $\therefore$ $\dfrac{y}{1-2y^2} = \infty$ or $1 - 2y^2 = 0$ or $y^2 = \dfrac{1}{2}$ $\therefore$ $y = \pm\dfrac{1}{\sqrt{2}}$

At $y = \dfrac{1}{\sqrt{2}}$ then $2x = \log y^2 - 2y^2$ $\therefore$ $2x = 2\log 2^{-\frac{1}{2}} - 2\times\dfrac{1}{2}$ or $2x = -(\log 2 + 1)$ $\therefore$ $x = -\dfrac{1}{2}(\log 2 + 1)$

At $y = -\dfrac{1}{\sqrt{2}}$ then $2x = \log y^2 - 2y^2$ $\therefore$ $2x = \log\left(-\dfrac{1}{\sqrt{2}}\right)^2 - 2\left(-\dfrac{1}{\sqrt{2}}\right)^2 = \log\dfrac{1}{2} - 1$ $\therefore$ $x = -\dfrac{1}{2}(\log 2 + 1)$

$$\text{Points are } \left[-\dfrac{1}{2}(\log 2 + 1), \dfrac{1}{\sqrt{2}}\right] \text{ and } \left[-\dfrac{1}{2}(\log 2 + 1), -\dfrac{1}{\sqrt{2}}\right] \text{ Ans.}$$

(17) (a) Normals at the points on the curve $y = x + \dfrac{1}{x}$ where the tangent makes an angle of $\dfrac{3\pi}{4}$ with $x -$ axis.

Solution: $-$ $y = x + \dfrac{1}{x}$ $\therefore$ $\dfrac{dy}{dx} = 1 - \dfrac{1}{x^2}$, $\tan\theta = \dfrac{3\pi}{4} = -1$

or $\dfrac{dy}{dx} = \tan\theta$ or $1 - \dfrac{1}{x^2} = -1$ or $x^2 - 1 = -x^2$ or $2x^2 = 1$ or $x^2 = \dfrac{1}{2}$ $\therefore$ $x = \pm\dfrac{1}{\sqrt{2}}$

At $x = \pm\dfrac{1}{\sqrt{2}}$ then $y = x + \dfrac{1}{x} = \dfrac{1}{\sqrt{2}} + \dfrac{1}{\frac{1}{\sqrt{2}}} = \dfrac{3}{\sqrt{2}}$ and $y = x + \dfrac{1}{x} = -\dfrac{1}{\sqrt{2}} - \dfrac{1}{\frac{1}{\sqrt{2}}} = -\dfrac{3}{\sqrt{2}}$

$$\text{Points are } \left(\dfrac{1}{\sqrt{2}}, \dfrac{3}{\sqrt{2}}\right), \left(-\dfrac{1}{\sqrt{2}}, -\dfrac{3}{\sqrt{2}}\right)$$

Normal is $y - \dfrac{3}{\sqrt{2}} = 1\left(x - \dfrac{1}{\sqrt{2}}\right)$ and $y + \dfrac{3}{\sqrt{2}} = 1\left(x + \dfrac{1}{\sqrt{2}}\right)$ or $x - y = -\sqrt{2}$ and $x - y = \sqrt{2}$ Ans.

(b) Find the points on the curve $2x^3 + y^2 - 24x = 0$ the tangents at which are parallel to $x -$ axis.

Solution: $-$ $2x^3 + y^2 - 24x = 0$ $\therefore$ $6x^2 + 2y\dfrac{dy}{dx} - 24 = 0$ or $\dfrac{dy}{dx} = \dfrac{24 - 6x^2}{2y}$

Tangent parallel to $x -$ axis $\dfrac{dy}{dx} = 0$ or $\dfrac{24 - 6x^2}{2y} = 0$ or $24 - 6x^2 = 0$ or $x^2 = 4$ $\therefore$ $x = \pm 2$

At $x = 2$ then $y^2 = 24(2) - 2(2)^3 = 48 - 16 = 32$ $\therefore$ $y = \pm 4\sqrt{2}$ point $x = 2, y = \pm 4\sqrt{2}$

At $x = -2$ then $y^2 = 24(-2) - 2(-2)^3 = -48 + 16 = -32$ $\therefore$ $y = \pm 4\sqrt{2}i$ point $x = -2, y = \pm 4\sqrt{2}i$

$$\text{Points are } \left(2, \pm 4\sqrt{2}\right), \left(-2, \pm 4\sqrt{2}i\right) \text{ Ans.}$$

Above Question: $-$ Tangent parallel to $y -$ axis $\dfrac{dy}{dx} = \infty$ or $\dfrac{24 - 6x^2}{2y} = \infty$ $\therefore$ $2y = 0$ or $y = 0$

At $y = 0$ then $2x^3 + y^2 - 24x = 0$ or $2x^3 - 24x = 0$ $\therefore$ $2x(x^2 - 12) = 0$ or $x = 0$ and $x = \pm 2\sqrt{3}$

$$\text{Points are } (0,0), \left(\pm 2\sqrt{3}, 0\right) \text{ Ans.}$$

(18) (a) Find the equation of the tangent and normal to the curve $y = e^{xy} + \sin^{-1}(\sin^2 x)$ at $x = 0$.

Solution: $-$ Given, $y = e^{xy} + \sin^{-1}(\sin^2 x)$ at $x = 0$, $y = 1 + 0 = 1$

or $\dfrac{dy}{dx} = e^{xy}.\left(x\dfrac{dy}{dx} + y.1\right) + \dfrac{1}{\sqrt{1 - \sin^4 x}}.2\sin x\cos x = \dfrac{ye^{xy} + \dfrac{2\sin x\cos x}{\sqrt{1 - \sin^4 x}}}{1 - xe^{xy}}$ at a point $(0,1)$.

or $\left(\dfrac{dy}{dx}\right)_{(0,1)} = \dfrac{1 + 0}{1 - 0} = 1$ slope of the tangent is 1 and slope of normal is -1.

Equation of the tangent and normal is $y - 1 = 1(x - 0)$ and $y - 1 = -1(x - 0)$ or $y - x = 1$ and $x + y = 1$ Ans.

(b) Find the equation of the tangent and normal to the curve $y = e^{\sin x} + a^{\cos x} + \log(1 + x)$.

Solution: – Given, $y = e^{\sin x} + a^{\cos x} + \log(1 + x)$ at $x = 0$, $y = e^0 + a^1 + \log 1 = 1 + a$

$\dfrac{dy}{dx} = e^{\sin x}. \cos x + a^{\cos x}. \log a (-\sin x) + \dfrac{1}{1+x}$ at a point $(0, 1 + a)$ or $\left(\dfrac{dy}{dx}\right)_{(0, 1+a)} = 1 + a \log a . 0 + 1 = 2$

slope of the tangent is 2 and slope of the normal is $-\dfrac{1}{2}$

then equation of the tangent and normal are $\therefore$ $y - (1 + a) = 2(x - 0)$ and $y - (1 + a) = -\dfrac{1}{2}(x - 0)$

or $y - (1 + a) = 2x$ and $2y - 2(1 + a) = -x$ or $y - 2x = 1 + a$ and $x + 2y = 2(1 + a)$ Ans.

(19) Find the angle of intersection of the curve: – (a) $x - y = \sqrt{2}\, a$ and $x + y = a$ (b) $x^2 + y^2 = 3a^2$ and $x^2 - y^2 = a^2$

Solution: – (a) $x - y = \sqrt{2}\, a$ and $x + y = a$ $\therefore$ $1 - \dfrac{dy}{dx} = 0$ and $1 + \dfrac{dy}{dx} = 0$ or $\dfrac{dy}{dx} = 1 = m_1$ and $\dfrac{dy}{dx} = -1 = m_2$

If θ be the angle between the curves, then $\tan\theta = \dfrac{m_1 - m_2}{1 + m_1 m_2} = \dfrac{1 + 1}{1 - 1} = \dfrac{2}{0} = \infty$ $\therefore$ $\cot\theta = 0$ or $\theta = \dfrac{\pi}{2}$ Ans.

(b) $x^2 + y^2 = 3a^2$ and $x^2 - y^2 = a^2$ $\therefore$ $2x + 2y\dfrac{dy}{dx} = 0$ and $2x - 2y\dfrac{dy}{dx} = 0$ or $\dfrac{dy}{dx} = -\dfrac{x}{y} = m_1$ and $\dfrac{dy}{dx} = \dfrac{x}{y} = m_2$

If θ be the angle between the curve then $\tan\theta = \dfrac{m_1 - m_2}{1 + m_1 m_2} = \dfrac{-2xy}{y^2 - x^2}$ for points of intersection

or $x^2 = 2a^2, y^2 = a^2$ or $x = \sqrt{2}\, a$, $y = a$ $\therefore$ $\tan\theta = \dfrac{-2\sqrt{2}\, a.a}{a^2 - 2a^2} = \dfrac{-2\sqrt{2}\, a^2}{-a^2} = 2\sqrt{2}$ $\therefore$ $\tan\theta = 2\sqrt{2}$ or $\theta = \tan^{-1}(2\sqrt{2})$ Ans.

(20) Find the angle of intersection of the curve: – (a) $x^2 y = 1$ and $xy = 1$ (b) $y^2 = x^2 + 4$ and $y^2 = 2x + 3$

(c) $x^2 + 2x = y^2$ and $x^2 + 1 = y^2$

Solution: – (a) $x^2 y = 1 \ldots\ldots\ldots$ (i) and $xy = 1 \ldots\ldots\ldots\ldots$ (ii)

Differentiating, $x^2 \dfrac{dy}{dx} + y.2x = 0$ and $x\dfrac{dy}{dx} + y.1 = 0$ or $\dfrac{dy}{dx} = -\dfrac{2y}{x} = m_1$ and $\dfrac{dy}{dx} = -\dfrac{y}{x} = m_2$

Solve the equatin (i) and (ii), we get $x = 1$, $y = 1$ then $m_1 = -\dfrac{2y}{x} = -2$, $m_2 = -\dfrac{y}{x} = -1$

If θ be the angle between the curves then $\tan\theta = \dfrac{m_1 - m_2}{1 + m_1 m_2} = \dfrac{-2 + 1}{1 + 2} = -\dfrac{1}{3}$ or $\tan\theta = -\dfrac{1}{3}$ $\therefore$ $\theta = \tan^{-1}\left(-\dfrac{1}{3}\right)$ Ans.

(b) $y^2 = x^2 + 4 \ldots\ldots\ldots\ldots\ldots$ (i) and $y^2 = 2x + 3 \ldots\ldots\ldots\ldots\ldots$ (ii)

Differentiating, $2y\dfrac{dy}{dx} = 2x$ and $2y\dfrac{dy}{dx} = 2$ or $\dfrac{dy}{dx} = \dfrac{x}{y} = m_1$ (say) and $\dfrac{dy}{dx} = \dfrac{1}{y} = m_2$ (say)

Solve the equation (i) and (ii), we get $y^2 = x^2 + 4$ and $y^2 = 2x + 3$

or $x^2 + 4 = 2x + 3$ or $x^2 - 2x + 1 = 0$ or $(x + 1)^2 = 0$

if $x = 1$ and $y^2 = x^2 + 4$ or $y^2 = 5$ $\therefore$ $y = \pm\sqrt{5}$ then $m_1 = \dfrac{x}{y} = \dfrac{1}{\sqrt{5}}$, $m_2 = \dfrac{1}{y} = \dfrac{1}{\sqrt{5}}$

If θ be the angle between the curves then $\tan\theta = \dfrac{m_1 - m_2}{1 + m_1 m_2} = \dfrac{\dfrac{1}{\sqrt{5}} - \dfrac{1}{\sqrt{5}}}{1 + \dfrac{1}{\sqrt{5}}.\dfrac{1}{\sqrt{5}}} = 0$ or $\tan\theta = 0$ $\therefore$ $\theta = 0$ Ans.

(c) $x^2 + 2x = y^2 \ldots\ldots\ldots\ldots\ldots$ (i) and $x^2 + 1 = y^2 \ldots\ldots\ldots\ldots\ldots\ldots$ (ii)

Differentiating, $2x + 2 = 2y\dfrac{dy}{dx}$ and $2x = 2y\dfrac{dy}{dx}$ or $\dfrac{dy}{dx} = \dfrac{x+1}{y} = m_1$ and $\dfrac{dy}{dx} = \dfrac{x}{y} = m_2$

Solve the equation (i) and (ii), we get $x^2 + 2x = y^2$ and $x^2 + 1 = y^2$ $\therefore x^2 + 2x = x^2 + 1$

or $2x = 1$ $\therefore x = \dfrac{1}{2}$ At $x = \dfrac{1}{2}$ then $x^2 + 1 = y^2$ or $y^2 = \dfrac{1}{4} + 1 = \dfrac{5}{4}$ $\therefore y = \dfrac{\sqrt{5}}{2}$

If θ be the angle between the curves then $\tan\theta = \dfrac{m_1 - m_2}{1 + m_1 m_2} = \dfrac{\dfrac{x+1}{y} - \dfrac{x}{y}}{1 + \dfrac{x+1}{y}\cdot\dfrac{x}{y}} = \dfrac{\dfrac{x+1-x}{y}}{\dfrac{y^2 + x^2 + x}{y^2}}$

$$\therefore \tan\theta = \dfrac{y}{y^2 + x^2 + x} = \dfrac{\dfrac{\sqrt{5}}{2}}{\dfrac{1}{4} + \dfrac{5}{4} + \dfrac{1}{2}} = \dfrac{\dfrac{\sqrt{5}}{2}}{\dfrac{1+5+2}{4}} = \dfrac{\sqrt{5}}{4} \quad \therefore \theta = \tan^{-1}\left(\dfrac{\sqrt{5}}{4}\right) \quad \text{Ans.}$$

(22) (a) The curves $x^2 + y^2 = a^2$ and $\dfrac{x}{a} + \dfrac{y}{b} = 1$ are intersect orthogonally then prove that $bx = -ay$.

Solution: – Given, $x^2 + y^2 = a^2$ and $\dfrac{x}{a} + \dfrac{y}{b} = 1$ Intersect orthogonally, $\left(\dfrac{dy}{dx}\right)_I \times \left(\dfrac{dy}{dx}\right)_{II} = -1$ or $m_1.m_2 = -1$

$x^2 + y^2 = a^2$ and $\dfrac{x}{a} + \dfrac{y}{b} = 1$ Differentiating, $2x + 2y\dfrac{dy}{dx} = 0$ and $\dfrac{1}{a} + \dfrac{1}{b}\dfrac{dy}{dx} = 0$ $\therefore \dfrac{dy}{dx} = -\dfrac{x}{y} = m_1$ and $\dfrac{dy}{dx} = -\dfrac{b}{a} = m_2$

or $\left(\dfrac{dy}{dx}\right)_I \times \left(\dfrac{dy}{dx}\right)_{II} = -1$ or $-\dfrac{x}{y} \times -\dfrac{b}{a} = -1$ or $\dfrac{bx}{ay} = -1$ $\therefore bx = -ay$ Proved.

(b) The curves $x^2 y = a(x+1)$ and $x^2 = by + 1$ are intersect orthogonally then prove that $x(a - b) = -2a$ and also prove that $\dfrac{a}{b} = \dfrac{x}{x+2}$.

Solution: – Given, $x^2 y = a(x+1) \ldots\ldots\ldots\ldots..$ (i) and $x^2 = by + 1 \ldots\ldots\ldots\ldots\ldots..$ (ii)

Differentiating, $2x\dfrac{dy}{dx} + 2xy = a$ and $2x = b\dfrac{dy}{dx}$ or $\dfrac{dy}{dx} = \dfrac{a - 2xy}{2x} = m_1$ and $\dfrac{dy}{dx} = \dfrac{2x}{b} = m_2$

We know that, $m_1.m_2 = -1$ or $\dfrac{a - 2xy}{2x} \times \dfrac{2x}{b} = -1$ or $a + b = 2xy \ldots\ldots\ldots\ldots.$ (A)

Equation (i), $x^2 y = a(x+1)$ or $xy = \dfrac{a(x+1)}{x}$ then put value xy in equation (A) we get

$\therefore a + b = 2xy$ or $a + b = 2\dfrac{a(x+1)}{x}$ or $ax + bx = 2ax + 2a$ or $2ax - ax - bx = -2a$

or $ax - bx = -2a$ $\therefore x(a - b) = -2a$ Proved.

Again, $ax + bx = 2ax + 2a$ or $ax - 2ax - 2a = -bx$ or $-(ax + 2a) = -bx$ or $a(x+2) = bx$ $\therefore \dfrac{a}{b} = \dfrac{x}{x+2}$ Proved.

(c) Two curves $\left(\dfrac{x}{b}\right)^3 + \left(\dfrac{y}{a}\right)^3 = 1$ and $x^3 y^3 = ab$ are touch at the any point then prove that $\dfrac{y}{x} = \dfrac{b}{a}$.

Solution: – Given, $\left(\dfrac{x}{b}\right)^3 + \left(\dfrac{y}{a}\right)^3 = 1$ and $x^3 y^3 = ab$

Differentiating, $3\left(\dfrac{x}{b}\right)^2 \cdot\dfrac{1}{b} + 3\left(\dfrac{y}{a}\right)^2 \cdot\dfrac{1}{a}\cdot\dfrac{dy}{dx} = 0$ and $x^3.3y^2\dfrac{dy}{dx} + y^3.3x^2 = 0$

$\therefore \dfrac{dy}{dx} = -\dfrac{3\left(\dfrac{x}{b}\right)^2 \cdot\dfrac{1}{b}}{3\left(\dfrac{y}{a}\right)^2 \cdot\dfrac{1}{a}} = -\dfrac{a^3}{b^3}\cdot\dfrac{x^2}{y^2}$ and $\dfrac{dy}{dx} = -\dfrac{y^3.3x^2}{x^3.3y^2} = -\dfrac{y}{x}$

Two curve are touch then we know that $\left(\dfrac{dy}{dx}\right)_I = \left(\dfrac{dy}{dx}\right)_{II}$ or $-\dfrac{a^3}{b^3}\cdot\dfrac{x^2}{y^2} = -\dfrac{y}{x}$ $\therefore \dfrac{y^3}{x^3} = \dfrac{a^3}{b^3}$ $\therefore \dfrac{y}{x} = \dfrac{a}{b}$ Proved.

Exercise – A6

(1) (a) Find the tangent and normal to the curve $y = 3x + 2e^x$ at the point $x = 0$.

(b) Find the equation of the tangent and normal to the curve $y^2 = \sqrt{xy} + \dfrac{2}{\sqrt{x}}$ at the point $x = 1$.

(c) Find the equation of the tangent and normal to the curve $y(x + 1)(x - 2) + x - 7 = 0$ at the point where it cuts the axis of x.

(2) (a) Find the equation of the normal to the curve $y = (1 + x)\tan^{-1}(\tan^2 x)$ at $x = 0$.

(b) The equation of the normal to the curve $x^3 - y^3 = xy$ at the point $(2,2)$.

(c) The equation of the tangent to the curve $e^{xy} = x + y$ at the point $(1,1)$.

(3) (a) on the ellipse $16x^2 + 25y^2 = 1$ the points at which the tangents are perpendicular to the line $5x = 4y$.

(b) If the tangent to the curve $y = f(x)$ at the point $(2,3)$ makes an angle $\dfrac{\pi}{3}$ with the positive x – axis then $f'(2)$ is equal to … … ….

(c) The straight line $2x - 3y = 5a$ will be a tangent to the ellipse $\dfrac{x^2}{4} + \dfrac{y^2}{5} = 1$ then find the value of a.

(4) (a) Find the equation of the tangent and normal to the curve $x^2y^2 = x^2 + y^2$ at the point $x = \sin\theta$, $y = \cos\theta$ and angle $\theta = \dfrac{\pi}{4}$.

(b) Find the equation of the tangent and normal to the curve $x^2 + y^2 = a^2$ at the point $x = a\tan^2\theta$, $y = a\sec^2\theta$ and angle $\theta = \dfrac{\pi}{4}$.

(c) The equation of the tangent and normal to the curve $y = e^{x+y} + \log x$ at the point $(1,0)$.

(5) (a) Tangents to $y = (x^2 - 1)(x + 2)$ at the points where the curve cuts the y – axis.

(b) Find the equation of the tangent to the curve $y = x^3 - 2$ of the point $(1, -1)$.

(c) Find the equation of the tangent and normal to the curve $y^2 = 2xy - 3$ at the point $y = 1$.

(6) (a) The point on the curve $y^2 + 3x = 12y$ where the tangent is vertical.

(b) The points on the curve $y^3 + 3x^2 = 12y$ where the tangent is horizantal.

(c) The points on the curves $2y^3 - 5x^2 = 24y$ where the tangent is vertical.

Answer

(1) (a) Tangent $5x - y + 2 = 0$, Normal $x + 5y = 10$

(b) Point are $(1, -1), (1,2)$, Tangent $2x - 5y = 7$ and $2x + 7y = 16$, Normal $5x + 2y = 3$ and $7x - 2y = 3$

(c) Tangent $x + 40y = 7$ and Normal $40x - y = 280$

(2) (a) Point $(x = 0, y = 0)$, $\boxed{x + y = 0}$ (b) $7x + 5y = 24$ (c) $x + y = 2$

(3) (a) Points $\left(\dfrac{1}{4\sqrt{2}}, \dfrac{1}{5\sqrt{2}}\right), \left(-\dfrac{1}{4\sqrt{2}}, -\dfrac{1}{5\sqrt{2}}\right)$ (b) $\sqrt{3}$ (c) $a = \pm\dfrac{2\sqrt{14}}{5}$

(4) (a) $x\sin^3\theta + y\cos^3\theta = \sin^4\theta + \cos^4\theta$ and $4x\cos^3\theta - 4y\sin^3\theta = \sin 4\theta$

(b) $x\sin^2\theta\cos^2\theta - y\cos^2\theta = a(\sin^4\theta - 1)$ and $x\cos^2\theta + y\sin^2\theta\cos^2\theta = 2a\sin^2\theta$ At $\theta = \dfrac{\pi}{4}$, $x - 2y = -3a$ and $2x + y = 4a$

(c) $(e + 1)x - (1 - e)y = e + 1$ and $(e - 1)x - (e + 1)y = e - 1$

---

(5) (a) $x + y = -2$ (b) $3x - y = 2$ (6) (a) Point $(12, 6)$ (b) Points are $(0, 0)$ & $(0, \pm 2\sqrt{3})$ (c) Points are $\left(\pm \dfrac{4\sqrt{2}}{\sqrt{5}}, -2 \right)$ or $\left(\pm 4\sqrt{\dfrac{2}{5}}, -2 \right)$

Differentiation

(**1**) **Basic formulae**: − (a) Definition: − Differential coefficient (or derivative) of a function.

If $y = f(x)$, then we define $\dfrac{dy}{dx} = \lim\limits_{\Delta x \to 0} \dfrac{f(x + \Delta x) - f(x)}{\Delta x}$ we may define $\dfrac{dy}{dx}$ at $x = a$ as

$$\left(\frac{dy}{dx} \right)_{x=a} = \lim_{x \to a} \frac{f(x) - f(a)}{x - a} \quad \text{Also} \quad \left(\frac{dy}{dx} \right)_{x=a} = \lim_{\Delta x \to 0} \frac{f(a + \Delta x) - f(a)}{\Delta x}$$

In evaluating the limits the following should be noted

(i) $\lim\limits_{\theta \to 0} \dfrac{\sin \theta}{\theta} = 1$ (ii) Limit of sum = sum of limits (iii) Limit of product = product of limits (iv) $\lim\limits_{h \to 0} (1 + h)^{\frac{1}{h}} = e$ and $\lim\limits_{h \to 0} \left(1 + \dfrac{h}{2} \right)^{\frac{1}{h}} = e^{\frac{1}{2}}$

(**2**) **Fundamental Theorems**: − Let $u, v, w, \ldots \ldots \ldots$ be functions of x whose derivative exist.

(i) $\dfrac{d(k)}{dx} = 0$ where k is a constant.   (ii) $\dfrac{d(ku)}{dx} = k \dfrac{du}{dx}$ (iii) $\dfrac{d(u \pm v)}{dx} = \dfrac{du}{dx} \pm \dfrac{dv}{dx}$ sum or difference (iv) $\dfrac{d(uv)}{dx} = u \dfrac{dv}{dx} + v \dfrac{du}{dx}$ Product

(v) $\dfrac{d \left(\frac{u}{v} \right)}{dx} = \dfrac{v \frac{du}{dx} - u \frac{dv}{dx}}{v^2}$ Quotient   (vi) If $y = f(t)$ and $t = \phi(x)$ then $\dfrac{dy}{dx} = \dfrac{dy}{dt} \times \dfrac{dt}{dx}$ (function of a function)

(vii) If $u = f(y)$, then $\dfrac{du}{dx} = \dfrac{du}{dy} \cdot \dfrac{dy}{dx} \ldots \ldots \ldots \ldots (A)$ $f'(y) = \dfrac{du}{dy}$

put value of $\dfrac{du}{dy}$ in equation (A), we get $\therefore \dfrac{du}{dx} = \dfrac{du}{dy} \cdot \dfrac{dy}{dx}$ $\therefore \dfrac{du}{dx} = f'(y) \cdot \dfrac{dy}{dx}$

(viii) $\dfrac{dy}{dx} \cdot \dfrac{dx}{dy} = 1$ or $\dfrac{dy}{dx} = \dfrac{1}{\left(\frac{dx}{dy} \right)}$

(ix) Differentiation of one function with respect to another function. Let $y = f(x)$, $z = \phi(x)$ and we have to differentiable

$f(x)$ w.r.t $\phi(x)$ so that we have to find the value of $\dfrac{dy}{dz} = \dfrac{dy}{dx} + \dfrac{dz}{dx}$ or $\dfrac{dy}{dz} = \dfrac{dy}{dx} \cdot \dfrac{dx}{dz}$

(x) Logarithmic differentiation: −If $y = [f_1(x)]^{f_2(x)}$ or $y = f_1(x) f_2(x) f_3(x) \ldots \ldots$ or $y = \dfrac{f_1(x) f_2(x) \ldots \ldots}{\phi_1(x) \phi_2(x) \ldots \ldots}$

Then it will be convenient to take log of both sides before performing differentiation.

(**3**) (i) **Parametric equation**: − If $x = f(t)$, $y = \phi(t)$ then $\dfrac{dy}{dx} = \dfrac{dy}{dt} + \dfrac{dx}{dt}$

(ii) Implicit function: − $f(x, y) = c$, Differentiate each term with respect to x, we get $\dfrac{d[\phi(y)]}{dx} = \dfrac{d[\phi(y)]}{dx} \cdot \dfrac{dy}{dx}$

Example: − $x^4 + y^4 - 5axy = 0$, Differentiating w.r.t x, we get

or $4x^3 + 4y^3 \dfrac{dy}{dx} - 5a \left(y \cdot 1 + x \dfrac{dy}{dx} \right) = 0$ or $4x^3 + 4y^3 \dfrac{dy}{dx} - 5ay - 5ax \dfrac{dy}{dx} = 0$ or $\dfrac{dy}{dx} (4y^3 - 5ax) = 5ay - 4x^3$ $\therefore \dfrac{dy}{dx} = \dfrac{5ay - 4x^3}{4y^3 - 5ax}$ Ans.

Partial Differentiation: − If $f(x, y) = c$ then we can find $\dfrac{dy}{dx}$ by the help of partial differentiation as under $\dfrac{dy}{dx} = -\dfrac{f_x}{f_y}$

where f_x is differential coefficient of $f(x, y)$ w.r.t. x treating y as constant. f_y is differentiation of $f(x, y)$ w.r.t. y treating x as constant.

Example: − If $f(x, y) = x^4 + y^4 - 5axy = 0$ then $f_x = 4x^3 - 5ay$, $f_y = 4y^3 - 5ax$

$$\therefore \frac{dy}{dx} = -\frac{f_x}{f_y} = -\frac{4x^3 - 5ay}{4y^3 - 5ax} \quad \therefore \frac{dy}{dx} = \frac{5ay - 4x^3}{4y^3 - 5ax} \quad \text{Ans.}$$

Some Standard Results

(i) $y = x^n$, $\dfrac{dy}{dx} = nx^{n-1}$ or $y = u^n$, $\dfrac{dy}{dx} = nu^{n-1}\dfrac{du}{dx}$

Note: $-\ y = \sqrt{x}$ then $\dfrac{dy}{dx} = \dfrac{1}{2\sqrt{x}}$ (Important result), $y = \dfrac{1}{x^n} = x^{-n} = -nx^{-n-1} = -nx^{-(n+1)} = -\dfrac{n}{x^{n+1}}$

(ii) $y = e^x$ then $\dfrac{dy}{dx} = e^x$ and $y = e^u$ then $\dfrac{dy}{dx} = e^u\dfrac{du}{dx}$ (iii) $y = \log x$ then $\dfrac{dy}{dx} = \dfrac{1}{x}$ and $y = \log u$ then $\dfrac{dy}{dx} = \dfrac{1}{u}\dfrac{du}{dx}$

(iv) $y = a^x$ then $\dfrac{dy}{dx} = a^x \log a$ and $y = a^u$ then $\dfrac{dy}{dx} = a^u \log a \dfrac{du}{dx}$ (v) $y = \sin x$ then $\dfrac{dy}{dx} = \cos x$ and $y = \sin u$ then $\dfrac{dy}{dx} = \cos u \dfrac{du}{dx}$

(vi) $y = \cos x$ then $\dfrac{dy}{dx} = -\sin x$ and $y = \cos u$ then $\dfrac{dy}{dx} = -\sin u \dfrac{du}{dx}$

(vii) $y = \tan x$ then $\dfrac{dy}{dx} = \sec^2 x$ and $y = \tan u$ then $\dfrac{dy}{dx} = \sec^2 u \dfrac{du}{dx}$

(viii) $y = \cot x$ then $\dfrac{dy}{dx} = -\csc^2 x$ and $y = \cot u$ then $\dfrac{dy}{dx} = -\csc^2 x \dfrac{du}{dx}$

(ix) $y = \sec x$ then $\dfrac{dy}{dx} = \sec x \tan x$ (x) $y = \csc x$ then $\dfrac{dy}{dx} = -\csc x \cot x$ (xi) $y = \sin^{-1} x$ then $\dfrac{dy}{dx} = \dfrac{1}{\sqrt{1 - x^2}}$

(xii) $y = \cos^{-1} x$ then $\dfrac{dy}{dx} = -\dfrac{1}{\sqrt{1 - x^2}}$ (xiii) $y = \tan^{-1} x$ then $\dfrac{dy}{dx} = \dfrac{1}{1 + x^2}$ (xiv) $y = \cot^{-1} x$ then $\dfrac{dy}{dx} = -\dfrac{1}{1 + x^2}$

(xv) $y = \sec^{-1} x$ then $\dfrac{dy}{dx} = \dfrac{1}{|x|\sqrt{x^2 - 1}}$, $|x| > 0$ (xvi) $y = \csc^{-1} x$ then $\dfrac{dy}{dx} = -\dfrac{1}{|x|\sqrt{x^2 - 1}}$, $|x| > 0$

Trigonometrical Expansions and Formulae be Remembered

(i) $\sin x = x - \dfrac{x^3}{3!} + \dfrac{x^5}{5!} - \cdots \ldots \ldots \ldots \ldots \ldots \ldots \ldots$

$\cos x = 1 - \dfrac{x^2}{2!} + \dfrac{x^4}{4!} - \dfrac{x^6}{6!} + \cdots \ldots \ldots \ldots \ldots \ldots \ldots \ldots$

$\tan x = x - \dfrac{x^3}{3} + \dfrac{2}{15}x^5 + \cdots \ldots \ldots \ldots \ldots \ldots \ldots \ldots \ldots$

(ii) (a) $\tan^{-1} x - \tan^{-1} y = \tan^{-1} \dfrac{x - y}{1 + xy}$ (b) $\tan^{-1} x + \tan^{-1} y = \tan^{-1} \dfrac{x + y}{1 - xy}$ (c) $2 \tan^{-1} x = \tan^{-1} \dfrac{2x}{1 - x^2}$

(d) $\sin^{-1} x \pm \sin^{-1} y = \sin^{-1}\left[x\sqrt{1 - y^2} \pm y\sqrt{1 - x^2}\right]$ (e) $\cos^{-1} x \pm \cos^{-1} y = \cos^{-1}\left[xy \mp \sqrt{1 - x^2}\sqrt{1 - y^2}\right]$

(iii) $\sin^{-1} x + \cos^{-1} x = \tan^{-1} x + \cot^{-1} x = \sec^{-1} x + \csc^{-1} x = \dfrac{\pi}{2}$

(iv) $\sin^{-1} x = \csc^{-1}\left(\dfrac{1}{x}\right)$, $\tan^{-1} x = \cot^{-1}\left(\dfrac{1}{x}\right)$ (v) $\sin^{-1}(\sin \theta) = \theta$, $\tan^{-1}(\tan \theta) = \theta$, $\cos^{-1}(\cos \theta) = \theta$

Example: $-\ \sin^{-1}(\cos \theta) = \sin^{-1} \sin\left(\dfrac{\pi}{2} - \theta\right) = \dfrac{\pi}{2} - \theta$, $\cos^{-1}(\sin \theta) = \cos^{-1} \cos\left(\dfrac{\pi}{2} - \theta\right) = \dfrac{\pi}{2} - \theta$

Important Results of Trigonometry

(vi) $\dfrac{1-\cos x}{1+\cos x} = \dfrac{2\sin^2\frac{x}{2}}{2\cos^2\frac{x}{2}} = \tan^2\dfrac{x}{2}$, $\dfrac{1+\cos x}{1-\cos x} = \cot^2\dfrac{x}{2}$, $\dfrac{1-\cos x}{\sin x} = \dfrac{2\sin^2\frac{x}{2}}{2\sin\frac{x}{2}\cos\frac{x}{2}} = \dfrac{\sin\frac{x}{2}}{\cos\frac{x}{2}} = \tan\left(\dfrac{x}{2}\right)$

or $\dfrac{1+\cos x}{\sin x} = \dfrac{2\cos^2\frac{x}{2}}{2\sin\frac{x}{2}\cos\frac{x}{2}} = \dfrac{\cos\frac{x}{2}}{\sin\frac{x}{2}} = \cot\left(\dfrac{x}{2}\right)$

or $\sqrt{1\pm\sin x} = \left[\cos^2\dfrac{x}{2} + \sin^2\dfrac{x}{2} \pm 2\sin\dfrac{x}{2}\cos\dfrac{x}{2}\right]^{\frac{1}{2}} = \left[\left\{\cos\dfrac{x}{2} \pm \sin\dfrac{x}{2}\right\}^2\right]^{\frac{1}{2}} = \cos\left(\dfrac{x}{2}\right) \pm \sin\left(\dfrac{x}{2}\right)$

$\tan A \pm \tan B = \dfrac{\sin A\cos B \pm \cos A\sin B}{\cos A\cos B} = \dfrac{\sin(A\pm B)}{\cos A\cos B}$ or $\tan\left(\dfrac{\pi}{4}+\theta\right) = \dfrac{1+\tan\theta}{1-\tan\theta}$, $\tan\left(\dfrac{\pi}{4}-\theta\right) = \dfrac{1-\tan\theta}{1+\tan\theta}$

(vii) $\cos x = 2\cos^2\left(\dfrac{x}{2}\right) - 1 = 1 - 2\sin^2\left(\dfrac{x}{2}\right) = \cos^2\left(\dfrac{x}{2}\right) - \sin^2\left(\dfrac{x}{2}\right)$.

(viii) $\sin x = 2\sin\dfrac{x}{2}\cos\dfrac{x}{2} = \dfrac{2\tan\left(\frac{x}{2}\right)}{1+\tan^2\left(\frac{x}{2}\right)}$ (ix) $\cos x = \dfrac{1-\tan^2\left(\frac{x}{2}\right)}{1+\tan^2\left(\frac{x}{2}\right)} = 1 - 2\sin^2\left(\dfrac{x}{2}\right) = 2\cos^2\left(\dfrac{x}{2}\right) - 1$

(x) $\tan x = \dfrac{2\tan\left(\frac{x}{2}\right)}{1-\tan^2\left(\frac{x}{2}\right)}$ (xi) $\sin 3x = 3\sin x - 4\sin^3 x$ (xii) $\cos 3x = 4\cos^3 x - 3\cos x$

Substitution for differentiation of algebric function

$\sqrt{a^2 - x^2}$, put $x = a\sin\theta$ or $\sqrt{a^2 + x^2}$, put $x = a\tan\theta$

$\sqrt{x^2 - a^2}$, put $x = a\sec\theta$ or $\sqrt{\left(\dfrac{a+x}{a-x}\right)}$, put $x = a\cos 2\theta$

(xiii) $\cos 2x = 2\cos^2 x - 1 = 1 - 2\sin^2 x = \cos^2 x - \sin^2 x$ (xiv) $\sin 2x = 2\sin x\cos x$

Differentiate from first Principles: −

(1) $\sin x$, Let $y = \sin x = f(x)$, $y + \delta y = \sin(x+\delta x)$ or $\delta y = \sin(x+\delta x) - y$ $\therefore$ $\dfrac{\delta y}{\delta x} = \dfrac{\sin(x+\delta x) - \sin x}{\delta x}$

$\dfrac{dy}{dx} = \lim_{\delta x\to 0}\dfrac{\sin(x+\delta x) - \sin x}{\delta x} = \lim_{\delta x\to 0}\dfrac{2\cos\left(x+\frac{\delta x}{2}\right).\sin\left(\frac{\delta x}{2}\right)}{\delta x} = \lim_{\delta x\to 0}\cos\left(x+\dfrac{\delta x}{2}\right).\left(\dfrac{\sin\left(\frac{\delta x}{2}\right)}{\frac{\delta x}{2}}\right) = \cos x . 1$

or $\dfrac{dy}{dx} = \cos x$ Ans. $\left[\because \sin C - \sin D = 2\cos\left(\dfrac{C+D}{2}\right).\sin\left(\dfrac{C-D}{2}\right)\right]$ $\left[\because \lim_{\theta\to 0}\dfrac{\sin\theta}{\theta} = 1\right]$

(2) $\cos x$, Ans: − $\dfrac{dy}{dx} = \dfrac{d(\cos x)}{dx} = -\sin x$ (Do yourself)

(3) $\tan x$, Let $y = \tan x$, $y + \delta y = \tan(x+\delta x)$ $\therefore$ $\delta y = \tan(x+\delta x) - y = \tan(x+\delta x) - \tan x$

$$\therefore \dfrac{\delta y}{\delta x} = \dfrac{\tan(x+\delta x) - \tan x}{\delta x}$$

or $\dfrac{dy}{dx} = \lim_{\delta x\to 0}\dfrac{\tan(x+\delta x) - \tan x}{\delta x}$ (change to sin and cos) $= \lim_{\delta x\to 0}\dfrac{\frac{\sin(x+\delta x)}{\cos(x+\delta x)} - \frac{\sin x}{\cos x}}{\delta x} = \lim_{\delta x\to 0}\dfrac{\cos x.\sin(x+\delta x) - \sin x.\cos(x+\delta x)}{\delta x.\cos(x+\delta x).\cos x}$

$= \lim_{\delta x\to 0}\dfrac{\sin(x+\delta x - x)}{\delta x.\cos(x+\delta x).\cos x}$

or $\dfrac{dy}{dx} = \lim_{\delta x\to 0}\left(\dfrac{\sin\delta x}{\delta x}\right).\dfrac{1}{\cos(x+\delta x).\cos x} = 1.\dfrac{1}{\cos^2 x} = \sec^2 x$ Ans. $[\because \sin x.\cos y - \sin y.\cos x = \sin(x-y)]$

(4) $\cot x$, Ans: $-\dfrac{dy}{dx} = \dfrac{d(\cot x)}{dx} = -\cosec^2 x$ (Do yourself)

(5) $\sec x$, Let $y = \sec x$, $y + \delta y = \sec(x + \delta x)$ or $\delta y = \sec(x + \delta x) - y = \sec(x + \delta x) - \sec x$

$$\therefore \frac{\delta y}{\delta x} = \frac{\sec(x + \delta x) - \sec x}{\delta x} \qquad \therefore \frac{dy}{dx} = \lim_{\delta x \to 0} \frac{\sec(x + \delta x) - \sec x}{\delta x} \text{ (change to cos)}$$

$$\text{or } \frac{dy}{dx} = \lim_{\delta x \to 0} \frac{\left(\frac{1}{\cos(x + \delta x)} - \frac{1}{\cos x}\right)}{\delta x} = \lim_{\delta x \to 0} \frac{\cos x - \cos(x + \delta x)}{\delta x . \cos x . \cos(x + \delta x)} = \lim_{\delta x \to 0} \frac{2\sin\left(\frac{x + x + \delta x}{2}\right) . \sin\left(\frac{x + \delta x - x}{2}\right)}{\delta x . \cos x . \cos(x + \delta x)} = \lim_{\delta x \to 0} \frac{2\sin\left(x + \frac{\delta x}{2}\right) . \sin\frac{\delta x}{2}}{\delta x . \cos x . \cos(x + \delta x)}$$

$$= \lim_{\delta x \to 0} \frac{\sin\left(x + \frac{\delta x}{2}\right)}{\cos x . \cos(x + \delta x)} \left(\frac{\sin\frac{\delta x}{2}}{\frac{\delta x}{2}}\right) = \frac{\sin x . 1}{\cos x . \cos x} = \tan x . \sec x$$

$$\therefore \frac{dy}{dx} = \tan x . \sec x \quad \text{Ans.} \qquad \left[\because \cos C - \cos D = 2\sin\left(\frac{C + D}{2}\right) . \sin\left(\frac{D - C}{2}\right)\right]$$

(6) $\cosec x$, Ans: $-\dfrac{dy}{dx} = \dfrac{d(\cosec x)}{dx} = -\cot x . \cosec x$ Ans. (Do yourself)

(7) e^x, Let $y = e^x$, $y + \delta y = e^{x + \delta x}$ or $\delta y = e^{x + \delta x} - y$ or $\delta y = e^{x + \delta x} - e^x$

$$\therefore \frac{\delta y}{\delta x} = \frac{e^{x + \delta x} - e^x}{\delta x} \quad \therefore \frac{dy}{dx} = \lim_{\delta x \to 0} \frac{e^{x + \delta x} - e^x}{\delta x} = \lim_{\delta x \to 0} \frac{e^x . e^{\delta x} - e^x}{\delta x} = \lim_{\delta x \to 0} \frac{e^x\left(e^{\delta x} - 1\right)}{\delta x} = \lim_{\delta x \to 0} e^x\left(\frac{e^{\delta x} - 1}{\delta x}\right) = e^x . 1$$

$$\therefore \frac{dy}{dx} = e^x \quad \text{Ans.} \qquad \left[\because \lim_{x \to 0} \frac{e^x - 1}{x} = 1\right]$$

(8) $\log x$, Let $y = \log x$, $y + \delta y = \log(x + \delta x)$ or $\delta y = \log(x + \delta x) - y$ or $\delta y = \log(x + \delta x) - \log x$

$$\therefore \frac{\delta y}{\delta x} = \frac{\log(x + \delta x) - \log x}{\delta x}, \quad \frac{dy}{dx} = \lim_{\delta x \to 0} \frac{\log(x + \delta x) - \log x}{\delta x} = \lim_{\delta x \to 0} \frac{\log\left(\frac{x + \delta x}{x}\right)}{\delta x} = \lim_{\delta x \to 0} \frac{\log\left(1 + \frac{\delta x}{x}\right)}{\delta x}$$

$$\frac{dy}{dx} = \lim_{\delta x \to 0} \left(\frac{\frac{\delta x}{x} - \frac{1}{2}\left(\frac{\delta x}{x}\right)^2 + \cdots\ldots\ldots}{\delta x}\right) = \lim_{\delta x \to 0} \frac{\delta x\left[\frac{1}{x} - \frac{1}{2} . \frac{\delta x}{x^2} + \cdots\ldots\ldots\right]}{\delta x} = \lim_{\delta x \to 0} \left[\frac{1}{x} - \frac{1}{2} . \frac{\delta x}{x^2} + \cdots\ldots\ldots\right] = \frac{1}{x}$$

$$\therefore \frac{dy}{dx} = \frac{1}{x} \quad \text{Ans.} \qquad \left[\because \log(1 + x) = x - \frac{1}{2}x^2 + \frac{1}{3}x^3 - \frac{1}{4}x^4 + \cdots\ldots\ldots\ldots\ldots\ldots\right]$$

(9) $\cos^{-1} x$, Let $y = \cos^{-1} x$ $\therefore \cos y = x$ or $x = \cos y$ $\therefore x + \delta x = \cos(y + \delta y)$

$$\therefore \delta x = \cos(y + \delta y) - \cos y \quad \text{or } \frac{\delta x}{\delta y} = \frac{\cos(y + \delta y) - \cos y}{\delta y} \quad \text{or } \frac{\delta y}{\delta x} = \frac{\delta y}{\cos(y + \delta y) - \cos y}$$

$$\frac{dy}{dx} = \lim_{\delta y \to 0} \frac{\delta y}{\cos(y + \delta y) - \cos y} = \lim_{\delta y \to 0} \frac{\delta y}{2\sin\left(\frac{y + \delta y + y}{2}\right) . \sin\left(\frac{y - y - \delta y}{2}\right)} = \lim_{\delta y \to 0} \frac{\delta y}{2\sin\left(y + \frac{\delta y}{2}\right) . \left[-\sin\left(\frac{\delta y}{2}\right)\right]}$$

$$= \lim_{\delta y \to 0} \frac{\delta y}{2\sin\left(y + \frac{\delta y}{2}\right) . \left[-\left(\frac{\sin\frac{\delta y}{2}}{\frac{\delta y}{2}}\right) . \frac{\delta y}{2}\right]} = \lim_{\delta y \to 0} \frac{1}{\sin\left(y + \frac{\delta y}{2}\right) . (-1)} = -\frac{1}{\sin y}$$

$$\therefore \frac{dy}{dx} = -\frac{1}{\sqrt{1 - \cos^2 y}} \quad \text{put } \cos y = x \text{ then } \frac{dy}{dx} = -\frac{1}{\sqrt{1 - x^2}} \quad \text{Ans.} \qquad [\because \sin^2 x + \cos^2 x = 1]$$

(10) $\sqrt{\cos x}$, Let $y = \sqrt{\cos x}$ $\therefore y + \delta y = \sqrt{\cos(x + \delta x)}$ or $\delta y = \sqrt{\cos(x + \delta x)} - y$

$$\text{or } \delta y = \sqrt{\cos(x + \delta x)} - \sqrt{\cos x} \quad \therefore \frac{\delta y}{\delta x} = \frac{\sqrt{\cos(x + \delta x)} - \sqrt{\cos x}}{\delta x}$$

$$\frac{dy}{dx} = \lim_{\delta x \to 0} \frac{\sqrt{\cos(x+\delta x)} - \sqrt{\cos x}}{\delta x} = \lim_{\delta x \to 0} \frac{\left(\sqrt{\cos(x+\delta x)} - \sqrt{\cos x}\right).\left(\sqrt{\cos(x+\delta x)} + \sqrt{\cos x}\right)}{\delta x.\left(\sqrt{\cos(x+\delta x)} + \sqrt{\cos x}\right)} = \lim_{\delta x \to 0} \frac{\cos(x+\delta x) - \cos x}{\delta x.\left(\sqrt{\cos(x+\delta x)} + \sqrt{\cos x}\right)}$$

$$= \lim_{\delta x \to 0} \frac{2\sin\left(x+\frac{\delta x}{2}\right).\sin\left(-\frac{\delta x}{2}\right)}{\delta x.\left(\sqrt{\cos(x+\delta x)} + \sqrt{\cos x}\right)}$$

$$\frac{dy}{dx} = \lim_{\delta x \to 0} \frac{\sin\left(x+\frac{\delta x}{2}\right)}{\left(\sqrt{\cos(x+\delta x)} + \sqrt{\cos x}\right)} . \lim_{\delta x \to 0} -\left(\frac{\sin\frac{\delta x}{2}}{\frac{\delta x}{2}}\right) = \frac{\sin x.(-1)}{2\sqrt{\cos x}} = -\frac{\sin x}{2\sqrt{\cos x}} \quad \text{Ans.}$$

$$\left[\because \sin(-\theta) = -\sin\theta \;,\; \sin C + \sin D = 2\sin\left(\frac{C+D}{2}\right)\cos\left(\frac{C-D}{2}\right) \;,\; \cos C + \cos D = 2\cos\left(\frac{C+D}{2}\right)\cos\left(\frac{C-D}{2}\right)\right]$$

(11) $e^{\cos x}$, Let $y = e^{\cos x}$ $\therefore \log y = \log_e e^{\cos x}$ or $\log y = \cos x \log_e e$ or $\log y = \cos x$

or $\log(y+\delta y) = \cos(x+\delta x)$ or $\log(y+\delta y) - \log y = \cos(x+\delta x) - \cos x$

or $\dfrac{\log(y+\delta y) - \log y}{\delta x} = \dfrac{\cos(x+\delta x) - \cos x}{\delta x}$ or $\left(\dfrac{\log(y+\delta y) - \log y}{\delta x}\right)\dfrac{\delta y}{\delta x} = \dfrac{\cos(x+\delta x) - \cos x}{\delta x}$

or $\dfrac{\log\left(1+\frac{\delta y}{y}\right)}{\delta y}.\dfrac{\delta y}{\delta x} = \dfrac{2\sin\left(x+\frac{\delta x}{2}\right)\sin\left(-\frac{\delta x}{2}\right)}{\delta x}$ $\therefore \lim_{\delta y \to 0} \dfrac{\log\left(1+\frac{\delta y}{y}\right)}{\delta y}.\dfrac{dy}{dx} = \lim_{\delta x \to 0} \sin\left(x+\frac{\delta x}{2}\right).-\left(\dfrac{\sin\frac{\delta x}{2}}{\frac{\delta x}{2}}\right)$

or $\lim_{\delta y \to 0} \dfrac{dy}{dx}\left[\dfrac{\frac{\delta y}{y} - \frac{1}{2}\left(\frac{\delta y}{y}\right)^2 + \cdots\cdots\cdots}{\delta y}\right] = \sin x\,(-1)$ or $\dfrac{1}{y}\dfrac{dy}{dx} = -\sin x$ $\therefore \dfrac{dy}{dx} = -y\sin x$

$$\therefore \frac{dy}{dx} = -e^{\cos x}\sin x \quad \text{Ans.} \quad [\text{ Put } y = e^{\cos x}]$$

(12) $x \log x$, Let $y = x \log x$ $\therefore y+\delta y = (x+\delta x)\log(x+\delta x)$ or $\delta y = (x+\delta x)\log(x+\delta x) - y$

$$\frac{dy}{dx} = \lim_{\delta x \to 0} \frac{(x+\delta x)\log(x+\delta x) - x\log x}{\delta x} = \lim_{\delta x \to 0} \frac{x\log(x+\delta x) + \delta x\log(x+\delta x) - x\log x}{\delta x}$$

$$\frac{dy}{dx} = \lim_{\delta x \to 0} \left\{\frac{x[\log(x+\delta x) - \log x]}{\delta x} + \frac{\delta x\log(x+\delta x)}{\delta x}\right\} = \lim_{\delta x \to 0} \frac{x\log\left(1+\frac{\delta x}{x}\right)}{\delta x} + \lim_{\delta x \to 0} \frac{\delta x\log(x+\delta x)}{\delta x}$$

$$\frac{dy}{dx} = x.\lim_{\delta x \to 0} \frac{\log\left(1+\frac{\delta x}{x}\right)}{\delta x} + \lim_{\delta x \to 0} \log(x+\delta x) = x.\lim_{\delta x \to 0} \frac{\frac{\delta x}{x} - \frac{1}{2}\left(\frac{\delta x}{x}\right)^2 + \cdots\cdots\cdots}{\delta x} + \log x = x.\lim_{\delta x \to 0} \frac{\delta x\left[\frac{1}{x} - \frac{1}{2}.\frac{\delta x}{x^2} + \cdots\cdots\cdots\right]}{\delta x} + \log x$$

$$= x.\frac{1}{x} + \log x = 1 + \log x \quad \text{Ans.}$$

(13) $x^2 e^x$, Let $y = x^2 e^x$ $\therefore y+\delta y = (x+\delta x)^2\, e^{x+\delta x}$ $\therefore \delta y = (x+\delta x)^2\, e^{x+\delta x} - x^2 e^x$

$$\frac{dy}{dx} = \lim_{\delta x \to 0} \frac{(x+\delta x)^2\, e^{x+\delta x} - x^2 e^x}{\delta x} = \lim_{\delta x \to 0} \frac{(x^2 + 2x\delta x + \delta x^2)e^{x+\delta x} - x^2 e^x}{\delta x}$$

$$\frac{dy}{dx} = \lim_{\delta x \to 0} \frac{x^2 e^{x+\delta x} + 2x\delta x e^{x+\delta x} + \delta x^2 e^{x+\delta x} - x^2 e^x}{\delta x} = \lim_{\delta x \to 0} \frac{x^2 e^{x+\delta x} - x^2 e^x}{\delta x} + \lim_{\delta x \to 0} \frac{2x\delta x e^{x+\delta x}}{\delta x} + \lim_{\delta x \to 0} \frac{\delta x^2 e^{x+\delta x}}{\delta x}$$

$$\frac{dy}{dx} = \lim_{\delta x \to 0} \frac{x^2\left(e^{x+\delta x} - e^x\right)}{\delta x} + 2xe^x + 0 = \lim_{\delta x \to 0} \frac{x^2 e^x\left(e^{\delta x} - 1\right)}{\delta x} + 2xe^x = x^2 e^x + 2xe^x$$

$$\therefore \frac{dy}{dx} = xe^x(x+2) \quad \text{Ans.} \qquad \left[\because \lim_{x \to 0} \frac{e^x - 1}{x} = 1\right]$$

Solved Example

(1) Find the differential coefficient of the following:— (a) $y = \tan^{-1}x + x$ (b) $y = \tan^{-1}(x+1) - \tan^{-1}x$

(c) $\sin^{-1} y = \cos^{-1} x + x$ (d) $y = \dfrac{3x^2 - 6x + 7}{x}$ (e) $y = \dfrac{x^2 + 2x + 3}{x^2}$ (f) $y = \dfrac{(x+1)^2}{x}$ (g) $y = \dfrac{2x}{a} - \dfrac{2b}{x}$

(h) $y = 2\sqrt{x} + \dfrac{1}{\sqrt{x}}$ (i) $y = 3x^{\frac{4}{3}} - x^{\frac{1}{3}} + 2x^{-\frac{1}{3}} - 3x^{-\frac{2}{3}}$ (j) $y = \tan x + \cot x$ (k) $y = \sec x + \csc x$ (l) $y = \tan x - \sec^2 x$

Solution: − (a) $y = \tan^{-1} x + x$ $\therefore$ $\dfrac{dy}{dx} = \dfrac{1}{1+x^2} + 1 = \dfrac{1 + 1 + x^2}{1 + x^2} = \dfrac{2 + x^2}{1 + x^2}$ Ans.

(b) $y = \tan^{-1}(x+1) - \tan^{-1} x$ $\therefore$ $\dfrac{dy}{dx} = \dfrac{1}{1+(x+1)^2} - \dfrac{1}{1+x^2} = \dfrac{1}{1+x^2+2x+1} - \dfrac{1}{1+x^2} = \dfrac{1}{x^2+2x+2} - \dfrac{1}{1+x^2}$

$\therefore$ $\dfrac{dy}{dx} = \dfrac{1+x^2-x^2-2x-2}{(x^2+2x+2)(1+x^2)} = \dfrac{-2x-1}{(x^2+2x+2)(1+x^2)} = -\dfrac{2x+1}{(x^2+2x+2)(1+x^2)}$ Ans.

(c) $\sin^{-1} y = \cos^{-1} x + x$ or $\dfrac{1}{\sqrt{1-y^2}}\dfrac{dy}{dx} = \dfrac{1}{-\sqrt{1-x^2}} + 1$ $\therefore$ $\dfrac{dy}{dx} = \sqrt{1-y^2}\left(\dfrac{\sqrt{1-x^2}-1}{\sqrt{1-x^2}}\right) = \sqrt{1-\sin^2(\cos^{-1}x+x)}\left(\dfrac{\sqrt{1-x^2}-1}{\sqrt{1-x^2}}\right)$

$= \cos(\cos^{-1}x + x)\left(\dfrac{\sqrt{1-x^2}-1}{\sqrt{1-x^2}}\right)$ Ans.

(d) $y = \dfrac{3x^2 - 6x + 7}{x}$ $\left[\text{formula,}\quad y = \dfrac{f(x)}{g(x)}, \quad \dfrac{dy}{dx} = \dfrac{g(x).f'(x) - f(x).g'(x)}{\left(g(x)\right)^2}\right]$

Here $f(x) = 3x^2 - 6x + 7$ and $g(x) = x$ then $f'(x) = 6x - 6$ and $g'(x) = 1$

$\therefore$ $\dfrac{dy}{dx} = \dfrac{x.(6x-6) - (3x^2-6x+7).1}{x^2} = \dfrac{6x^2 - 6x - 3x^2 + 6x - 7}{x^2} = \dfrac{3x^2 - 7}{x^2} = 3 - \dfrac{7}{x^2}$ Ans.

(e) $y = \dfrac{x^2 + 2x + 3}{x^2}$, Let $f(x) = x^2 + 2x + 3$ and $g(x) = x^2$ then $f'(x) = 2x + 2$ and $g'(x) = 2x$

use formula , $\dfrac{dy}{dx} = \dfrac{g(x).f'(x) - f(x).g'(x)}{\left(g(x)\right)^2} = \dfrac{x^2.(2x+2) - (x^2+2x+3).2x}{(x^2)^2}$

$\dfrac{dy}{dx} = \dfrac{2x^3 + 2x^2 - 2x^3 - 4x^2 - 6x}{x^4} = \dfrac{-2x^2 - 6x}{x^4} = \dfrac{-2x(x+3)}{x^4} = \dfrac{-2(x+3)}{x^3} = -2\left[\dfrac{1}{x^2} + \dfrac{3}{x^3}\right]$ Ans.

(f) $y = \dfrac{(x+1)^2}{x}$, Let $f(x) = (x+1)^2$ and $g(x) = x$ then $f'(x) = 2(x+1)$ and $g'(x) = 1$

use formula , $\dfrac{dy}{dx} = \dfrac{g(x).f'(x) - f(x).g'(x)}{\left(g(x)\right)^2} = \dfrac{x.2(x+1) - (x+1)^2.1}{x^2} = \dfrac{2x^2 + 2x - (x^2 + 2x + 1)}{x^2}$

$\dfrac{dy}{dx} = \dfrac{2x^2 + 2x - x^2 - 2x - 1}{x^2} = \dfrac{x^2 - 1}{x^2} = \dfrac{x^2}{x^2} - \dfrac{1}{x^2} = 1 - \dfrac{1}{x^2}$ Ans.

(g) $y = \dfrac{2x}{a} - \dfrac{2b}{x} = \dfrac{2x^2 - 2ab}{ax}$, Let $f(x) = 2x^2 - 2ab$ and $g(x) = ax$ then $f'(x) = 4x$ and $g'(x) = a$

use formula, $\dfrac{dy}{dx} = \dfrac{g(x).f'(x) - f(x).g'(x)}{\left(g(x)\right)^2} = \dfrac{ax.4x - (2x^2 - 2ab).a}{(ax)^2} = \dfrac{4ax^2 - 2ax^2 + 2a^2b}{a^2x^2} = \dfrac{2ax^2 + 2a^2b}{a^2x^2}$

$\therefore$ $\dfrac{dy}{dx} = \dfrac{2a(x^2 + ab)}{a^2x^2} = \dfrac{2(x^2 + ab)}{ax^2} = \dfrac{2}{a} + \dfrac{2b}{x^2}$ Ans.

(h) $y = 2\sqrt{x} + \dfrac{1}{\sqrt{x}} = \dfrac{2x+1}{\sqrt{x}}$, Let $f(x) = 2x+1$ and $g(x) = \sqrt{x}$ then $f'(x) = 2$ and $g'(x) = \dfrac{1}{2\sqrt{x}}$

use formula, $\dfrac{dy}{dx} = \dfrac{g(x).f'(x) - f(x).g'(x)}{\left(g(x)\right)^2} = \dfrac{\sqrt{x}.2 - (2x+1).\dfrac{1}{2\sqrt{x}}}{\left(\sqrt{x}\right)^2} = \dfrac{\dfrac{4x - 2x - 1}{2\sqrt{x}}}{x} = \dfrac{2x-1}{2x\sqrt{x}}$ Ans.

(i) $y = 3x^{\frac{4}{3}} - x^{\frac{1}{3}} + 2x^{-\frac{1}{3}} - 3x^{-\frac{2}{3}}$ $\left[\text{formula}, \quad y = x^n, \quad \dfrac{dy}{dx} = nx^{n-1}\right]$

$$\frac{dy}{dx} = 3.\frac{4}{3}x^{\frac{4}{3}-1} - \frac{1}{3}x^{\frac{1}{3}-1} + 2.\frac{-1}{3}x^{-\frac{1}{3}-1} - 3.\frac{-2}{3}x^{-\frac{2}{3}-1} = 4x^{\frac{1}{3}} - \frac{1}{3}x^{-\frac{2}{3}} - \frac{2}{3}x^{-\frac{4}{3}} + 2x^{-\frac{5}{3}} \quad \text{Ans.}$$

(j) $y = \tan x + \cot x$ $\therefore$ $\dfrac{dy}{dx} = \sec^2 x - \mathrm{cosec}^2 x = \dfrac{1}{\cos^2 x} - \dfrac{1}{\sin^2 x} = \dfrac{\sin^2 x - \cos^2 x}{\sin^2 x \cos^2 x} = \dfrac{-(\cos^2 x - \sin^2 x).4}{(2 \sin x \cos x)^2}$

$$\frac{dy}{dx} = -\frac{4 \cos 2x}{\sin^2 2x} = -4 \cot 2x \,.\, \mathrm{cosec}\, 2x \quad \text{Ans.}$$

IInd Method: $- \; y = \tan x + \cot x = \dfrac{\sin x}{\cos x} + \dfrac{\cos x}{\sin x} = \dfrac{\sin^2 x + \cos^2 x}{\sin x \cos x} = \dfrac{1}{\sin x \cos x}$

$$\left[\text{use formula} \;\; y = \frac{f(x)}{g(x)}, \; \frac{dy}{dx} = \frac{g(x).f'(x) - f(x).g'(x)}{\left(g(x)\right)^2} \;\; \text{and} \;\; y = f(x)g(x), \; \frac{dy}{dx} = f(x)g'(x) + g(x)f'(x)\right]$$

Here Let $f(x) = 1$ and $g(x) = \sin x \cos x$ then $f'(x) = 0$ and $g'(x) = \sin x\,(-\sin x) + \cos x .\cos x = \cos 2x$

$$\frac{dy}{dx} = \frac{\sin x \cos x . 0 - 1.\cos 2x}{(\sin x \cos x)^2} = -\frac{\cos 2x . 4}{(2 \sin x \cos x)^2} = -\frac{4 \cos 2x}{\sin^2 2x} = -4 \cot 2x . \mathrm{cosec}\, 2x \quad \text{Ans.}$$

(k) $y = \sec x + \mathrm{cosec}\, x = \dfrac{1}{\cos x} + \dfrac{1}{\sin x} = \dfrac{\sin x + \cos x}{\sin x \cos x}$

Let $f(x) = \sin x + \cos x$ and $g(x) = \sin x \cos x$

$f'(x) = \cos x - \sin x$ and $g'(x) = \sin x \dfrac{d(\cos x)}{dx} + \cos x \dfrac{d(\sin x)}{dx} = -\sin^2 x + \cos^2 x = \cos^2 x - \sin^2 x = \cos 2x$

$$\frac{dy}{dx} = \frac{g(x).f'(x) - f(x).g'(x)}{\left(g(x)\right)^2} = \frac{\sin x \cos x\,(\cos x - \sin x) - (\sin x + \cos x).\cos 2x}{(\sin x \cos x)^2}$$

$$= \frac{\sin x \cos x\,(\cos x - \sin x) - (\sin x + \cos x)(\cos^2 x - \sin^2 x)}{\sin^2 x \cos^2 x}$$

$\dfrac{dy}{dx} = \dfrac{\sin x \cos^2 x - \sin^2 x \cos x - \sin x \cos^2 x + \sin^3 x - \cos^3 x + \sin^2 x \cos x}{\sin^2 x \cos^2 x} = \dfrac{\sin^3 x - \cos^3 x}{\sin^2 x \cos^2 x} = \dfrac{\sin^3 x}{\sin^2 x \cos^2 x} - \dfrac{\cos^3 x}{\sin^2 x \cos^2 x} = \dfrac{\sin x}{\cos^2 x} - \dfrac{\cos x}{\sin^2 x}$

$$= \tan x \sec x - \cot x \,\mathrm{cosec}\, x \quad \text{Ans.}$$

or Direct Differentiate, $\dfrac{dy}{dx} = \dfrac{d(\sec x)}{dx} + \dfrac{d(\mathrm{cosec}\, x)}{dx} = \tan x \sec x + (-\cot x \,\mathrm{cosec}\, x) = \tan x \sec x - \cot x \,\mathrm{cosec}\, x \quad \text{Ans.}$

(l) $y = \tan x - \sec^2 x$ $\therefore$ $\dfrac{dy}{dx} = \sec^2 x - 2 \sec x \sec x \tan x = \sec^2 x\,(1 - 2 \tan x) = (1 + \tan^2 x)(1 - 2 \tan x)$

$\dfrac{dy}{dx} = 1 + \tan^2 x - 2 \tan x - 2 \tan^3 x = (1 + \tan^2 x)\left(1 - 2\dfrac{\sin x}{\cos x}\right) = \sec^2 x\left(\dfrac{\cos x - 2 \sin x}{\cos x}\right) = \dfrac{1}{\cos^2 x}\left(\dfrac{\cos x - 2 \sin x}{\cos x}\right)$

$$\therefore \; \frac{dy}{dx} = \frac{1}{\cos^3 x}(\cos x - 2 \sin x) \quad \text{Ans.}$$

IInd Method: $- \; y = \tan x - \sec^2 x = \dfrac{\sin x}{\cos x} - \dfrac{1}{\cos^2 x} = \dfrac{\sin x \cos x - 1}{\cos^2 x}$

$$\frac{dy}{dx} = \frac{\cos^2 x . \dfrac{d(\sin x \cos x - 1)}{dx} - (\sin x \cos x - 1).\dfrac{d(\cos^2 x)}{dx}}{(\cos^2 x)^2} = \frac{\cos^2 x\,(\cos^2 x - \sin^2 x) - (\sin x \cos x - 1).2 \cos x\,(-\sin x)}{\cos^4 x}$$

$\dfrac{dy}{dx} = \dfrac{\cos^4 x - \sin^2 x \cos^2 x + 2 \sin^2 x \cos^2 x - 2 \sin x \cos x}{\cos^4 x} = 1 + \tan^2 x - 2 \tan x \sec^2 x = \sec^2 x - 2 \tan x \sec^2 x$

$$\frac{dy}{dx} = \sec^2 x\,(1 - 2 \tan x) = \frac{1}{\cos^3 x}(\cos x - 2 \sin x) \quad \text{Ans.}$$

(2) Find the differential coefficient of the following: $-$ (a) $y = \tan^{-1} x$ (b) $y = \sin^{-1} x$ (c) $y = \cos^{-1} x$

(d) $y = \cot^{-1} x$ (e) $y = \sec^{-1} x$ (f) $y = \csc^{-1} x$ (g) $y = x \sin^{-1} x$ (h) $y = \dfrac{\cos^{-1} x}{x^2}$ (i) $y = \dfrac{\tan^{-1} x}{1 + x^2}$

(j) $y = \sqrt{1 - x^2}\, \sin^{-1} x$ (k) $y = x\sqrt{1 + x}$ (l) $y = x\sqrt{1 + x} + (x + 1)\sqrt{x - 3}$

Solution: – (a) $y = \tan^{-1} x$ or $\tan y = x$ or $\sec^2 y \dfrac{dy}{dx} = 1$ $\therefore \dfrac{dy}{dx} = \dfrac{1}{\sec^2 y} = \dfrac{1}{1 + \tan^2 y} = \dfrac{1}{1 + x^2}$ Ans.

(b) $y = \sin^{-1} x$ or $\sin y = x$ $\therefore \cos y \dfrac{dy}{dx} = 1$ or $\dfrac{dy}{dx} = \dfrac{1}{\cos y} = \dfrac{1}{\sqrt{1 - \sin^2 y}} = \dfrac{1}{\sqrt{1 - x^2}}$ Ans.

(c) $y = \cos^{-1} x$ $\therefore \dfrac{dy}{dx} = \dfrac{1}{-\sqrt{1 - x^2}}$ Ans.   (Do yourself), same as above question.

(d) $y = \cot^{-1} x$ or $\cot y = x$ or $-\csc^2 y \dfrac{dy}{dx} = 1$ or $\dfrac{dy}{dx} = \dfrac{1}{-\csc^2 y} = -\dfrac{1}{\sqrt{1 + \cot^2 y}} = -\dfrac{1}{\sqrt{1 + x^2}}$ Ans.

(e) $y = \sec^{-1} x$ or $\sec y = x$ or $\tan y \sec y \dfrac{dy}{dx} = 1$ or $\dfrac{dy}{dx} = \dfrac{1}{\tan y \sec y} = \dfrac{1}{\sqrt{\sec^2 y - 1}\,.\,\sec y} = \dfrac{1}{x\sqrt{x^2 - 1}}$ Ans.

(f) $y = \csc^{-1} x$ or $\csc y = x$ or $-\cot y \csc y \dfrac{dy}{dx} = 1$ or $\dfrac{dy}{dx} = -\dfrac{1}{\cot y \csc y} = -\dfrac{1}{\csc y \sqrt{\csc^2 y - 1}}$ $\therefore \dfrac{dy}{dx} = -\dfrac{1}{x\sqrt{x^2 - 1}}$ Ans.

(g) $y = x \sin^{-1} x$, Let $f(x) = x$ and $g(x) = \sin^{-1} x$ then $f'(x) = 1$ and $g'(x) = \dfrac{1}{\sqrt{1 - x^2}}$

$\left[\text{use formula, } y = f(x).g(x) \text{ then } \dfrac{dy}{dx} = f(x).g'(x) + g(x).f'(x)\right]$ $\therefore \dfrac{dy}{dx} = x.\dfrac{1}{\sqrt{1 - x^2}} + \sin^{-1} x.1 = \dfrac{x + \sin^{-1} x.\sqrt{1 - x^2}}{\sqrt{1 - x^2}}$ Ans.

(h) $y = \dfrac{\cos^{-1} x}{x^2}$, Let $f(x) = \cos^{-1} x$ and $g(x) = x^2$ then $f'(x) = -\dfrac{1}{\sqrt{1 - x^2}}$ and $g'(x) = 2x$

$\left[\text{use formula } y = \dfrac{f(x)}{g(x)} , \quad \dfrac{dy}{dx} = \dfrac{g(x).f'(x) - f(x).g'(x)}{(g(x))^2}\right]$

$\therefore \dfrac{dy}{dx} = \dfrac{x^2.\left(-\dfrac{1}{\sqrt{1 - x^2}}\right) - \cos^{-1} x.2x}{(x^2)^2} = \dfrac{-\left(x^2 + 2x \cos^{-1} x \sqrt{1 - x^2}\right)}{x^4} = -\dfrac{x + 2 \cos^{-1} x \sqrt{1 - x^2}}{x^3}$ Ans.

(i) $y = \dfrac{\tan^{-1} x}{1 + x^2}$, Let $f(x) = \tan^{-1} x$ and $g(x) = 1 + x^2$ then $f'(x) = \dfrac{1}{1 + x^2}$ and $g'(x) = 2x$

use above formula, $\dfrac{dy}{dx} = \dfrac{(1 + x^2)\dfrac{1}{1 + x^2} - \tan^{-1} x.2x}{(1 + x^2)^2} = \dfrac{1 - 2x \tan^{-1} x}{(1 + x^2)^2}$ Ans.

(j) $y = \sqrt{1 - x^2}\, \sin^{-1} x$, Let $f(x) = \sqrt{1 - x^2}$ and $g(x) = \sin^{-1} x$ then $f'(x) = \dfrac{1}{2\sqrt{1 - x^2}}(-2x)$ and $g'(x) = \dfrac{1}{\sqrt{1 - x^2}}$

$\left[\text{use formula } y = f(x).g(x), \quad \dfrac{dy}{dx} = f(x)g'(x) + g(x)f'(x)\right]$ $\therefore \dfrac{dy}{dx} = \sqrt{1 - x^2}.\dfrac{1}{\sqrt{1 - x^2}} + \sin^{-1} x.\left(-\dfrac{x}{\sqrt{1 - x^2}}\right) = \dfrac{\sqrt{1 - x^2} - x \sin^{-1} x}{\sqrt{1 - x^2}}$ Ans.

(k) $y = x\sqrt{1 + x}$, $\dfrac{dy}{dx} = \dfrac{3x + 2}{2\sqrt{1 + x}}$ Ans. (Same as above question)

(l) $y = x\sqrt{1 + x} + (x + 1)\sqrt{x - 3}$, Let $u = x\sqrt{1 + x}$ and $v = (x + 1)\sqrt{x - 3}$ then $y = u + v$ or $\dfrac{dy}{dx} = \dfrac{du}{dx} + \dfrac{dv}{dx}$

Now $u = x\sqrt{1 + x}$, $\dfrac{du}{dx} = x.\dfrac{1}{2\sqrt{1 + x}} + \sqrt{1 + x}.1 = \dfrac{3x + 2}{2\sqrt{1 + x}}$

and $v = (x + 1)\sqrt{x - 3}$, $\dfrac{dv}{dx} = (x + 1).\dfrac{1}{2\sqrt{x - 3}} + \sqrt{x - 3}.1 = \dfrac{3x - 5}{2\sqrt{x - 3}}$

$$\therefore \frac{dy}{dx} = \frac{du}{dx} + \frac{dv}{dx} = \frac{3x+2}{2\sqrt{1+x}} + \frac{3x-5}{2\sqrt{x-3}} = \frac{(3x+2)\sqrt{x-3} + (3x-5)\sqrt{1+x}}{2\sqrt{x+1}\sqrt{x-3}} \quad \text{Ans.}$$

(3) Find the differential coefficient of the following: — (a) $y = ax^3 + bx^2 + cx + d$ (b) $y = x^4 - \dfrac{4}{3}x^3 + 2x^2 - 4x + 5$

(c) $y = 3x^{-\frac{2}{3}} + 2x^{\frac{1}{2}} - x$ (d) $y = \sqrt{\dfrac{1}{x+1}}$ (e) $y = \sqrt{x - 2x^2}$ (f) $y = x^{-3} + 2\log x - 5e^x$ (g) $y = \tan x + \cot x - e^x$

(h) $y = x^5 + 3\log x - \operatorname{cosec} x$ (i) $y^2 = x^2 + 2x + 1$ (j) $\log y = \sin x + \cos x$ (k) $\sqrt{y} = x^2 + 2x$ (l) $\sin y = e^x$

(m) $\sqrt{\sin y} = (1+x)^{\frac{1}{2}}$ (n) $\cos^2 y = x + 1$ (o) $\sin x + y^2 = \tan x + 1$

Solution: — (a) $y = ax^3 + bx^2 + cx + d$, $\dfrac{dy}{dx} = 3ax^2 + 2bx + c$ Ans. $\left[\text{formula, } y = x^n \text{ then } \dfrac{dy}{dx} = nx^{n-1}\right]$

(b) $y = x^4 - \dfrac{4}{3}x^3 + 2x^2 - 4x + 5$, $\dfrac{dy}{dx} = 4x^3 - 4x^2 + 4x - 4 = 4(x^3 - x^2 + x - 1)$ Ans.

(c) $y = 3x^{-\frac{2}{3}} + 2x^{\frac{1}{2}} - x$, $\dfrac{dy}{dx} = -\dfrac{2}{3}3x^{-\frac{5}{3}} + 2\dfrac{1}{2}x^{-\frac{1}{2}} - 1 = -\dfrac{2}{x^{\frac{5}{3}}} + \dfrac{1}{x^{\frac{1}{2}}} - 1$ Ans.

(d) $y = \sqrt{\dfrac{1}{x+1}} = \dfrac{1}{\sqrt{x+1}}$, $\dfrac{dy}{dx} = \dfrac{\sqrt{x+1}.0 - 1.\dfrac{1}{2\sqrt{x+1}}}{(\sqrt{x+1})^2} = \dfrac{-1}{2(x+1)\sqrt{x+1}} = -\dfrac{1}{2(x+1)^{\frac{3}{2}}}$ Ans.

$$\left[\text{use formula } y = \dfrac{f(x)}{g(x)}, \quad \dfrac{dy}{dx} = \dfrac{g(x).f'(x) - f(x).g'(x)}{(g(x))^2}\right]$$

(e) $y = \sqrt{x - 2x^2}$ squaring both side $\therefore y^2 = x - 2x^2$ or $2y\dfrac{dy}{dx} = 1 - 4x$ $\therefore \dfrac{dy}{dx} = \dfrac{1-4x}{2y} = \dfrac{1-4x}{2\sqrt{x-2x^2}}$ Ans.

(f) $y = x^{-3} + 2\log x - 5e^x$, $\dfrac{dy}{dx} = -3x^{-4} + \dfrac{2}{x} - 5e^x = -\dfrac{3}{x^4} + \dfrac{2}{x} - 5e^x$ Ans.

(g) $y = \tan x + \cot x - e^x$, $\dfrac{dy}{dx} = \sec^2 x - \operatorname{cosec}^2 x - e^x = \dfrac{1}{\cos^2 x} - \dfrac{1}{\sin^2 x} - e^x = \dfrac{\sin^2 x - \cos^2 x}{\sin^2 x \cos^2 x} - e^x$

or $\dfrac{dy}{dx} = \dfrac{-4\cos 2x}{(2\sin x \cos x)^2} - e^x = \dfrac{-4\cos 2x}{\sin^2 2x} - e^x = -4\cot 2x \operatorname{cosec} 2x - e^x$ Ans.

(h) $y = x^5 + 3\log x - \operatorname{cosec} x$, $\dfrac{dy}{dx} = 5x^4 + \dfrac{3}{x} - (-\cot x \operatorname{cosec} x) = 5x^4 + \dfrac{3}{x} + \cot x \operatorname{cosec} x$ Ans.

(i) $y^2 = x^2 + 2x + 1$, $2y\dfrac{dy}{dx} = 2x + 2$ $\therefore \dfrac{dy}{dx} = \dfrac{2(x+1)}{2y} = \dfrac{x+1}{\sqrt{x^2+2x+1}} = \dfrac{x+1}{\sqrt{(x+1)^2}} = \dfrac{x+1}{x+1} = 1$ Ans.

IInd Method: — $y^2 = (x+1)^2$ $\therefore y = \sqrt{(x+1)^2} = x+1$ $\therefore \dfrac{dy}{dx} = 1 + 0 = 1$ Ans.

(j) $\log y = \sin x + \cos x$ $\therefore y = e^{\sin x + \cos x}$ or $\dfrac{1}{y}\dfrac{dy}{dx} = \cos x - \sin x$ $\therefore \dfrac{dy}{dx} = y(\cos x - \sin x)$

$$\dfrac{dy}{dx} = e^{\sin x + \cos x}(\cos x - \sin x) \text{Ans.}$$

(k) $\sqrt{y} = x^2 + 2x$ squaring both side, we get $y = (x^2 + 2x)^2$ or $y = x^4 + 4x^3 + 4x^2$

or $\dfrac{dy}{dx} = 4x^3 + 12x^2 + 8x = 4x(x^2 + 3x + 2) = 4x(x^2 + 2x + x + 2) = 4x[x(x+2) + 1(x+2)] = 4x(x+1)(x+2)$ Ans.

(l) $\sin y = e^x$, $\cos y \dfrac{dy}{dx} = e^x$ $\therefore$ $\dfrac{dy}{dx} = \dfrac{e^x}{\cos y} = \dfrac{e^x}{\sqrt{1 - \sin^2 y}} = \dfrac{e^x}{\sqrt{1 - e^{2x}}}$ or $\dfrac{e^x\sqrt{1 + (e^x)^2}}{\sqrt{1 - (e^{2x})^2}}$ Ans.

(m) $\sqrt{\sin y} = (1 + x)^{\frac{1}{2}}$ squaring both side we have $\sin y = 1 + x$ $\therefore$ $\cos y \dfrac{dy}{dx} = 1$

$$\text{or } \dfrac{dy}{dx} = \dfrac{1}{\cos y} = \dfrac{1}{\sqrt{1 - \sin^2 y}} \quad \therefore \dfrac{dy}{dx} = \dfrac{1}{\sqrt{1 - (1 + x)^2}} = \dfrac{1}{\sqrt{1 - 1 - 2x - x^2}} = \dfrac{1}{\sqrt{-x(x + 2)}} \quad \text{Ans.}$$

(n) $\cos^2 y = x + 1$, $2\cos y (-\sin y)\dfrac{dy}{dx} = 1$ or $-2\sin y \cos y \dfrac{dy}{dx} = 1$ $\therefore$ $\dfrac{dy}{dx} = -\dfrac{1}{2\sin y \cos y}$

or $\cos^2 y = x + 1$ or $\cos y = \sqrt{x + 1}$, $\sin y = \sqrt{1 - \cos^2 y} = \sqrt{1 - x - 1} = \sqrt{-x}$ or $\dfrac{dy}{dx} = -\dfrac{1}{2\sin y \cos y} = -\dfrac{1}{2\sqrt{-x}\sqrt{x + 1}}$ Ans.

(o) $\sin x + y^2 = \tan x + 1$ or $y^2 = \tan x - \sin x + 1$ $\therefore$ $2y\dfrac{dy}{dx} = \sec^2 x - \cos x$ $\therefore$ $\dfrac{dy}{dx} = \dfrac{\sec^2 x - \cos x}{2y}$

$$y^2 = \tan x - \sin x + 1 \quad \text{or } y = \sqrt{\tan x - \sin x + 1} \quad \therefore \dfrac{dy}{dx} = \dfrac{\sec^2 x - \cos x}{2\sqrt{\tan x - \sin x + 1}} \quad \text{Ans.}$$

(4) (a) If $y = x^4 - 3x^3 + 2x^2 + 3x - 5$ then find $\dfrac{dy}{dx}$ at $x = 3$.

(b) If $y = \sqrt{\sin x + \cos x}$, find $\dfrac{dy}{dx}$ at $x = \dfrac{\pi}{4}$. (c) If $y = |x|^2 - 3|x| + 4$ then find $\dfrac{dy}{dx}$ at $x = 2$ and $x = -1$.

(d) If $y = |x|^3 - 2|x|^2 + 3|x| + 5$ then find $\dfrac{dy}{dx}$ at $x = 2$. (e) If $y = \log|x + 1|$ then find $\dfrac{dy}{dx}$ at $x = 1$.

(f) If $y = x^3 + \sin x - \dfrac{3}{2}\log x$ then find $\dfrac{dy}{dx}$, when $x = \dfrac{\pi}{3}$. (g) If $y = |x - 1| + |x - 3|$ then find $\dfrac{dy}{dx}$.

(h) If $y = \log e^x + e^x$ then find $\dfrac{dy}{dx}$, when $x = \log_e 4$. (i) If $x + y = e^{x+y} + e^x + e^y$ then find $\dfrac{dy}{dx}$.

Solution: $-$ (a) $y = x^4 - 3x^3 + 2x^2 + 3x - 5$, $\dfrac{dy}{dx} = 4x^3 - 9x^2 + 4x + 3$ at $x = 3$

$$\left(\dfrac{dy}{dx}\right)_{x=3} = 4(3)^3 - 9(3)^2 + 4(3) + 3 = 108 - 81 + 12 + 3 = 123 - 81 = 42 \quad \text{Ans.}$$

(b) $y = \sqrt{\sin x + \cos x}$, $\dfrac{dy}{dx} = \dfrac{1}{2\sqrt{\sin x + \cos x}} \cdot (\cos x - \sin x) = \dfrac{(\cos x - \sin x)}{2\sqrt{\sin x + \cos x}}$ at $x = \dfrac{\pi}{4}$

$$\left(\dfrac{dy}{dx}\right)_{x=\frac{\pi}{4}} = \dfrac{(\cos x - \sin x)}{2\sqrt{\sin x + \cos x}} = \dfrac{\left(\cos\frac{\pi}{4} - \sin\frac{\pi}{4}\right)}{2\sqrt{\sin\frac{\pi}{4} + \cos\frac{\pi}{4}}} = 0 \quad \text{Ans.}$$

(c) $y = |x|^2 - 3|x| + 4$ $\therefore x = 0$ then $x > 0$, x is positive and $x < 0$, x is negative

$$\text{or } y = \begin{cases} x^2 - 3x + 4, & x > 0 \\ x^2 + 3x + 4, & x < 0 \end{cases} \quad \therefore \dfrac{dy}{dx} = \begin{cases} 2x - 3, & x > 0 \\ 2x + 3, & x < 0 \end{cases}$$

At $x = 2$, $\left(\dfrac{dy}{dx}\right)_{x=2} = \begin{cases} 2 \times 2 - 3, & x > 0 \\ 2 \times 2 + 3, & x < 0 \end{cases} = \begin{cases} 4 - 3, & x > 0 \\ 4 + 3, & x < 0 \end{cases} = \begin{cases} 1, & x > 0 \\ 7, & x < 0 \end{cases}$ Ans.

At $x = -1$, $\left(\dfrac{dy}{dx}\right)_{x=-1} = \begin{cases} 2(-1) - 3, & x > 0 \\ 2(-1) + 3, & x < 0 \end{cases} = \begin{cases} -2 - 3, & x > 0 \\ -2 + 3, & x < 0 \end{cases} = \begin{cases} -5, & x > 0 \\ 1, & x < 0 \end{cases}$ Ans.

(d) $y = |x|^3 - 2|x|^2 + 3|x| + 5$ at $x = 2$ is a positive integer. or $\dfrac{dy}{dx} = 3x^2 - 4x + 3$ at $x = 2$ $\therefore \left(\dfrac{dy}{dx}\right)_{x=2} = 12 - 8 + 3 = 7$ Ans.

(e) $y = \log|x + 1|$ $\therefore x + 1 = 0$ or $x = -1$ $\therefore y = \begin{cases} \log(x + 1), & x > -1 \\ \log -(x + 1), & x < -1 \end{cases}$ *it is not possible.*

$$\therefore \frac{dy}{dx} = \frac{1}{x + 1} \text{ at } x = 1 \quad \therefore \left(\frac{dy}{dx}\right)_{x=1} = \frac{1}{1 + 1} = \frac{1}{2} \quad \text{Ans.}$$

(f) $y = x^3 + \sin x - \dfrac{3}{2}\log x$, $\quad \dfrac{dy}{dx} = 3x^2 + \cos x - \dfrac{3}{2} \cdot \dfrac{1}{x} = 3x^2 + \cos x - \dfrac{3}{2x}$

At $x = \dfrac{\pi}{3}$ then $\dfrac{dy}{dx} = 3\left(\dfrac{\pi}{3}\right)^2 + \cos\dfrac{\pi}{3} - \dfrac{3}{2\frac{\pi}{3}} = \dfrac{3\pi^2}{9} + \dfrac{1}{2} - \dfrac{9}{2\pi} = \dfrac{\pi^2}{3} + \dfrac{1}{2} - \dfrac{9}{2\pi} = \dfrac{2\pi^3 + 3\pi - 27}{6\pi}$ Ans.

(g) $y = |x - 1| + |x - 3|$, Put $x - 1 = 0$ and $x - 3 = 0$ then $x = 1, 3$

$-\infty$	$+$ ve	1	$-$ ve	3	$+$ ve	∞

or $y = \begin{cases} (x - 1) + (x - 3) \text{ if } x < 1 \text{ and } x \geq 3 \\ -(x - 1) - (x - 3) \text{ if } 1 \leq x < 3 \end{cases} = \begin{cases} 2x - 4 \text{ if } x < 1 \text{ and } x \geq 3 \\ -2x + 4 \text{ if } 1 \leq x < 3 \end{cases}$

$\therefore \dfrac{dy}{dx} = \begin{cases} 2 \text{ if } x < 1, \ x \geq 3 \\ -2 \text{ if } 1 \leq x < 3 \end{cases}$ or $\dfrac{dy}{dx} = 2$ and -2 $\quad \therefore \dfrac{dy}{dx} = \{2, -2\}$ Ans.

(h) $y = \log e^x + e^x$, $\dfrac{dy}{dx} = \dfrac{1}{e^x} + e^x$ At $x = \log_e 4$ then $\dfrac{dy}{dx} = \dfrac{1}{e^{\log_e 4}} + e^{\log_e 4} = \dfrac{1}{4} + 4 = \dfrac{1 + 16}{4} = \dfrac{17}{4}$ Ans.

(i) $x + y = e^{x+y} + e^x + e^y$, $1 + \dfrac{dy}{dx} = e^{x+y}\left(1 + \dfrac{dy}{dx}\right) + e^x + e^y\dfrac{dy}{dx}$ or $\dfrac{dy}{dx}(1 - e^y - e^{x+y}) = e^{x+y} + e^x - 1$

$$\dfrac{dy}{dx} = \dfrac{e^{x+y} + e^x - 1}{1 - e^y - e^{x+y}} = \dfrac{-(1 - e^y - e^{x+y})}{1 - e^y - e^{x+y}} = -1 \text{ Ans.}$$

IInd Method: $- \log(x + y) = \log e^{x+y} + \log e^x + \log e^y = x + y + x + y = 2x + 2y$

or $\dfrac{1}{x + y}\left(1 + \dfrac{dy}{dx}\right) = 2 + 2\dfrac{dy}{dx}$ or $\dfrac{1}{x + y}\dfrac{dy}{dx} - 2\dfrac{dy}{dx} = 2 - \dfrac{1}{x + y}$ or $\dfrac{dy}{dx}\left(\dfrac{1}{x + y} - 2\right) = \dfrac{2x + 2y - 1}{x + y}$

$$\therefore \dfrac{dy}{dx} = \dfrac{\dfrac{2x + 2y - 1}{x + y}}{\dfrac{1 - 2x - 2y}{x + y}} = \dfrac{2x + 2y - 1}{1 - 2x - 2y} = \dfrac{-(1 - 2x - 2y)}{1 - 2x - 2y} = -1 \text{ Ans.}$$

(5) Find the differential coefficient of the following: $-$

(a) If $y = (x^2 - 2x + 3)(x^2 + 2)$ find $\dfrac{dy}{dx}$. (b) $y = \dfrac{1 + x^3}{1 - x^3}$ (c) $y = \dfrac{1}{x\tan x}$ (d) $y = \dfrac{\tan x - \cot x}{\tan x + \cot x}$ (e) $y = \dfrac{1 + \sin x}{1 - \sin x}$ (f) $y = \sqrt{\dfrac{1 - x}{1 + x}}$

(g) $y = \dfrac{x\sin x}{a + x}$ (h) $y = \dfrac{x\cot x}{\tan x + \cot x}$ (i) $y = x^3\sin x$ (j) $y = (x^2 - 1)\sec x$ (k) $y = \dfrac{1 + \tan x}{1 - \tan x}$ (l) $y = \dfrac{x^2 - \sec x}{1 + \tan x}$ (m) $y = \dfrac{x^2\cos x}{1 - \tan x}$

Solution: $-$ (a) $y = (x^2 - 2x + 3)(x^2 + 2)$, Let $f(x) = x^2 - 2x + 3$ and $g(x) = x^2 + 2$

$f'(x) = 2x - 2$ and $g'(x) = 2x$ $\quad \left[\text{use formula, } y = f(x)g.(x) \text{ then } \dfrac{dy}{dx} = f(x)g'(x) + g(x)f'(x)\right]$

$\dfrac{dy}{dx} = (x^2 - 2x + 3).2x + (x^2 + 2).(2x - 2) = 2x^3 - 4x^2 + 6x + 2x^3 + 4x - 2x^2 - 4 = 4x^3 - 6x^2 + 10x - 4$ Ans.

(b) $y = \dfrac{1 + x^3}{1 - x^3}$, Let $f(x) = 1 + x^3$ and $g(x) = 1 - x^3$ then $f'(x) = 3x^2$ and $g'(x) = -3x^2$

$$\left[\text{use formula } y = \dfrac{f(x)}{g(x)}, \quad \dfrac{dy}{dx} = \dfrac{g(x).f'(x) - f(x).g'(x)}{\left(g(x)\right)^2}\right]$$

$$\frac{dy}{dx} = \frac{(1-x^3).3x^2 - (1+x^3).(-3x^2)}{(1-x^3)^2} = \frac{3x^2 - 3x^5 + 3x^2 + 3x^3}{(1-x^3)^2} = \frac{6x^2}{(1-x^3)^2} \quad \text{Ans.}$$

(c) $y = \dfrac{1}{x\tan x}$, Let $f(x) = 1$ and $g(x) = x\tan x$ then $f'(x) = 0$ and $g'(x) = x.\sec^2 x + \tan x.1$

use above formula $\dfrac{dy}{dx} = \dfrac{x\tan x.0 - 1(x\sec^2 x + \tan x)}{x^2\tan^2 x} = \dfrac{-(x\sec^2 x + \tan x)}{x^2\tan^2 x} = \dfrac{-x(1+\tan^2 x) - \tan x}{x^2\tan^2 x} = \dfrac{-x - x\tan^2 x - \tan x}{x^2\tan^2 x}$

$$\frac{dy}{dx} = \frac{-x}{x^2\tan^2 x} - \frac{x\tan^2 x}{x^2\tan^2 x} - \frac{\tan x}{x^2\tan^2 x} = -\frac{1}{x\tan^2 x} - \frac{1}{x} - \frac{1}{x^2\tan x} = -\frac{\cos^2 x}{x\sin^2 x} - \frac{\cos x}{x^2\sin x} - \frac{1}{x} \quad \text{Ans.}$$

IInd Method: $-$ $y = \dfrac{1}{x\tan x} = \dfrac{\cos x}{x\sin x}$, Let $f(x) = \cos x$ and $g(x) = x\sin x$ then $f'(x) = -\sin x$ and $g'(x) = x\cos x + \sin x$

use above formula, $\dfrac{dy}{dx} = \dfrac{x\sin x(-\sin x) - \cos x(x\cos x + \sin x)}{x^2\sin^2 x} = \dfrac{-x\sin^2 x - x\cos^2 x - \sin x\cos x}{x^2\sin^2 x} = \dfrac{-x(\sin^2 x + \cos^2 x) - \sin x\cos x}{x^2\sin^2 x}$

$$= \frac{-x - \sin x\cos x}{x^2\sin^2 x} = -\frac{x}{x^2\sin^2 x} - \frac{\sin x\cos x}{x^2\sin^2 x} = -\frac{1}{x\sin^2 x} - \frac{\cos x}{x^2\sin x} = -\frac{\operatorname{cosec}^2 x}{x} - \frac{\cot x}{x^2} \quad \text{Ans.}$$

(d) $y = \dfrac{\tan x - \cot x}{\tan x + \cot x}$, Let $f(x) = \tan x - \cot x$ and $g(x) = \tan x + \cot x$ then $f'(x) = \sec^2 x + \operatorname{cosec}^2 x$ and $g'(x) = \sec^2 x - \operatorname{cosec}^2 x$

use above formula, $\dfrac{dy}{dx} = \dfrac{(\tan x + \cot x).(\sec^2 x + \operatorname{cosec}^2 x) - (\tan x - \cot x).(\sec^2 x - \operatorname{cosec}^2 x)}{(\tan x + \cot x)^2}$

$$\frac{dy}{dx} = \frac{(\tan x + \cot x).(\sec^2 x + \operatorname{cosec}^2 x)}{(\tan x + \cot x)^2} - \frac{(\tan x - \cot x).(\sec^2 x - \operatorname{cosec}^2 x)}{(\tan x + \cot x)^2}$$

$$= \frac{\sec^2 x + \operatorname{cosec}^2 x}{\tan x + \cot x} - \frac{(\tan x - \cot x).(\sec^2 x - \operatorname{cosec}^2 x)}{(\tan x + \cot x)^2} \quad (\text{change sin and cos})$$

$$\frac{dy}{dx} = 2\sin 2x \quad \text{Ans.}$$

IInd Method: $-$ $y = \dfrac{\tan x - \cot x}{\tan x + \cot x}$ (change sin and cos) or $y = -\cos 2x$ or $\dfrac{dy}{dx} = -(-\sin 2x).2 = 2\sin 2x$ Ans.

(e) $y = \dfrac{1+\sin x}{1-\sin x} = \dfrac{f(x)}{g(x)}$ (say) where $f(x) = 1+\sin x$, $f'(x) = \cos x$ and $g(x) = 1-\sin x$, $g'(x) = -\cos x$

or $\dfrac{dy}{dx} = \dfrac{g(x).f'(x) - f(x).g'(x)}{(g(x))^2} = \dfrac{(1-\sin x).\cos x - (1+\sin x).(-\cos x)}{(1-\sin x)^2} = \dfrac{\cos x - \sin x\cos x + \cos x + \sin x\cos x}{(1-\sin x)^2} = \dfrac{2\cos x}{(1-\sin x)^2}$ Ans.

(f) $y = \sqrt{\dfrac{1-x}{1+x}} = \dfrac{\sqrt{1-x}}{\sqrt{1+x}} = \dfrac{f(x)}{g(x)}$ (say) Here $f(x) = \sqrt{1-x}$, $f'(x) = \dfrac{1}{2\sqrt{1-x}}(-1)$ and $g(x) = \sqrt{1+x}$, $g'(x) = \dfrac{1}{2\sqrt{1+x}}$

$$\frac{dy}{dx} = \frac{g(x).f'(x) - f(x).g'(x)}{(g(x))^2} = \frac{\sqrt{1+x}.\left(-\dfrac{1}{2\sqrt{1-x}}\right) - \sqrt{1-x}.\left(\dfrac{1}{2\sqrt{1+x}}\right)}{(\sqrt{1+x})^2} = \frac{-\dfrac{\sqrt{1+x}}{2\sqrt{1-x}} - \dfrac{\sqrt{1-x}}{2\sqrt{1+x}}}{1+x} = \frac{-1-x-1+x}{2(1+x)\sqrt{(1-x)(1+x)}}$$

$$= -\frac{1}{(1+x)\sqrt{1-x^2}} \quad \text{Ans.}$$

(g) $y = \dfrac{x\sin x}{a+x} = \dfrac{f(x)}{g(x)}$ (say) Here $f(x) = x\sin x$, $f'(x) = x\dfrac{d(\sin x)}{dx} + \sin x\dfrac{d(x)}{dx} = x\cos x + \sin x$ and $g(x) = a+x$, $g'(x) = 1$

$$\therefore \frac{dy}{dx} = \frac{g(x).f'(x) - f(x).g'(x)}{(g(x))^2} = \frac{(a+x)(x\cos x + \sin x) - x\sin x.1}{(a+x)^2} = \frac{ax\cos x + a\sin x + x^2\cos x + x\sin x - x\sin x}{(a+x)^2}$$

$$= \frac{ax\cos x + a\sin x + x^2\cos x}{(a+x)^2} \quad \text{Ans.}$$

(h) $y = \dfrac{x\cot x}{\tan x + \cot x} = \dfrac{x\dfrac{\cos x}{\sin x}}{\dfrac{\sin x}{\cos x} + \dfrac{\cos x}{\sin x}} = \dfrac{\dfrac{x\cos x}{\sin x}}{\dfrac{\sin^2 x + \cos^2 x}{\sin x\cos x}} = x\cos^2 x$, $\dfrac{dy}{dx} = x\dfrac{d(\cos^2 x)}{dx} + \cos^2 x\dfrac{d(x)}{dx}$

$$\frac{dy}{dx} = x.\, 2\cos x\,(-\sin x) + \cos^2 x\,.\,1 = -2x\sin x\cos x + \cos^2 x = -x\sin 2x + \cos^2 x = -x\sin 2x + \frac{1 + \cos 2x}{2}$$

$$= \frac{-2x\sin 2x + \cos 2x + 1}{2} \quad \text{Ans.}$$

(i) $y = x^3 \sin x$, $\dfrac{dy}{dx} = x^3$(differential coefficient(d. c) of $\sin x$) + $\sin x$ (differential coefficient(d. c) of x^3)

$$\therefore \frac{dy}{dx} = x^3\cos x + \sin x\,.\,3x^2 = x^2(x\cos x + 3\sin x) \quad \text{Ans.}$$

(j) $y = (x^2 - 1)\sec x = f(x)\,.\,g(x)$, $\dfrac{dy}{dx} = f(x)g'(x) + g(x)f'(x)$ where $f(x) = x^2 - 1$, $f'(x) = 2x$ and $g(x) = \sec x$, $g'(x) = \tan x \sec x$

$$\therefore \frac{dy}{dx} = (x^2 - 1).\tan x\sec x + \sec x\,.\,2x = \sec x\,(x^2\tan x - \tan x + 2x) \quad \text{Ans.}$$

(k) $y = \dfrac{1 + \tan x}{1 - \tan x} = \dfrac{f(x)}{g(x)}$ (say) here $f(x) = 1 + \tan x$, $f'(x) = \sec^2 x$ and $g(x) = 1 - \tan x$, $g'(x) = -\sec^2 x$

or $\dfrac{dy}{dx} = \dfrac{g(x).f'(x) - f(x).g'(x)}{\big(g(x)\big)^2} = \dfrac{(1 - \tan x).\sec^2 x - (1 + \tan x).(-\sec^2 x)}{(1 - \tan x)^2}$

$$\therefore \frac{dy}{dx} = \frac{\sec^2 x - \tan x\sec^2 x + \sec^2 x + \tan x\sec^2 x}{(1 - \tan x)^2} = \frac{2\sec^2 x}{(1 - \tan x)^2} \quad \text{Ans.}$$

(l) $y = \dfrac{x^2 - \sec x}{1 + \tan x}$ (use above formula and solve this question) or $\dfrac{dy}{dx} = \dfrac{(2x - \tan x\sec x)}{1 + \tan x} - \dfrac{(x^2 - \sec x)\sec^2 x}{(1 + \tan x)^2}$ Ans.

(m) $y = \dfrac{x^2\cos x}{1 - \tan x}$ (same as above question), $\dfrac{dy}{dx} = \dfrac{(2x\cos x - x^2\sin x)}{(1 - \tan x)} + \dfrac{x^2\sec x}{(1 - \tan x)^2}$ Ans.

(6) Find the differential coefficient of the following: − (a) $y = x^2$ (b) $y = 3x^3$

(c) $y = x + x^3 + 2x^5$ (d) $y + 1 = x^3 + 2x^2$ (e) $y = \dfrac{1}{x^3}$ (f) $y = \sqrt{x + 1}$ (g) $y = \dfrac{1}{\sqrt{2x + 1}}$

(h) $\sqrt{y + 2} = x + 1$ (i) $y = \sin x + e^x$ (j) $\tan y = 2\sin x + x$ (k) $y = e^x\cos x + x\log x$ (l) $y = 2\sqrt{x} - \dfrac{1}{\sqrt{x}}$

Solution: − (a) $y = x^2$, $\dfrac{dy}{dx} = 2x$ Ans. $\left[\text{formula, } y = x^n,\ \dfrac{dy}{dx} = nx^{n-1}\ \text{or}\ \dfrac{d(x^n)}{dx} = nx^{n-1}\right]$

(b) $y = 3x^3$, $\dfrac{dy}{dx} = 9x^2$ Ans. (c) $y = x + x^3 + 2x^5$, $\dfrac{dy}{dx} = 1 + 3x^2 + 10x^4$ Ans.

(d) $y + 1 = x^3 + 2x^2$, $\dfrac{dy}{dx} = 3x^2 + 4x$ Ans. (e) $y = \dfrac{1}{x^3} = x^{-3}$, $\dfrac{dy}{dx} = -3x^{-4} = -\dfrac{3}{x^4}$ Ans.

(f) $y = \sqrt{x + 1}$ squaring both of sides $y^2 = x + 1$, $2y\dfrac{dy}{dx} = 1$ $\therefore \dfrac{dy}{dx} = \dfrac{1}{2y} = \dfrac{1}{2\sqrt{x + 1}}$ Ans.

(g) $y = \dfrac{1}{\sqrt{2x + 1}} = \dfrac{1}{(2x + 1)^{\frac{1}{2}}} = (2x + 1)^{-\frac{1}{2}}$, $\dfrac{dy}{dx} = -\dfrac{1}{2}.2\,(2x + 1)^{-\frac{1}{2}-1} = -(2x + 1)^{-\frac{3}{2}}$ Ans.

$$\left[\text{Use formula } y = (ax)^n,\ \frac{dy}{dx} = a^n.nx^{n-1}\ \text{and}\ y = (ax + b)^n,\ \frac{dy}{dx} = na(ax + b)^{n-1}\right]$$

(h) $\sqrt{y + 2} = x + 1$ squaring both of sides $y + 2 = (x + 1)^2$ or $y + 2 = x^2 + 2x + 1$ $\therefore \dfrac{dy}{dx} = 2x + 2 = 2(x + 1)$ Ans.

(i) $y = \sin x + e^x$, $\dfrac{dy}{dx} = \cos x + e^x$ Ans.

(j) $\tan y = 2 \sin x + x$, $\sec^2 y \dfrac{dy}{dx} = 2 \cos x + 1$ $\therefore \dfrac{dy}{dx} = \dfrac{1 + 2 \cos x}{\sec^2 y} = \dfrac{1 + 2 \cos x}{\sec^2\{\tan(2 \sin x + x)\}}$ **Ans.**

(k) $y = e^x \cos x + x \log x$, $\dfrac{dy}{dx} = e^x \dfrac{d(\cos x)}{dx} + \cos x \dfrac{d(e^x)}{dx} + x \dfrac{d(\log x)}{dx} + \log x \dfrac{d(x)}{dx}$

$$\dfrac{dy}{dx} = -e^x \sin x + e^x \cos x + x\dfrac{1}{x} + \log x . 1 \quad \therefore \dfrac{dy}{dx} = e^x(\cos x - \sin x) + (1 + \log x) \quad \text{Ans.}$$

(l) $y = 2\sqrt{x} - \dfrac{1}{\sqrt{x}} = \dfrac{2x - 1}{\sqrt{x}}$ or $\dfrac{dy}{dx} = \dfrac{g(x).f'(x) - f(x).g'(x)}{\left(g(x)\right)^2} = \dfrac{\sqrt{x}\dfrac{d(2x-1)}{dx} - (2x-1)\dfrac{d(\sqrt{x})}{dx}}{\left(\sqrt{x}\right)^2} = \dfrac{\sqrt{x}.2 - (2x-1).\dfrac{1}{2\sqrt{x}}}{x}$

$$\dfrac{dy}{dx} = \dfrac{\dfrac{4x - 2x + 1}{2\sqrt{x}}}{x} \quad \therefore \dfrac{dy}{dx} = \dfrac{2x+1}{2x\sqrt{x}} = \dfrac{2x+1}{2(x)^{\frac{3}{2}}} \quad \text{Ans.}$$

Exercise – A7

(1) Find the differential coefficient of the following: –

(a) $y = \sin(ax^2 + bx + c)$ (b) $y = \cos(2x - 3)$ (c) $y = \sqrt{x^2 - 2x + 3}$ (d) $y = \sin\sqrt{2x^2 - 3}$

(e) $y = \sqrt{\sin(2 + 3x^2)}$ (f) $y = \cos\sqrt{\sqrt{x} + 3}$ (g) $y = \sqrt{\cos(x - 2)}$ (h) $y = \sin\sqrt{\tan x}$

(i) $y = \tan\sqrt{\sin\sqrt{x} + \cos\sqrt{x}}$ (j) $y = \cos^2\sqrt{\sin x + 3}$ (k) $y = \sqrt{\sin\sqrt{\cos\sqrt{x-1}}}$ (l) $y = \log\sqrt{\sin x + \cos x}$

(m) $y = \cot(x - 2)$ (n) $y = \sqrt{\cot\sqrt{2x+1}}$ (o) $y = \sec\sqrt{\text{cosec}(x^2 + 1)}$ (p) $y = \text{cosec}(\sec\theta + \tan\theta)$

(q) $y = \sqrt{\sec(\tan\theta + 1)}$ (r) $y = \sqrt{\text{cosec}(\sin\theta + \sec\theta)}$

(2) (a) $y = \sin^{-1}(\sec x + \tan x)$ (b) $y = \cos^{-1}(\cot\theta - \text{cosec}\,\theta)$ (c) $y = \tan^{-1}(\sin\theta - \cos\theta)$

(d) $y = \cot^{-1}(\sec\theta + \text{cosec}\,\theta)$ (e) $y = \sec^{-1}(\tan\theta - \theta)$ (f) $y = \text{cosec}^{-1}(\sin\theta + \tan\theta)$ (g) $y = \sin^{-1}\sqrt{x^2 + 1}$

(h) $y = \sin^{-1}\sqrt{x^2 - 1} + \cos^{-1}\sqrt{x - 1}$ (i) $y = \tan^{-1}(x^2 - 5)$ (j) $y = \sin\sqrt{\cos\sqrt{x-1}}$ (k) $y = \cos\sqrt{\sin\sqrt{x}}$

(l) $y = \tan\sqrt{\cot\sqrt{x^2 + 1}}$ (m) $y = \cot\sqrt{\cos\sqrt{1 + \sin\theta}}$ (n) $y = \sec\sqrt{\text{cosec}\sqrt{x + 1}}$ (o) $y = \text{cosec}\sqrt{\sin\sqrt{1 + \cos\theta}}$

(p) $y = \dfrac{\tan x}{x} - \sqrt{1 + x^2}$ (q) $y = \sin^2(x^2 - 1)$ (r) $y = \sin^2\left(\sqrt{x^2 - x + 1}\right)$ (s) $y = \cos^2(1 - x^2)$

(t) $y = \cos^2\sqrt{x^3 - 1}$ (u) $y = \tan^2(ax^3 + bx^2 + cx + d)$ (v) $y = \tan^2\sqrt{1 - ax^2}$ (w) $y = \sqrt{\tan(2x + 1)}$

(3) (a) $y = x\sin x^3$ (b) $y = x^2\cos(x^3 + 1)$ (c) $y = x^3\tan x^2$ (d) $y = (x - 1)\sec x^3$ (e) $y = \sqrt{x}\cot x^2$

(f) $y = (x^2 + 1)\text{cosec}\sqrt{x + 1}$ (g) $y = \dfrac{\sin(x^2 + 1)}{x^3}$ (h) $y = \dfrac{\tan\sqrt{x + 1}}{x}$ (i) $y = \sin\left[\dfrac{1 + x}{1 - x}\right]$

(j) $y = \sin\sqrt{x^2 + ax + 1} \times \cos\sqrt{ax^2 + 1}$ (k) $y = \tan(ax^2 + bx + 1) + \cot\sqrt{ax^2 + bx + c}$ (l) $y = x^2\log\sqrt{x - 3}$

(m) $y = \sin^3\sqrt{5 - 2x + x^2}$ (n) $y = (x - 1)^2.(3x^2 - 1)$ (o) $y = \cos(\log x^2)$ (p) $y = \sin(e^x + 1)$

(q) $y = \tan[\log(x - 5)]$ (r) $y = \log[\log(3 - x)]$ (s) $y = e^{\sqrt{x}}.\log x$ (t) $y = e^{2x}.x^3$ (u) $y = \sqrt{1 + \tan\theta}$ (v) $y = \sqrt{\sin\sqrt{x}}$

(4) (a) $y = \dfrac{\tan\left(\dfrac{1-x}{1+x}\right)}{x} + \dfrac{\cot\left(\dfrac{1+x}{1-x}\right)}{x}$ (b) $y = \cos\sqrt{\sin\sqrt{ax+b}}$ (c) $y = \dfrac{\sqrt{\tan x}}{\sqrt{x^2-3}}$ (d) $y = \dfrac{x^2+\sin x^3}{x^3+\cos x^2}$ (e) $y = x^{\frac{3}{2}}\cos(ax^2+bx+c)$

(5) Find the differentiate the following: – (a) $y = \tan^{-1}\left(\dfrac{\sqrt{x}-1}{1+\sqrt{x}}\right)$ (b) $y = \tan^{-1}\left(\dfrac{\sin x+\cos x}{1-\sin x\cos x}\right)$ (c) $y = \tan^{-1}\left(\dfrac{6x}{1-x^2}\right)$

(d) $y = \tan^{-1}\left(\dfrac{x+3}{1+x^2}\right) + \tan^{-1}\left(\dfrac{2x+1}{2-3x}\right)$ (e) $y = \cot^{-1}\left(\dfrac{\sqrt{1+x}-\sqrt{1-x}}{\sqrt{1-x}+\sqrt{1+x}}\right)$ (f) $y = \sin^{-1}\dfrac{1}{\sqrt{1+x^2}} + \tan^{-1}\dfrac{x}{\sqrt{1-x^2}}$

(6) (a) $y = \sin^{-1}\sqrt{x} + \sin^{-1}\sqrt{1-x}$ (b) $y = \sin^{-1}(x^2-1) - \sin^{-1}(2x^2)$ (c) $y = \cos^{-1}\dfrac{1}{1+x} + \cos^{-1}(1-x)$

(d) $y = \cos^{-1}\dfrac{x}{\sqrt{1+x}} - \cos^{-1}\dfrac{1}{\sqrt{1-x}}$ (e) $y = \sin^{-1}\left[\sqrt{x}.\sqrt{2-x} + \sqrt{x-1}.\sqrt{1-x}\right]$ (f) $y = \cos^{-1}\left[\sqrt{x}.\dfrac{1}{\sqrt{1+x^2}} - \sqrt{1-x}.\sqrt{\dfrac{x^2}{1+x^2}}\right]$

(7) (a) Differentiate $\tan^{-1}\dfrac{\sqrt{1-x^2}}{x}$ with respect to $\tan^{-1}x$ (b) Differentiate $\sin^{-1}\sqrt{\dfrac{1+x}{1-x}}$ with respect to $\sqrt{x}$

(c) Differentiate $x^{\cos^{-1}\sqrt{x}}$ with respect to $\cos^{-1}\sqrt{x}$ (d) Differentiate $\cot^{-1}\dfrac{x}{\sqrt{1+x^2}}$ with respect to $\csc^{-1}\dfrac{1}{x^2+1}$

(8) (a) Find the differential coefficient of $\log_{(1-x)}\cos^{-1}(1-x)$ with respect to $2^{2(1-x)}$ and also find its value at $x = \dfrac{1}{2}$.

(b) If $\sqrt{1+4x^2} - \sqrt{1-9y^2} = a(2x-3y)$ then find $\dfrac{dy}{dx}$. (c) If $y = \tan^{-1}\dfrac{1}{1+x} + \cot^{-1}(1+x)$ find $\dfrac{d^2y}{dx^2}$.

(d) If $x = a(1-\cos t)$, $y = a(1+\sin t)$ find $\dfrac{dy}{dx}$ and independent of t. (e) If $x = a\left[\sin t - \log\left(\tan\dfrac{t}{2}\right)\right]$, $y = a\cos t$ find $\dfrac{dy}{dx}$ at $t = \dfrac{\pi}{4}$.

(9) (a) If $x = t^2 + \dfrac{1}{t^2}$ and $y = t^2 - \dfrac{1}{t^2}$ find $\dfrac{d^2y}{dx^2}$ and $\dfrac{dy}{dx}$ is independent of t. (b) If $x = \dfrac{1-t^2}{t^3}$, $y = \dfrac{3}{2t^3} - \dfrac{2}{t^2}$ find $\dfrac{dy}{dx}$ at $t = 2$.

(c) If $x = 2\sin t + \sin 2t$, $y = 2\cos t + \cos 2t$ find the value of $\dfrac{dy}{dx}$ at $x = \dfrac{\pi}{2}$.

(d) If $x = \sin^{-1}\dfrac{1}{\sqrt{1+t^2}}$, $y = \cos^{-1}\dfrac{t}{\sqrt{1+t^2}}$ show that $\dfrac{dy}{dx}$ is independent of t.

(10) (a) If $x = \dfrac{\sin^2 t}{\sqrt{\cos 2t}}$, $y = \dfrac{\cos^2 t}{\sqrt{\sin 2t}}$ find $\dfrac{dy}{dx}$ at $t = \dfrac{\pi}{4}$ (b) If $x = \tan\theta.\sqrt{\cos 2\theta}$, $y = \cot\theta.\sqrt{\sin 2\theta}$ find $\dfrac{dy}{dx}$ at $\theta = \dfrac{\pi}{6}$

(c) If $x^3 - 2xy + y^3 = a^3$ find $\dfrac{d^2y}{dx^2}$. (d) If $x = \cos^{-1}\left(2t\sqrt{1-t^2}\right)$ and $y = \dfrac{\pi}{2} + \sin^{-1}t$ then find the value of $\dfrac{d^2y}{dx^2}$ at $t = \dfrac{\pi}{6}$.

(11) (a) $y = (\cos^{-1}x)^x$ (b) $y = x^{x^2}$ (c) $y = (\sin x)^{\log x}$ (d) $y = (\cos x)^{\log x}$ (e) $y = (\tan x)^{\log x}$ (f) $y = (\log x)^{\log(\tan x)}$

(g) $y = (\sin x)^{\cos x} + (\tan x)^{\cot x}$ (h) $y = (\cot x)^{\cos x}$ (i) $y = (\tan x)^{\sin x}$ (j) $y = e^{x\cos x^2} + (\cot x)^x$

(12) (a) $y = e^{x+x^2+x^3+\cdots\ldots\ldots\infty}$ (b) $y = e^{(\sin x+\cos x)}$ (c) $y = \sqrt{\sin x\sqrt{\sin x\sqrt{\sin\ldots\ldots\ldots\ldots\infty}}}$

(d) $y = \sqrt{\sin x + \sqrt{\sin x + \sqrt{\sin x + \sqrt{\ldots\ldots\ldots\ldots\infty}}}}$ (e) $y = e^{\sin x+\sin^2 x+\sin^3 x+\cdots\ldots\ldots\infty}$ (f) $y = (\sin x)^{(\sin x)^{\sin x\cdots\ldots\infty}}$

(g) $y = (\log x)^{(\log x)^{\log x\cdots\ldots\infty}}$ (h) $y = e^{(\sqrt{x})^{e^{(\sqrt{x})^{\cdots\cdots\infty}}}}$ (i) $y = \log(\log(\log(\cos x)))$ (j) $y = x^x + x^{\sin x}$ (k) $y = a^x + a^{\sin x}$

(l) $e^y = e^{x-y}$ (m) $x^m y^n = (x-y)^{m-n}$ (n) $x^y + y^x = (x-y)^{x-y}$

(13) (a) If $y = x^2 \log\left(\dfrac{a+x}{a-x}\right)$ then $\dfrac{dy}{dx} = 2x\left(\dfrac{ax}{a^2 - x^2} + \log\left(\dfrac{a+x}{a-x}\right)\right)$ (b) If $y = \left(x - \sqrt{x^2 - a^2}\right)^n$ then $\dfrac{dy}{dx}$.

(c) $y = \dfrac{1 - \sqrt{x-1}}{1 + \sqrt{x-1}} + \dfrac{\sqrt{x-1}}{\sqrt{x}}$, find $\dfrac{dy}{dx}$.   (d) $\cos y = a\cos(x+y)$, find $\dfrac{dy}{dx}$.   (e) $y = 1 + \dfrac{Ax^2}{x-A} + \dfrac{Bx}{x-B} + \dfrac{C}{x-C}$, find $\dfrac{dy}{dx}$.

(f) $y = \tan^{-1}\left(\dfrac{\sqrt{a^2 + b^2}\,\cos x}{b - a\cos x}\right)$, find $\dfrac{dy}{dx}$.   (g) $y\sqrt{x^2 + 1} = \log\left(\sqrt{x^2 + 1} - x\right)$, find $\dfrac{dy}{dx}$.

(h) If $y = 2\tan^{-1}\left(\dfrac{x}{\sqrt{1 - x^2}}\right) + \log\left(\dfrac{1 + 2\sqrt{x} + x}{1 - 2\sqrt{x} + x}\right)$ then find $\dfrac{dy}{dx}$.

(14) (a) $y = x + \dfrac{1}{x + \dfrac{1}{x + \dfrac{1}{x + \cdots \,.... \,\infty}}}$ (b) $y = \sin x + \dfrac{1}{\sin x + \dfrac{1}{\sin x + \cdots \,...... \,\infty}}$

(c) If $x^2 + y^2 + 2xy + x + y + 5 = 0$ then find the value of $\dfrac{d^2 y}{dx^2}$.   (d) $y = (\log_{\sin x}\cos x)(\log_{\tan x}\cot x) + \cos^{-1}\dfrac{x}{1+x}$ at $x = \dfrac{\pi}{4}$.

(e) $f(x) = \log_x \cos x^2 + \cos x^2$, find $\dfrac{dy}{dx}$ with respect to $\sqrt{x+1}$ (f) If $y = (\cos^{-1} x + \sin^{-1} x)x$ then prove that $\dfrac{d^2 y}{dx^2} = 0$ and $\dfrac{dy}{dx} = \dfrac{\pi}{2}$.

(g) If $Ax^2 + By^2 = 1$ then prove that $By^2 y'' - Axy' + Ay = 0$ or $(1 - Ax^2)y'' - Axy' + Ay = 0$

(h) If $y = (\cos^{-1} x)^3$ then prove that $\dfrac{d^2 y}{dx^2} + \dfrac{x}{1 - x^2}\dfrac{dy}{dx} - \dfrac{6}{1 - x^2}y^{\frac{1}{3}} = 0$.   (i) If $y = e^{a\cos^{-1} x}$ then prove that $\left(\sqrt{1 - x^2}\right)y_1 + ay = 0$.

(j) If $y = \tan^{-1}\sqrt{x - 1}$ then prove that $y_2 + \dfrac{1}{x}y_1 = 0$ or $xy_2 = -y_1$.

(15) (a) $y = \cos x.\cos 3x.\cos 5x$ (b) $y = 8\sin x.\sin 2x.\sin 4x$ (c) $y = \cos 5x.\sin 3x$ (d) $y = \sin 2x + \cos 3x$

(e) $y = 2\tan 2\theta + 3\cot 2\theta$ (f) $y = \sec 2\theta.\csc 3\theta$ (g) $y = \cos 3\theta + \sec 2\theta$ (h) $y = \sin 2\theta + \csc 3\theta$ (i) $y = \log_2 \log_2 \log_2 x^2$

(j) $y = 4\log_e \log_e \log_e \log_e x$ (k) $f(x) = \log_{x^3} \log x^2$ then find $f'(x)$ at $x = e$.

(16) (a) $y = \sin^m x.\cos^n x$ (b) $y = \sin^n x.\cos mx$ (c) $y = f(f(\log x))$ where $f(x) = \log x$.

(d) $y = f(f(\sin x))$, where $f(x) = \sin x$ then prove that $\dfrac{dy}{dx} = \cos x.\cos(\sin x).\cos(\sin(\sin x))$.

(e) If $x\sqrt{1 - y} - y\sqrt{1 - x} = 0$ then $\dfrac{dy}{dx} = -\dfrac{1}{(x-1)^2}$.

(f) If $\cos^{-1}\left(\dfrac{x^2 + y^2}{x^2 - y^2}\right) = \log a$ then show that $\dfrac{dy}{dx} = \dfrac{y}{x}$ or $xy' - y = 0$ or $x\dfrac{dy}{dx} - y = 0$.

(g) If $\sqrt{x + y} + \sqrt{2x - y} = c$ then show that $\dfrac{d^2 y}{dx^2}$.

(17) (a) Prove that the value of $6 + \log_{\frac{3}{2}}\left[\dfrac{1}{3\sqrt{2}}\sqrt{4 - \dfrac{1}{3\sqrt{2}}\sqrt{4 - \dfrac{1}{3\sqrt{2}}\sqrt{4 - \dfrac{1}{3\sqrt{2}}\sqrt{\cdots \,......}}}}\right]$ is 4.

Answer

Solution: $-$ (1) (a) $y = \sin(ax^2 + bx + c)$,   Let $u = ax^2 + bx + c$, $\dfrac{du}{dx} = 2ax + b$

Now, $y = \sin u$, $\dfrac{dy}{du} = \cos u$ $\therefore \dfrac{dy}{dx} = \dfrac{dy}{du} \times \dfrac{du}{dx} = \cos u.(2ax+b) = (2ax+b)\cos(ax^2+bx+c)$ Ans.

IInd Method:$- \ y = \sin(ax^2+bx+c)$

or $\dfrac{dy}{dx} = \cos(ax^2+bx+c).\dfrac{d(ax^2+bx+c)}{dx} = \cos(ax^2+bx+c)(2ax+b) = (2ax+b)\cos(ax^2+bx+c)$ Ans.

(b) Do yourself. $y = \cos(2x-3)$, Let $u = 2x-3$ $\therefore \dfrac{dy}{dx} = -2\sin(2x-3)$ Ans.

(c) $y = \sqrt{x^2-2x+3}$, Let $u = x^2-2x+3$ $\therefore \dfrac{du}{dx} = 2x-2$ or $\left[\text{use formula, } y = \sqrt{f(x)} \ \therefore \dfrac{dy}{dx} = \dfrac{1}{2\sqrt{f(x)}}.f'(x)\right]$

or $y = \sqrt{u}$ $\therefore \dfrac{dy}{du} = \dfrac{1}{2\sqrt{u}}$ $\therefore \dfrac{dy}{dx} = \dfrac{dy}{du} \times \dfrac{du}{dx} = \dfrac{1}{2\sqrt{u}}(2x-2) = \dfrac{2(x-1)}{2\sqrt{x^2-2x+3}} = \dfrac{(x-1)}{\sqrt{x^2-2x+3}}$ Ans.

(d) $y = \sin\sqrt{2x^2-3}$, Let $u = 2x^2-3$ $\therefore \dfrac{du}{dx} = 4x$ or $y = \sin\sqrt{u}$, Let $v = \sqrt{u}$ $\therefore \dfrac{dv}{du} = \dfrac{1}{2\sqrt{u}}$ or $y = \sin v$ $\therefore \dfrac{dy}{dv} = \cos v$

$\therefore \dfrac{dy}{dx} = \dfrac{dy}{dv} \times \dfrac{dv}{du} \times \dfrac{du}{dx} = \cos v.\dfrac{1}{2\sqrt{u}}.4x = \dfrac{2x\cos\sqrt{2x^2-3}}{\sqrt{2x^2-3}}$ Ans.

Direct:$- \ y = \sin\sqrt{2x^2-3}$, $\dfrac{dy}{dx} = \cos\sqrt{2x^2-3}.\dfrac{d(\sqrt{2x^2-3})}{dx} = \cos\sqrt{2x^2-3}.\dfrac{1}{2\sqrt{2x^2-3}}.4x = \dfrac{2x\cos\sqrt{2x^2-3}}{\sqrt{2x^2-3}}$ Ans.

(e) $y = \sqrt{\sin(2+3x^2)}$, Let $u = 2+3x^2$ $\therefore \dfrac{du}{dx} = 6x$ or $y = \sqrt{\sin u}$, Let $v = \sin u$ $\therefore \dfrac{dv}{du} = \cos u$

$y = \sqrt{v}$, $\dfrac{dy}{dv} = \dfrac{1}{2\sqrt{v}}$ $\therefore \dfrac{dy}{dx} = \dfrac{dy}{dv} \times \dfrac{dv}{du} \times \dfrac{du}{dx} = \dfrac{1}{2\sqrt{v}}.\cos u.6x = \dfrac{3x\cos(2+3x^2)}{\sqrt{\sin u}} = \dfrac{3x\cos(2+3x^2)}{\sqrt{\sin(2+3x^2)}}$ Ans.

(f) $y = \cos\sqrt{\sqrt{x}+3}$, Let $u = \sqrt{x}$ $\therefore \dfrac{du}{dx} = \dfrac{1}{2\sqrt{x}}$ or $y = \cos\sqrt{u+3}$, Let $v = u+3$ $\therefore \dfrac{dv}{du} = 1$

$y = \cos\sqrt{v}$, Let $w = \sqrt{v}$ $\therefore \dfrac{dw}{dv} = \dfrac{1}{2\sqrt{v}} = \dfrac{1}{2\sqrt{u+3}} = \dfrac{1}{2\sqrt{\sqrt{x}+3}}$

$y = \cos w$, $\dfrac{dy}{dw} = -\sin w = -\sin\sqrt{v} = -\sin\sqrt{u+3} = -\sin\sqrt{\sqrt{x}+3}$

$\therefore \dfrac{dy}{dx} = \dfrac{dy}{dw} \times \dfrac{dw}{dv} \times \dfrac{dv}{du} \times \dfrac{du}{dx} = -\sin\sqrt{\sqrt{x}+3}.\dfrac{1}{2\sqrt{\sqrt{x}+3}}.1.\dfrac{1}{2\sqrt{x}} = -\dfrac{1}{4}\dfrac{\sin\sqrt{\sqrt{x}+3}}{\sqrt{\sqrt{x}+3}.\sqrt{x}}$ Ans.

IInd Method:$- \ y = \cos\sqrt{\sqrt{x}+3}$, $\dfrac{dy}{dx} = -\sin\sqrt{\sqrt{x}+3}.\dfrac{d(\sqrt{\sqrt{x}+3})}{dx}.\dfrac{d(\sqrt{x})}{dx} = -\sin\sqrt{\sqrt{x}+3}.\dfrac{1}{2\sqrt{\sqrt{x}+3}}.\dfrac{1}{2\sqrt{x}} = -\dfrac{1}{4}.\dfrac{\sin\sqrt{\sqrt{x}+3}}{\sqrt{\sqrt{x}+3}.\sqrt{x}}$ Ans.

(g) solve same as above question. $y = \sqrt{\cos(x-2)}$, $\dfrac{dy}{dx} = -\dfrac{1}{2}.\dfrac{\sin(x-2)}{\sqrt{\cos(x-2)}}$ Ans.

(h) $y = \sin\sqrt{\tan x}$, Let $u = \tan x$ $\therefore \dfrac{du}{dx} = \sec^2 x$

$y = \sin\sqrt{u}$, Let $v = \sqrt{u}$ $\therefore \dfrac{dv}{du} = \dfrac{1}{2\sqrt{u}} = \dfrac{1}{2\sqrt{\tan x}}$ or $y = \sin v$ $\therefore \dfrac{dy}{dv} = \cos v = \cos\sqrt{u} = \cos\sqrt{\tan x}$

$\therefore \dfrac{dy}{dx} = \dfrac{dy}{dv} \times \dfrac{dv}{du} \times \dfrac{du}{dx} = \cos\sqrt{\tan x}.\dfrac{1}{2\sqrt{\tan x}}.\sec^2 x = \dfrac{\sec^2 x.\cos\sqrt{\tan x}}{2\sqrt{\tan x}} = \dfrac{\cos\sqrt{\tan x}}{2\cos^2 x.\sqrt{\tan x}}$ Ans.

(i) $y = \tan\sqrt{\sin\sqrt{x}+\cos\sqrt{x}}$, Let $u = \sqrt{x}$ $\therefore \dfrac{du}{dx} = \dfrac{1}{2\sqrt{x}}$

Now, $y = \tan \sqrt{\sin u + \cos u}$, Let $v = \sin u + \cos u$ $\therefore \dfrac{dv}{du} = \cos u - \sin u = \cos \sqrt{x} - \sin \sqrt{x}$ put $u = \sqrt{x}$

Now, $y = \tan \sqrt{v}$, Let $w = \sqrt{v}$ $\therefore \dfrac{dw}{dv} = \dfrac{1}{2\sqrt{v}} = \dfrac{1}{2\sqrt{\sin u + \cos u}} = \dfrac{1}{2\sqrt{\sin \sqrt{x} + \cos \sqrt{x}}}$

put $v = \sin u + \cos u$, $u = \sqrt{x}$

Now, $y = \tan w$ $\therefore \dfrac{dy}{dw} = \sec^2 w = \sec^2 \sqrt{v} = \sec^2 \sqrt{\sin u + \cos u} = \sec^2 \sqrt{\sin \sqrt{x} + \cos \sqrt{x}}$

$$\left(\text{put } w = \sqrt{v}, \ v = \sin u + \cos u \text{ and } u = \sqrt{x} \right)$$

Then, $\dfrac{dy}{dx} = \dfrac{dy}{dw} \times \dfrac{dw}{dv} \times \dfrac{dv}{du} \times \dfrac{du}{dx} = \sec^2 \sqrt{\sin \sqrt{x} + \cos \sqrt{x}} \cdot \dfrac{1}{2\sqrt{\sin \sqrt{x} + \cos \sqrt{x}}} \cdot \left(\cos \sqrt{x} - \sin \sqrt{x} \right) \cdot \dfrac{1}{2\sqrt{x}}$

$\therefore \dfrac{dy}{dx} = \dfrac{1}{4} \cdot \dfrac{\left(\cos \sqrt{x} - \sin \sqrt{x} \right)}{\cos^2 \sqrt{\sin \sqrt{x} + \cos \sqrt{x}} \cdot \sqrt{\sin \sqrt{x} + \cos \sqrt{x}} \cdot \sqrt{x}}$ Ans.

(j) $y = \cos^2 \sqrt{\sin x + 3}$, Let $u = \sin x + 3$ $\therefore \dfrac{du}{dx} = \cos x$

Now, $y = \cos^2 \sqrt{u}$, Let $v = \sqrt{u}$ $\therefore \dfrac{dv}{du} = \dfrac{1}{2\sqrt{u}} = \dfrac{1}{2\sqrt{\sin x + 3}}$ (put $u = \sin x + 3$)

Now, $y = \cos^2 v = (\cos v)^2$, Let $w = \cos v$ $\therefore \dfrac{dw}{dv} = -\sin v = -\sin \sqrt{u}$

$$\dfrac{dw}{dv} = -\sin \sqrt{\sin x + 3} \quad (\text{put } v = \sqrt{u} \text{ and } u = \sin x + 3)$$

Now, $y = w^2$ $\therefore \dfrac{dy}{dw} = 2w = 2 \cos v = 2 \cos \sqrt{u} = 2 \cos \sqrt{\sin x + 3}$ (put $w = \cos v, v = \sqrt{u}$ and $u = \sin x + 3$)

Then, $\dfrac{dy}{dx} = \dfrac{dy}{dw} \times \dfrac{dw}{dv} \times \dfrac{dv}{du} \times \dfrac{du}{dx} = 2 \cos \sqrt{\sin x + 3} \cdot \left(-\sin \sqrt{\sin x + 3} \right) \cdot \dfrac{1}{2\sqrt{\sin x + 3}} \cdot \cos x$

$\therefore \dfrac{dy}{dx} = \dfrac{-2 \sin \sqrt{\sin x + 3} \cdot \cos \sqrt{\sin x + 3} \cdot \cos x}{2\sqrt{\sin x + 3}} = -\dfrac{\sin\{2(\sqrt{\sin x + 3})\} \cdot \cos x}{2\sqrt{\sin x + 3}}$ Ans. [$\sin 2x = 2 \sin x \cos x$]

(k) Do yourself. $y = \sqrt{\sin \sqrt{\cos \sqrt{x - 1}}}$, $\dfrac{dy}{dx} = -\dfrac{1}{8} \cdot \dfrac{\sin \sqrt{x - 1} \cdot \cos \sqrt{\cos \sqrt{x - 1}}}{\sqrt{\sin \sqrt{\cos \sqrt{x - 1}}} \cdot \sqrt{\cos \sqrt{x - 1}} \cdot \sqrt{x - 1}}$ Ans.

(l) Do yourself. $y = \log \sqrt{\sin x + \cos x}$, $\dfrac{dy}{dx} = \dfrac{1}{2} [\sec 2x - \tan 2x]$ Ans.

(m) $y = \cot(x - 2)$, Let $u = x - 2$ $\therefore \dfrac{du}{dx} = 1$

Now, $y = \cot u = \dfrac{\cos u}{\sin u} = \dfrac{f(x)}{g(x)}$ (say) $\therefore \dfrac{dy}{du} = \dfrac{g(x) \cdot f'(x) - f(x) \cdot g'(x)}{\left(g(x) \right)^2} = \dfrac{\sin u \cdot \dfrac{d(\cos u)}{du} - \cos u \cdot \dfrac{d(\sin u)}{du}}{(\sin u)^2}$

$\therefore \dfrac{dy}{du} = \dfrac{\sin u \cdot (-\sin u) - \cos u \cdot \cos u}{(\sin u)^2} = \dfrac{-\sin^2 u - \cos^2 u}{\sin^2 u} = \dfrac{-(\sin^2 u + \cos^2 u)}{\sin^2 u} = -\dfrac{1}{\sin^2 u} = -\csc^2 u$

$$\therefore \dfrac{dy}{dx} = \dfrac{dy}{du} \times \dfrac{du}{dx} = -\csc^2 u \cdot 1 = -\csc^2 (x - 2) \quad \text{Ans.} \quad (\text{put } u = x - 2)$$

(n) $y = \sqrt{\cot \sqrt{2x + 1}}$, Let $u = \sqrt{2x + 1}$ $\therefore \dfrac{du}{dx} = \dfrac{1}{2\sqrt{2x + 1}} \cdot 2 = \dfrac{1}{\sqrt{2x + 1}}$

Now, $y = \sqrt{\cot u}$, $\dfrac{dy}{du} = \dfrac{1}{2\sqrt{\cot u}} . \dfrac{d(\cot u)}{du} = \dfrac{1}{2\sqrt{\cot u}} . (-\cosec^2 u) = -\dfrac{\cosec^2 \sqrt{2x+1}}{2\sqrt{\cot \sqrt{2x+1}}}$ (put $u = \sqrt{2x+1}$)

$$\therefore \dfrac{dy}{dx} = \dfrac{dy}{du} \times \dfrac{du}{dx} = -\dfrac{\cosec^2 \sqrt{2x+1}}{2\sqrt{\cot \sqrt{2x+1}}} . \dfrac{1}{\sqrt{2x+1}} = -\dfrac{\cosec^2 \sqrt{2x+1}}{2\sqrt{2x+1}.\sqrt{\cot \sqrt{2x+1}}} \quad \text{Ans.}$$

(o) $y = \sec\sqrt{\cosec(x^2+1)}$, Let $u = x^2 + 1$ $\quad \therefore \dfrac{du}{dx} = 2x$

$y = \sec\sqrt{\cosec u}$, Let $v = \sqrt{\cosec u}$ $\quad \therefore \dfrac{dv}{du} = \dfrac{1}{2\sqrt{\cosec u}} . \dfrac{d(\cosec u)}{du} = \dfrac{1}{2\sqrt{\cosec u}} . (-\cot u . \cosec u)$

or $\dfrac{dv}{du} = -\dfrac{\cot u . \cosec u}{2\sqrt{\cosec u}} = -\dfrac{\cot(x^2+1) . \cosec(x^2+1)}{2\sqrt{\cosec(x^2+1)}}$ (put $u = x^2 + 1$)

$y = \sec v$ $\quad \therefore \dfrac{dy}{dv} = \tan v . \sec v = \tan\sqrt{\cosec u} . \sec\sqrt{\cosec u} = \tan\sqrt{\cosec(x^2+1)} . \sec\sqrt{\cosec(x^2+1)}$ (put $v = \sqrt{\cosec u}$ and $u = x^2 + 1$)

$\therefore \dfrac{dy}{dx} = \dfrac{dy}{dv} \times \dfrac{dv}{du} \times \dfrac{du}{dx} = \tan\sqrt{\cosec(x^2+1)} . \sec\sqrt{\cosec(x^2+1)} . -\dfrac{\cot(x^2+1) . \cosec(x^2+1)}{2\sqrt{\cosec(x^2+1)}} . 2x$

$\dfrac{dy}{dx} = -\tan\sqrt{\cosec(x^2+1)} . \sec\sqrt{\cosec(x^2+1)} . \dfrac{\cot(x^2+1) . \sqrt{\cosec(x^2+1)} . \sqrt{\cosec(x^2+1)}}{\sqrt{\cosec(x^2+1)}} . x$

$$= -\tan\sqrt{\cosec(x^2+1)} . \sec\sqrt{\cosec(x^2+1)} . \cot(x^2+1) . \sqrt{\cosec(x^2+1)} . x$$

put $\theta = x^2 + 1$, $\dfrac{dy}{dx} = -\tan\sqrt{\cosec\theta} . \sec\sqrt{\cosec\theta} . \cot\theta . \sqrt{\cosec\theta} . \sqrt{\theta-1}$ Ans. where $\theta = x^2 + 1, x = \sqrt{\theta-1}$

(p) Do yourself. Hint $- y = \cosec(\sec\theta + \tan\theta)$, $\dfrac{dy}{d\theta} = -\sec(\tan\theta + \sec\theta) . \cot(\tan\theta + \sec\theta) . \cosec(\tan\theta + \sec\theta)$ Ans.

(q) $y = \sqrt{\sec(1 + \tan\theta)}$, Let $u = 1 + \tan\theta$ $\quad \therefore \dfrac{du}{d\theta} = \sec^2\theta$

Now, $y = \sqrt{\sec u}$, Let $v = \sec u$ $\quad \therefore \dfrac{dv}{du} = \tan u . \sec u = \tan(1 + \tan\theta) . \sec(1 + \tan\theta)$ [put $u = 1 + \tan\theta$]

Now, $y = \sqrt{v}$, $\dfrac{dy}{dv} = \dfrac{1}{2\sqrt{v}} = \dfrac{1}{2\sqrt{\sec u}} = \dfrac{1}{2\sqrt{\sec(1 + \tan\theta)}}$ [put $v = \sec u$ and $u = 1 + \tan\theta$]

Then, $\dfrac{dy}{dx} = \dfrac{dy}{dv} \times \dfrac{dv}{du} \times \dfrac{du}{d\theta} = \dfrac{1}{2\sqrt{\sec(1 + \tan\theta)}} . \tan(1 + \tan\theta) . \sec(1 + \tan\theta) . \sec^2\theta = \dfrac{1}{2} . \sec^2\theta . \sqrt{\sec(1 + \tan\theta)} . \tan(1 + \tan\theta)$ Ans.

(r) Do yourself. $y = \sqrt{\cosec(\sin\theta + \sec\theta)}$, $\quad \dfrac{dy}{d\theta} = -\dfrac{\cot(\sin\theta + \sec\theta) . \cosec(\sin\theta + \sec\theta) . (\cos\theta + \tan\theta . \sec\theta)}{2\sqrt{\cosec(\sin\theta + \sec\theta)}}$ Ans.

(2) (a) $y = \sin^{-1}(\sec x + \tan x)$, Let $u = \sec x + \tan x$ $\quad \therefore \dfrac{du}{dx} = \sec x . \tan x + \sec^2 x = \sec x (\tan x + \sec x)$

Now, $y = \sin^{-1} u$ $\quad \therefore \dfrac{dy}{du} = \dfrac{1}{\sqrt{1 - u^2}} = \dfrac{1}{\sqrt{1 - (\sec x + \tan x)^2}}$ [put $u = \sec x + \tan x$]

Then, $\dfrac{dy}{dx} = \dfrac{dy}{du} \times \dfrac{du}{dx} = \dfrac{1}{\sqrt{1 - (\sec x + \tan x)^2}} . \sec x (\tan x + \sec x)$ Ans.

(b) Do yourself. $y = \cos^{-1}(\cot\theta - \cosec\theta)$, $\dfrac{dy}{dx} = \dfrac{(1 - \cos\theta)}{\sqrt{2}\sin\theta . \sqrt{\cos\theta (1 - \cos\theta)}}$ Ans.

(c) Do yourself. Ans: $-\dfrac{\sin\theta + \cos\theta}{2(1 - \sin\theta \cos\theta)}$ $\quad$ (d) Ans: $-\dfrac{(\cos\theta - \sin\theta)(1 + \sin\theta \cos\theta)}{1 + \sin\theta \cos\theta (2 + \sin\theta \cos\theta)}$

(e) Ans: $-\dfrac{(\sec^2\theta - 1)}{(\tan\theta - \theta)\sqrt{(\tan\theta - \theta)^2 - 1}}$ or $\dfrac{\tan^2\theta}{(\tan\theta - \theta)\sqrt{\tan^2\theta - 2\tan\theta - 1 + \theta^2}}$

(f) Ans: $-\dfrac{1}{(\sin\theta + \tan\theta)\sqrt{(\sin\theta + \tan\theta)^2 - 1}} \times (\cos\theta + \sec^2\theta)$

(g) $y = \sin^{-1}\sqrt{x^2 + 1}$, Let $u = \sqrt{x^2 + 1}$ $\therefore \dfrac{du}{dx} = \dfrac{1}{2\sqrt{x^2 + 1}} \cdot \dfrac{d(x^2)}{dx} = \dfrac{1}{2\sqrt{x^2 + 1}} \cdot 2x = \dfrac{x}{\sqrt{x^2 + 1}}$

Now, $y = \sin^{-1}u$ $\therefore \dfrac{dy}{du} = \dfrac{1}{\sqrt{1 - u^2}} = \dfrac{1}{\sqrt{1 - \left(\sqrt{x^2 + 1}\right)^2}} = \dfrac{1}{\sqrt{1 - x^2 - 1}} = \dfrac{1}{\sqrt{-x^2}} = \dfrac{1}{|x|}$ $\left[\text{put } u = \sqrt{x^2 + 1}\right]$

Then, $\dfrac{dy}{dx} = \dfrac{dy}{du} \times \dfrac{du}{dx} = \dfrac{1}{|x|} \cdot \dfrac{x}{\sqrt{x^2 + 1}} = \dfrac{x}{|x| \cdot \sqrt{x^2 + 1}}$ Ans.

(h) $y = \sin^{-1}\sqrt{x^2 - 1} + \cos^{-1}\sqrt{x - 1}$

Let $u = \sqrt{x^2 - 1}$, $\dfrac{du}{dx} = \dfrac{1}{2\sqrt{x^2 - 1}} \cdot 2x = \dfrac{x}{\sqrt{x^2 - 1}}$ and $v = \sqrt{x - 1}$, $\dfrac{dv}{dx} = \dfrac{1}{\sqrt{x - 1}}$

Now, $y = \sin^{-1}u + \cos^{-1}v$ $\therefore \dfrac{dy}{dx} = \dfrac{1}{\sqrt{1 - u^2}} \cdot \dfrac{du}{dx} - \dfrac{1}{\sqrt{1 - v^2}} \cdot \dfrac{dv}{dx}$

$\therefore \dfrac{dy}{dx} = \dfrac{1}{\sqrt{1 - \left(\sqrt{x^2 - 1}\right)^2}} \cdot \dfrac{x}{\sqrt{x^2 - 1}} - \dfrac{1}{\sqrt{1 - \left(\sqrt{x - 1}\right)^2}} \cdot \dfrac{1}{\sqrt{x - 1}} = \dfrac{x}{\sqrt{x^2 - 1} \cdot \sqrt{1 - x^2 + 1}} - \dfrac{1}{\sqrt{x - 1}} \cdot \dfrac{1}{\sqrt{1 - x + 1}}$

$= \dfrac{x}{\sqrt{x^2 - 1} \cdot \sqrt{2 - x^2}} - \dfrac{1}{\sqrt{x - 1} \cdot \sqrt{2 - x}}$ Ans.

(i) $y = \tan^{-1}\sqrt{x^2 - 5}$, Let $u = \sqrt{x^2 - 5}$ $\therefore \dfrac{du}{dx} = \dfrac{1}{2\sqrt{x^2 - 5}} \cdot \dfrac{d(x^2 - 5)}{dx} = \dfrac{1}{2\sqrt{x^2 - 5}} \cdot 2x = \dfrac{x}{\sqrt{x^2 - 5}}$

Now, $y = \tan^{-1}u$ $\therefore \dfrac{dy}{du} = \dfrac{1}{1 + u^2} = \dfrac{1}{1 + \left(\sqrt{x^2 - 5}\right)^2} = \dfrac{1}{1 + x^2 - 5} = \dfrac{1}{x^2 - 4}$ $\left[\text{Put } u = \sqrt{x^2 - 5}\right]$

Then, $\dfrac{dy}{dx} = \dfrac{dy}{du} \times \dfrac{du}{dx} = \dfrac{1}{x^2 - 4} \cdot \dfrac{x}{\sqrt{x^2 - 5}} = \dfrac{x}{(x - 2)(x + 2)\sqrt{x^2 - 5}}$ Ans.

(j) see question no. (1) (k), Ans: $-\dfrac{\sin\sqrt{x - 1} \cdot \cos\sqrt{\cos\sqrt{x - 1}}}{4\sqrt{x - 1} \cdot \sqrt{\cos\sqrt{x - 1}}}$ (k) Ans: $-\dfrac{1}{4} \cdot \dfrac{\sin\sqrt{\sin\sqrt{x}} \cdot \cos\sqrt{x}}{\sqrt{x} \cdot \sqrt{\sin\sqrt{x}}}$

(l) $y = \tan\sqrt{\cot\sqrt{x^2 + 1}}$, $\dfrac{dy}{dx} = -\dfrac{1}{2} \cdot \dfrac{x \cdot \sec^2\sqrt{\cot\sqrt{x^2 + 1}} \cdot \text{cosec}^2\sqrt{x^2 + 1}}{\sqrt{x^2 + 1} \cdot \sqrt{\cot\sqrt{x^2 + 1}}}$ Ans.

(m) $y = \cot\sqrt{\cos\sqrt{1 + \sin\theta}}$, $\dfrac{dy}{d\theta} = \dfrac{\cos\theta \cdot \sin\sqrt{1 + \sin\theta} \cdot \text{cosec}^2\sqrt{\cos\sqrt{1 + \sin\theta}}}{4\sqrt{1 + \sin\theta} \cdot \sqrt{\cos\sqrt{1 + \sin\theta}}}$ Ans.

(n) $y = \sec\sqrt{\text{cosec}\sqrt{x + 1}}$, $\dfrac{dy}{dx} = \dfrac{\tan\sqrt{\text{cosec}\sqrt{x + 1}} \cdot \cot\sqrt{x + 1} \cdot \sec\sqrt{\text{cosec}\sqrt{x + 1}} \cdot \text{cosec}\sqrt{x + 1}}{4\sqrt{x + 1} \cdot \sqrt{\text{cosec}\sqrt{x + 1}}}$ Ans.

(o) $y = \text{cosec}\sqrt{\sin\sqrt{1 + \cos\theta}}$, $\dfrac{dy}{d\theta} = \dfrac{(2x - x^2) \cdot \cos\sqrt{x} \cdot \cot\sqrt{\sin\sqrt{x}} \cdot \text{cosec}\sqrt{\sin\sqrt{x}}}{4\sqrt{x} \cdot \sqrt{\sin\sqrt{x}}}$ (where $1 + \cos\theta = x$) Ans.

(p) $y = \dfrac{\tan x}{x} - \sqrt{1 + x^2}$, $\dfrac{dy}{dx} = \dfrac{x\dfrac{d(\tan x)}{dx} - \tan x\dfrac{d(x)}{dx}}{x^2} - \dfrac{1}{2\sqrt{1 + x^2}} \cdot \dfrac{d(1 + x^2)}{dx} = \dfrac{x\sec^2 x - \tan x}{x^2} - \dfrac{1}{2\sqrt{1 + x^2}} \cdot 2x$

$\therefore \dfrac{dy}{dx} = \dfrac{x \cdot \sqrt{1 + x^2}[(1 + \tan^2 x) - \tan x] - x^3}{x^2\sqrt{1 + x^2}} = \dfrac{x \cdot \sqrt{1 + x^2}(\tan^2 x - \tan x + 1) - x^3}{x^2\sqrt{1 + x^2}}$ Ans.

(q) $y = \sin^2(x^2 - 1)$, Let $u = x^2 - 1$ $\therefore \dfrac{du}{dx} = 2x$

Now, $y = \sin^2 u = (\sin u)^2$, Let $v = \sin u$ $\therefore \dfrac{dv}{du} = \cos u = \cos(x^2 - 1)$ [Put $u = x^2 - 1$]

Now, $y = v^2$ $\therefore \dfrac{dy}{dv} = 2v = 2\sin u = 2\sin(x^2 - 1)$ [Put $v = \sin u$ and $u = x^2 - 1$]

Then, $\dfrac{dy}{dx} = \dfrac{dy}{dv} \times \dfrac{dv}{du} \times \dfrac{du}{dx} = 2\sin(x^2 - 1).\cos(x^2 - 1).2x = 2x\sin 2(x^2 - 1)$ Ans.

(r) $y = \sin^2\left(\sqrt{x^2 - x + 1}\right)$, Let $u = \sqrt{x^2 - x + 1}$ $\therefore \dfrac{du}{dx} = \dfrac{1}{2\sqrt{x^2 - x + 1}}.\dfrac{d(x^2 - x + 1)}{dx} = \dfrac{2x - 1}{2\sqrt{x^2 - x + 1}}$

Now, $y = \sin^2 u = (\sin u)^2$, Let $v = \sin u$ $\therefore \dfrac{dv}{du} = \cos u = \cos\left(\sqrt{x^2 - x + 1}\right)$ $\left[\text{Put } u = \sqrt{x^2 - x + 1}\right]$

Now, $y = v^2$ $\therefore \dfrac{dy}{dv} = 2v = 2\sin u = 2\sin\left(\sqrt{x^2 - x + 1}\right)$ $\left[\text{Put } v = \sin u \text{ and } u = \sqrt{x^2 - x + 1}\right]$

Then, $\dfrac{dy}{dx} = \dfrac{dy}{dv} \times \dfrac{dv}{du} \times \dfrac{du}{dx} = 2\sin\left(\sqrt{x^2 - x + 1}\right).\cos\left(\sqrt{x^2 - x + 1}\right).\dfrac{2x - 1}{2\sqrt{x^2 - x + 1}} = \dfrac{(2x - 1).\sin\left(2\sqrt{x^2 - x + 1}\right)}{2\sqrt{x^2 - x + 1}}$ Ans.

(s) $y = \cos^2(1 - x^2)$, Let $u = 1 - x^2$ $\therefore \dfrac{du}{dx} = -2x$

or $y = (\cos u)^2$, Let $v = \cos u$ $\therefore \dfrac{dv}{du} = -\sin u = -\sin(1 - x^2)$ [Put $u = 1 - x^2$]

$\therefore y = v^2$, $\dfrac{dy}{dv} = 2v = 2\cos u = 2\cos(1 - x^2)$ [Put $v = \cos u$ and $u = 1 - x^2$]

or $\dfrac{dy}{dx} = \dfrac{dy}{dv} \times \dfrac{dv}{du} \times \dfrac{du}{dx} = 2\cos(1 - x^2).[-\sin(1 - x^2)].(-2x) = 2x.\sin[2(1 - x^2)]$ Ans.

(t) $y = \cos^2\sqrt{x^3 - 1}$, $\dfrac{dy}{dx} = \dfrac{-3x^2.\sin\left[2\sqrt{x^3 - 1}\right]}{2\sqrt{x^3 - 1}}$ Ans. [See Question No. $-(2)(q)$.]

(u) $y = \tan^2(ax^3 + bx^2 + cx + d)$, Let $u = ax^3 + bx^2 + cx + d$ $\therefore \dfrac{du}{dx} = 3ax^2 + 2bx + c$

or $y = (\tan u)^2$, Let $v = \tan u$ $\therefore \dfrac{dv}{du} = \sec^2 u = \sec^2(ax^3 + bx^2 + cx + d)$ [Put $u = ax^3 + bx^2 + cx + d$]

or $y = v^2$, $\dfrac{dy}{dv} = 2v = 2\tan u = 2\tan(ax^3 + bx^2 + cx + d)$ [Put $v = \tan u$ and $u = ax^3 + bx^2 + cx + d$]

or $\dfrac{dy}{dx} = \dfrac{dy}{dv} \times \dfrac{dv}{du} \times \dfrac{du}{dx} = 2\tan(ax^3 + bx^2 + cx + d).\sec^2(ax^3 + bx^2 + cx + d).(3ax^2 + 2bx + c)$

$\therefore \dfrac{dy}{dx} = (3ax^2 + 2bx + c).[2\tan(ax^3 + bx^2 + cx + d) + 2\tan^3(ax^3 + bx^2 + cx + d)]$ Ans.

(v) $y = \tan^2\sqrt{1 - ax^2}$, Let $u = \sqrt{1 - ax^2}$ $\therefore \dfrac{du}{dx} = \dfrac{1}{2\sqrt{1 - ax^2}}.2ax = \dfrac{ax}{\sqrt{1 - ax^2}}$

$y = \tan^2 u = (\tan u)^2$, Let $v = \tan u$ $\therefore \dfrac{dv}{du} = \sec^2 u = \sec^2\sqrt{1 - ax^2}$ $\left[\text{Put } u = \sqrt{1 - ax^2}\right]$

$y = v^2$ $\therefore \dfrac{dy}{dv} = 2v = 2\tan u = 2\tan\sqrt{1 - ax^2}$ $\left[\text{Put } v = \tan u \text{ and } u = \sqrt{1 - ax^2}\right]$

or $\dfrac{dy}{dx} = \dfrac{dy}{dv} \times \dfrac{dv}{du} \times \dfrac{du}{dx} = 2\tan\sqrt{1 - ax^2}.\sec^2\sqrt{1 - ax^2}.\dfrac{ax}{\sqrt{1 - ax^2}} = \dfrac{-2ax\left(\tan\sqrt{1 - ax^2} + \tan^3\sqrt{1 - ax^2}\right)}{\sqrt{1 - ax^2}}$ Ans.

(w) $y = \sqrt{\tan(2x+1)}$, Let $u = 2x+1$ $\therefore \dfrac{du}{dx} = 2$

$y = \sqrt{\tan u}$, $\dfrac{dy}{du} = \dfrac{1}{2\sqrt{\tan u}} \cdot \dfrac{d(\tan u)}{du} = \dfrac{1}{2\sqrt{\tan u}} \cdot \sec^2 u = \dfrac{\sec^2(2x+1)}{2\sqrt{\tan(2x+1)}}$ [Put $u = 2x+1$]

or $\dfrac{dy}{dx} = \dfrac{dy}{du} \times \dfrac{du}{dx} = \dfrac{\sec^2(2x+1)}{2\sqrt{\tan(2x+1)}} \cdot 2 = \dfrac{\sec^2(2x+1)}{\sqrt{\tan(2x+1)}} = \dfrac{1 + \tan^2(2x+1)}{\sqrt{\tan(2x+1)}} = \dfrac{1}{\sqrt{\tan(2x+1)}} + \dfrac{\tan^2(2x+1)}{\sqrt{\tan(2x+1)}}$

$\qquad = [\tan(2x+1)]^{-\frac{1}{2}} + [\tan(2x+1)]^{\frac{3}{2}}$ Ans.

(3) (a) $y = x\sin x^3$, Let $u = x^3$ or $x = u^{\frac{1}{3}}$ $\therefore \dfrac{du}{dx} = 3x^2$

$y = u^{\frac{1}{3}} \cdot \sin u$, $\dfrac{dy}{du} = u^{\frac{1}{3}} \cdot \dfrac{d(\sin u)}{du} + \sin u \cdot \dfrac{d\left(u^{\frac{1}{3}}\right)}{du} = u^{\frac{1}{3}} \cdot \cos u + \sin u \cdot \dfrac{1}{3} u^{\frac{1}{3}-1} = u^{\frac{1}{3}} \cdot \cos u + \sin u \cdot \dfrac{u^{-\frac{2}{3}}}{3} = u^{\frac{1}{3}} \cdot \cos u + \dfrac{\sin u}{3u^{\frac{2}{3}}} = \dfrac{3u\cos u + \sin u}{3u^{\frac{2}{3}}}$

$\qquad = \dfrac{3x^3 \cos x^3 + \sin x^3}{3(x^3)^{\frac{2}{3}}} = \dfrac{3x^3 \cos x^3 + \sin x^3}{3x^2}$ [Put $u = x^3$]

$\therefore \dfrac{dy}{dx} = \dfrac{dy}{du} \times \dfrac{du}{dx} = \dfrac{3x^3 \cos x^3 + \sin x^3}{3x^2} \cdot 3x^2 = 3x^3 \cos x^3 + \sin x^3$ Ans.

(b) $y = x^2 \cos(x^3+1) = f(x) \cdot g(x)$,

or $\dfrac{dy}{dx} = f(x) \cdot g'(x) + g(x) \cdot f'(x) = x^2 \cdot \dfrac{d[\cos(x^3+1)]}{dx} + \cos(x^3+1) \cdot \dfrac{d(x^2)}{dx} = x^2 \cdot (-\sin(x^3+1)) \cdot 3x^2 + \cos(x^3+1) \cdot 2x$

$\qquad = 2x \cdot \cos(x^3+1) - 3x^2 \cdot \sin(x^3+1)$ Ans.

(c) $y = x^3 \tan x^2$, $\dfrac{dy}{dx} = x^3 \cdot \sec^2 x^2 \cdot 2x + \tan x^2 \cdot 3x^2 = 2x^4 \sec^2 x^2 + 3x^2 \tan x^2 = 2x^4(1 + \tan^2 x^2) + 3x^2 \tan x^2$

$\qquad = 2x^4 + 2x^4 \tan^2 x^2 + 3x^2 \tan x^2 = 2x^4 + x^2 \tan x^2 (3 + 2x^2 \tan x^2)$ Ans.

(d) $y = (x-1)\sec x^3$, Let $f(x) = x-1$, $f'(x) = 1$ and $g(x) = \sec x^3$, $g'(x) = \tan x^3 \cdot \sec x^3 \cdot 3x^2$

$\therefore \dfrac{dy}{dx} = f(x) \cdot g'(x) + g(x) \cdot f'(x) = (x-1) \cdot \tan x^3 \cdot \sec x^3 \cdot 3x^2 + \sec x^3 \cdot 1 = 3x^2(x-1) \cdot \dfrac{\sin x^3}{\cos x^3} \cdot \dfrac{1}{\cos x^3} + \dfrac{1}{\cos x^3}$

$\qquad \therefore \dfrac{dy}{dx} = \dfrac{3x^2(x-1)\sin x^3 + \cos x^3}{\cos^2 x^3} = \dfrac{3x^2 \sin x^3 (x-1) + \cos x^3}{\cos^2 x^3}$ Ans.

(e) $y = \sqrt{x} \cot x^2$, Let $f(x) = \sqrt{x}$, $f'(x) = \dfrac{1}{2\sqrt{x}}$ and $g(x) = \cot x^2 = \dfrac{\cos x^2}{\sin x^2}$ Put $z = x^2$, $\dfrac{dz}{dx} = 2x$

$g(x) = \dfrac{\cos x^2}{\sin x^2} = \dfrac{\cos z}{\sin z} = \dfrac{u}{v}$ (say) $u = \cos z$, $\dfrac{du}{dz} = -\sin z$ and $v = \sin z$, $\dfrac{dv}{dz} = \cos z$

$g'(x) = \dfrac{v \cdot \dfrac{du}{dz} \cdot \dfrac{dz}{dx} - u \cdot \dfrac{dv}{dz} \cdot \dfrac{dz}{dx}}{v^2} = \dfrac{\sin z \cdot (-\sin z)\dfrac{dz}{dx} - \cos z \cdot \cos z \cdot \dfrac{dz}{dx}}{(\sin z)^2} = \dfrac{-(\sin^2 z + \cos^2 z) \cdot \dfrac{dz}{dx}}{\sin^2 z} = -\dfrac{1 \cdot \dfrac{dz}{dx}}{\sin^2 z} = -\dfrac{dz}{dx} \cdot \mathrm{cosec}^2 z = -2x \cdot \mathrm{cosec}^2 x^2$

$\therefore \dfrac{dy}{dx} = f(x) \cdot g'(x) + g(x) \cdot f'(x) = -\sqrt{x} \cdot 2x \cdot \mathrm{cosec}^2 x^2 + \cot x^2 \cdot \dfrac{1}{2\sqrt{x}} = \dfrac{-2x\sqrt{x}}{\sin^2 x^2} + \dfrac{\cos x^2}{2\sqrt{x} \cdot \sin x^2} = \dfrac{-4x^2 + \sin x^2 \cdot \cos x^2}{2\sqrt{x} \cdot \sin^2 x^2} = \dfrac{-4x^2 + \dfrac{2\sin x^2 \cdot \cos x^2}{2}}{2\sqrt{x} \cdot \sin^2 x^2}$

$\qquad = \dfrac{-8x^2 + \sin 2x^2}{4\sqrt{x} \cdot \sin^2 x^2} = \dfrac{\sin 2x^2 - 8x^2}{4\sqrt{x} \cdot \sin^2 x^2}$ Ans.

(f) $y = (x^2+1)\mathrm{cosec}\sqrt{x+1}$, $\dfrac{dy}{dx} = \dfrac{4x\sqrt{x+1}\sin\sqrt{x+1} - (x^2+1)\cos\sqrt{x+1}}{2\sqrt{x+1} \cdot \sin^2\sqrt{x+1}}$ Ans.

(g) $y = \dfrac{\sin(x^2+1)}{x^3}$, $\dfrac{dy}{dx} = \dfrac{2\cos(x^2+1)}{x^2} - \dfrac{3\sin(x^2+1)}{x^4}$ Ans.   (h) $y = \dfrac{\tan\sqrt{x+1}}{x}$, $\dfrac{dy}{dx} = \dfrac{\sec^2\sqrt{x+1}}{2x\sqrt{x+1}} - \dfrac{\tan\sqrt{x+1}}{x^2}$ Ans.

(i) $y = \sin\left(\dfrac{1+x}{1-x}\right)$, Let $u = \dfrac{1+x}{1-x}$ $\therefore \dfrac{du}{dx} = \dfrac{(1-x).1-(1+x).(-1)}{(1-x)^2} = \dfrac{1-x+1+x}{(1-x)^2} = \dfrac{2}{(1-x)^2}$

or $y = \sin u$, $\dfrac{dy}{du} = \cos u = \cos\left(\dfrac{1+x}{1-x}\right)$ $\left[\text{Put } u = \dfrac{1+x}{1-x}\right]$ $\therefore \dfrac{dy}{dx} = \dfrac{dy}{du} \times \dfrac{du}{dx} = \cos\left(\dfrac{1+x}{1-x}\right).\dfrac{2}{(1-x)^2} = \dfrac{2\cos\left(\dfrac{1+x}{1-x}\right)}{(1-x)^2}$ Ans.

(j) $y = \sin\sqrt{x^2 + ax + 1}.\cos\sqrt{ax^2 + 1} = u.v$ (say)   Let $u = \sin\sqrt{x^2 + ax + 1}$ and $v = \cos\sqrt{ax^2 + 1}$

$\dfrac{du}{dx} = \cos\sqrt{x^2 + ax + 1}.\dfrac{1}{2\sqrt{x^2 + ax + 1}}.(2x + a)$ and $\dfrac{dv}{dx} = -\sin\sqrt{ax^2 + 1}.\dfrac{1}{2\sqrt{ax^2 + 1}}.2ax$

$\therefore \dfrac{dy}{dx} = u.\dfrac{dv}{dx} + v.\dfrac{du}{dx}$ or $\dfrac{dy}{dx} = -\sin\sqrt{x^2 + ax + 1}.\dfrac{ax.\sin\sqrt{ax^2 + 1}}{\sqrt{ax^2 + 1}} + \cos\sqrt{ax^2 + 1}.\dfrac{(2x + a).\cos\sqrt{x^2 + ax + 1}}{2\sqrt{x^2 + ax + 1}}$ Ans.

(k) $y = \tan(ax^2 + bx + 1) + \cot\sqrt{ax^2 + bx + c}$

$\dfrac{dy}{dx} = \sec^2(ax^2 + bx + 1).(2ax + b) + \left(-\text{cosec}^2\sqrt{ax^2 + bx + c}.\dfrac{1}{2\sqrt{ax^2 + bx + c}}.(2ax + b)\right)$

$\dfrac{dy}{dx} = \dfrac{(2ax + b)}{2\sqrt{ax^2 + bx + c}}\left[2\sqrt{ax^2 + bx + c}.\sec^2(ax^2 + bx + 1) - \text{cosec}^2\sqrt{ax^2 + bx + c}\right]$ Ans.

(l) $y = x^2\log\sqrt{x - 3}$, Let $u = x^2$, $\dfrac{du}{dx} = 2x$ and $v = \log\sqrt{x - 3}$, $\dfrac{dv}{dx} = \dfrac{1}{\sqrt{x - 3}}.\dfrac{1}{2\sqrt{x - 3}}.1 = \dfrac{1}{2(x - 3)}$

$\therefore \dfrac{dy}{dx} = u.\dfrac{dv}{dx} + v.\dfrac{du}{dx} = x^2.\dfrac{1}{2(x - 3)} + \log\sqrt{x - 3}.2x = x\left[\dfrac{x}{2(x - 3)} + \log(x - 3)\right]$ Ans.

(m) $y = \sin^3\sqrt{5 - 2x + x^2}$, Let $u = \sqrt{5 - 2x + x^2}$ $\therefore \dfrac{du}{dx} = \dfrac{1}{2\sqrt{5 - 2x + x^2}}.(2x - 2)$

$\therefore y = \sin^3 u = (\sin u)^3$, Let $v = \sin u$ $\therefore \dfrac{dv}{du} = \cos u = \cos\sqrt{5 - 2x + x^2}$ $\left[\text{Put } u = \sqrt{5 - 2x + x^2}\right]$

$\therefore y = v^3$, $\dfrac{dy}{dv} = 3v^2 = 3(\sin u)^2 = 3\left(\sin\sqrt{5 - 2x + x^2}\right)^2$ $\left[\text{Put } v = \sin u \text{ and } u = \sqrt{5 - 2x + x^2}\right]$

$\therefore \dfrac{dy}{dx} = \dfrac{dy}{dv} \times \dfrac{dv}{du} \times \dfrac{du}{dx} = 3\left(\sin\sqrt{5 - 2x + x^2}\right)^2.\cos\sqrt{5 - 2x + x^2}.\dfrac{1}{2\sqrt{5 - 2x + x^2}}.(2x - 2)$

$\qquad = \dfrac{3(x - 1).\sin^2\sqrt{5 - 2x + x^2}.\cos\sqrt{5 - 2x + x^2}}{\sqrt{5 - 2x + x^2}}$ Ans.

(n) $y = (x - 1)^2.(3x^2 - 1)$, Let $u = (x - 1)^2$ $\therefore \dfrac{du}{dx} = 2(x - 1)$ and $v = (3x^2 - 1)$ $\therefore \dfrac{dv}{dx} = 6x$

$\therefore \dfrac{dy}{dx} = u.v = u.\dfrac{dv}{dx} + v.\dfrac{du}{dx} = (x - 1)^2.6x + (3x^2 - 1).2(x - 1) = 2(x - 1)[3x(x - 1) + 3x^2 - 1] = 2(x - 1)[3x^2 - 3x + 3x^2 - 1]$

$\qquad = 2(x - 1)(6x^2 - 3x - 1)$ Ans.

(o) $y = \cos(\log x^2)$, Let $u = x^2$ $\therefore \dfrac{du}{dx} = 2x$ $\therefore y = \cos(\log u)$ Now, Let $v = \log u$ $\therefore \dfrac{dv}{du} = \dfrac{1}{u} = \dfrac{1}{x^2}$ $[\text{Put } u = x^2]$

$\therefore y = \cos v$, $\dfrac{dy}{dv} = -\sin v = -\sin(\log u) = -\sin(\log x^2)$ $[\text{Put } v = \log u \text{ and } u = x^2]$

$\qquad\qquad\qquad\qquad \therefore \dfrac{dy}{dx} = \dfrac{dy}{dv} \times \dfrac{dv}{du} \times \dfrac{du}{dx} = -\sin(\log x^2).\dfrac{1}{x^2}.2x = -\dfrac{2\sin(\log x^2)}{x}$ Ans.

(p) $y = \sin(e^x + 1)$, $\dfrac{dy}{dx} = \cos(e^x + 1).\dfrac{d(e^x + 1)}{dx} = e^x.\cos(e^x + 1)$ Ans.

(q) $y = \tan[\log(x - 5)]$, $\dfrac{dy}{dx} = \sec^2[\log(x - 5)] \cdot \dfrac{d(\log(x - 5))}{dx} = \sec^2[\log(x - 5)] \cdot \dfrac{1}{(x - 5)} \cdot \dfrac{d(x - 5)}{dx} = \dfrac{\sec^2[\log(x - 5)]}{(x - 5)}$

$$= \dfrac{1 + \tan^2[\log(x - 5)]}{(x - 5)} \quad \text{Ans.}$$

(r) $y = \log[\log(3 - x)]$, Let $u = 3 - x$ $\therefore \dfrac{du}{dx} = -1$

$\Rightarrow y = \log[\log u]$, Let $v = \log u$ $\therefore \dfrac{dv}{du} = \dfrac{1}{u} = \dfrac{1}{(3 - x)}$ [Put $u = 3 - x$]

$\Rightarrow y = \log v$ $\therefore \dfrac{dy}{dv} = \dfrac{1}{v} = \dfrac{1}{\log u} = \dfrac{1}{\log(3 - x)}$ [Put $v = \log u$ and $u = 3 - x$]

$\therefore \dfrac{dy}{dx} = \dfrac{dy}{dv} \times \dfrac{dv}{du} \times \dfrac{du}{dx} = \dfrac{1}{\log(3 - x)} \cdot \dfrac{1}{(3 - x)} \cdot (-1) = -\dfrac{1}{(3 - x) \cdot \log(3 - x)} = \dfrac{1}{(x - 3) \cdot \log(3 - x)}$ Ans.

(s) $y = e^{\sqrt{x}} \cdot \log x$, Let $u = e^{\sqrt{x}}$ $\therefore \dfrac{du}{dx} = e^{\sqrt{x}} \cdot \dfrac{1}{2\sqrt{x}}$ and $v = \log x$ $\therefore \dfrac{dv}{dx} = \dfrac{1}{x}$

$\Rightarrow y = u \cdot v$, $\dfrac{dy}{dx} = u \cdot \dfrac{dv}{dx} + v \cdot \dfrac{du}{dx} = e^{\sqrt{x}} \cdot \dfrac{1}{x} + \log x \cdot e^{\sqrt{x}} \cdot \dfrac{1}{2\sqrt{x}} = \dfrac{2\sqrt{x} \cdot e^{\sqrt{x}} + x \cdot e^{\sqrt{x}} \cdot \log x}{2x \cdot \sqrt{x}}$

$$\therefore \dfrac{dy}{dx} = \dfrac{\sqrt{x} \cdot e^{\sqrt{x}}(2 + \sqrt{x} \cdot \log x)}{2x \cdot \sqrt{x}} = \dfrac{e^{\sqrt{x}}(2 + \sqrt{x} \cdot \log x)}{2x} \quad \text{Ans.}$$

(t) $y = e^{2x} \cdot x^3$, $\dfrac{dy}{dx} = e^{2x} \cdot \dfrac{d(x^3)}{dx} + x^3 \cdot \dfrac{d(e^{2x})}{dx} = 3x^2 \cdot e^{2x} + x^3 \cdot e^{2x} \cdot 2 = x^2 \cdot e^{2x}(3 + 2x)$ Ans.

(u) $y = \sqrt{1 + \tan \theta}$, $\dfrac{dy}{d\theta} = \dfrac{1}{2\sqrt{1 + \tan \theta}} \cdot \dfrac{d(1 + \tan \theta)}{d\theta} = \dfrac{1}{2\sqrt{1 + \tan \theta}} \cdot \sec^2 \theta = \dfrac{1 + \tan^2 \theta}{2\sqrt{1 + \tan \theta}}$ Ans.

(v) $y = \sqrt{\sin \sqrt{x}}$, Let $u = \sqrt{x}$ $\therefore \dfrac{du}{dx} = \dfrac{1}{2\sqrt{x}}$

$\Rightarrow y = \sqrt{\sin u}$, $\dfrac{dy}{du} = \dfrac{1}{2\sqrt{\sin u}} \cdot \dfrac{d(\sin u)}{du} = \dfrac{1}{2\sqrt{\sin u}} \cdot \cos u = \dfrac{\cos \sqrt{x}}{2\sqrt{\sin \sqrt{x}}}$ [Put $u = \sqrt{x}$]

$\therefore \dfrac{dy}{dx} = \dfrac{dy}{du} \times \dfrac{du}{dx} = \dfrac{\cos \sqrt{x}}{2\sqrt{\sin \sqrt{x}}} \cdot \dfrac{1}{2\sqrt{x}} = \dfrac{\cos \sqrt{x}}{4\sqrt{x} \cdot \sqrt{\sin \sqrt{x}}} = \dfrac{\cos \sqrt{x}}{\sqrt{16x \cdot \sin \sqrt{x}}}$ Ans.

(4) (a) $y = \dfrac{\tan\left(\frac{1 - x}{1 + x}\right)}{x} + \dfrac{\cot\left(\frac{1 + x}{1 - x}\right)}{x}$, Let $u = \dfrac{\tan\left(\frac{1 - x}{1 + x}\right)}{x}$ and $v = \dfrac{\cot\left(\frac{1 + x}{1 - x}\right)}{x}$

$\Rightarrow u = \dfrac{\tan\left(\frac{1 - x}{1 + x}\right)}{x}$, $\dfrac{du}{dx} = \dfrac{x \cdot \sec^2\left(\frac{1 - x}{1 + x}\right) \cdot \frac{d}{dx}\left(\frac{1 - x}{1 + x}\right) - \tan\left(\frac{1 - x}{1 + x}\right) \cdot \frac{d(x)}{dx}}{x^2}$

or $\dfrac{d}{dx}\left(\dfrac{1 - x}{1 + x}\right) = \dfrac{(1 + x) \cdot (-1) - (1 - x) \cdot 1}{(1 + x)^2} = \dfrac{-1 - x - 1 + x}{(1 + x)^2} = -\dfrac{2}{(1 + x)^2}$

$\therefore \dfrac{du}{dx} = \dfrac{x \cdot \sec^2\left(\frac{1 - x}{1 + x}\right) \cdot \left(-\frac{2}{(1 + x)^2}\right) - \tan\left(\frac{1 - x}{1 + x}\right) \cdot 1}{x^2} = -\dfrac{2x \cdot \sec^2\left(\frac{1 - x}{1 + x}\right) + (1 + x)^2 \cdot \tan\left(\frac{1 - x}{1 + x}\right)}{x^2 \cdot (1 + x)^2}$

$\Rightarrow v = \dfrac{\cot\left(\frac{1 + x}{1 - x}\right)}{x}$, $\dfrac{d}{dx}\left(\dfrac{1 + x}{1 - x}\right) = \dfrac{(1 - x) \cdot 1 - (1 + x) \cdot (-1)}{(1 - x)^2} = \dfrac{1 - x + 1 + x}{(1 - x)^2} = \dfrac{2}{(1 - x)^2}$

or $\dfrac{dv}{dx} = \dfrac{-x \cdot \text{cosec}^2\left(\frac{1 + x}{1 - x}\right) \cdot \frac{2}{(1 - x)^2} - \cot\left(\frac{1 + x}{1 - x}\right) \cdot 1}{x^2} = -\dfrac{2x \cdot \text{cosec}^2\left(\frac{1 + x}{1 - x}\right) + (1 - x)^2 \cdot \cot\left(\frac{1 + x}{1 - x}\right)}{x^2 \cdot (1 - x)^2}$

$$\therefore \frac{dy}{dx} = u + v = \frac{du}{dx} + \frac{dv}{dx}$$

$$\frac{dy}{dx} = -\frac{2x.\sec^2\left(\frac{1-x}{1+x}\right) + (1+x)^2.\tan\left(\frac{1-x}{1+x}\right)}{x^2.(1+x)^2} - \frac{2x.\csc^2\left(\frac{1+x}{1-x}\right) + (1-x)^2.\cot\left(\frac{1+x}{1-x}\right)}{x^2.(1-x)^2}$$

$$= \frac{-2x\left[\sec^2\left(\frac{1-x}{1+x}\right) + (1+x)^2.\tan\left(\frac{1-x}{1+x}\right) + \csc^2\left(\frac{1+x}{1-x}\right) + (1-x)^2.\cot\left(\frac{1+x}{1-x}\right)\right]}{x^2.(1+x)^2} \quad \text{Ans.}$$

(b) $y = \cos\sqrt{\sin\sqrt{ax+b}}$, Let $u = \sqrt{ax+b}$ $\quad\therefore \frac{du}{dx} = \frac{1}{2\sqrt{ax+b}}.\frac{d}{dx}(ax+b) = \frac{a}{2\sqrt{ax+b}}$

$\Rightarrow y = \cos\sqrt{\sin u}$, Let $v = \sqrt{\sin u}$ $\quad\therefore \frac{dv}{du} = \frac{1}{2\sqrt{\sin u}}.\frac{d}{du}(\sin u) = \frac{\cos u}{2\sqrt{\sin u}} = \frac{\cos(ax+b)}{2\sqrt{\sin\sqrt{ax+b}}}$ [Put $u = \sqrt{ax+b}$]

$\Rightarrow y = \cos v$, $\frac{dy}{dv} = -\sin v = -\sin\sqrt{\sin u} = -\sin\sqrt{\sin\sqrt{ax+b}}$ [Put $v = \sqrt{\sin u}$ and $u = \sqrt{ax+b}$]

$$\therefore \frac{dy}{dx} = \frac{dy}{dv} \times \frac{dv}{du} \times \frac{du}{dx} = -\sin\sqrt{\sin\sqrt{ax+b}}.\frac{\cos(ax+b)}{2\sqrt{\sin\sqrt{ax+b}}}.\frac{a}{2\sqrt{ax+b}} = \frac{-a\sin\sqrt{\sin\sqrt{ax+b}}.\cos(ax+b)}{4.\sqrt{\sin\sqrt{ax+b}}.\sqrt{ax+b}} \quad \text{Ans.}$$

(c) $y = \frac{\sqrt{\tan x}}{\sqrt{x^2-3}} = \frac{u}{v}$ (say) $u = \sqrt{\tan x}$, $\frac{du}{dx} = \frac{1}{2\sqrt{\tan x}}.\sec^2 x$ and $v = \sqrt{x^2-3}$, $\frac{dv}{dx} = \frac{1}{2\sqrt{x^2-3}}.2x$

or $\frac{dy}{dx} = \frac{v.\frac{du}{dx} - u.\frac{dv}{dx}}{v^2} = \frac{\sqrt{x^2-3}.\frac{1}{2\sqrt{\tan x}}.\sec^2 x - \sqrt{\tan x}.\frac{1}{2\sqrt{x^2-3}}.2x}{\left(\sqrt{x^2-3}\right)^2} = \frac{\frac{(x^2-3).\sec^2 x - 2x.\tan x}{2\sqrt{\tan x}.\sqrt{x^2-3}}}{(x^2-3)} = \frac{(x^2-3).\sec^2 x - 2x.\tan x}{2.(x^2-3).\sqrt{x^2-3}.\sqrt{\tan x}}$ Ans.

(d) $y = \frac{x^2 + \sin x^3}{x^3 + \cos x^2} = \frac{u}{v}$ (say) $u = x^2 + \sin x^3$ and $v = x^3 + \cos x^2$ or $\frac{du}{dx} = 2x + \cos x^3.3x^2$, $\frac{dv}{dx} = 3x^2 - \sin x^2.2x$

or $\frac{dy}{dx} = \frac{v.\frac{du}{dx} - u.\frac{dv}{dx}}{v^2} = \frac{(x^3 + \cos x^2)(2x + 3x^2.\cos x^3) - (x^2 + \sin x^3)(3x^2 - 2x.\sin x^2)}{(x^3 + \cos x^2)^2}$

$$= \frac{x[(x^3 + \cos x^2)(2 + 3x\cos x^3) - (x^2 + \sin x^3)(3x - 2\sin x^2)]}{(x^3 + \cos x^2)^2} \quad \text{Ans.}$$

(e) $y = x^{\frac{3}{2}}.\cos(ax^2 + bx + c) = u.v$ (say), $u = x^{\frac{3}{2}}$ $\therefore \frac{du}{dx} = \frac{3}{2}x^{\frac{3}{2}-1} = \frac{3}{2}\sqrt{x}$ [$y = x^n$, $y' = nx^{n-1}$]

and $v = \cos(ax^2 + bx + c)$ $\therefore \frac{dv}{dx} = -\sin(ax^2 + bx + c).\frac{d}{dx}(ax^2 + bx + c) = -\sin(ax^2 + bx + c).(2ax + b)$

$$\therefore \frac{dy}{dx} = u.\frac{dv}{dx} + v.\frac{du}{dx} = x^{\frac{3}{2}}.[-\sin(ax^2 + bx + c).(2ax + b)] + \cos(ax^2 + bx + c).\frac{3}{2}\sqrt{x}$$

$$= \frac{3}{2}\sqrt{x}.\cos(ax^2 + bx + c) - x^{\frac{3}{2}}.(2ax + b).\sin(ax^2 + bx + c) \quad \text{Ans.}$$

(5) (a) $y = \tan^{-1}\left(\frac{\sqrt{x}-1}{1+\sqrt{x}}\right) = \tan^{-1}\sqrt{x} - \tan^{-1}1$ $\left[\text{formula}, \ \tan^{-1}\left(\frac{A-B}{1+AB}\right) = \tan^{-1}A - \tan^{-1}B\right]$

$\therefore \frac{dy}{dx} = \frac{1}{1+(\sqrt{x})^2}.\frac{d}{dx}(\sqrt{x}) - 0 = \frac{1}{1+x}.\frac{1}{2\sqrt{x}} = \frac{1}{2\sqrt{x}(1+x)}$ Ans. $\left[\text{formula}, \ y = \tan^{-1}x \ , \ \frac{dy}{dx} = \frac{1}{1+x^2}.\frac{d}{dx}(x)\right]$

IInd Method: $-$ $y = \tan^{-1}\left(\frac{\sqrt{x}-1}{1+\sqrt{x}}\right)$, Let $u = \frac{\sqrt{x}-1}{1+\sqrt{x}} = \frac{f(x)}{g(x)}$ (say) $f'(x) = \frac{1}{2\sqrt{x}}$ and $g'(x) = \frac{1}{2\sqrt{x}}$

or $\frac{du}{dx} = \frac{g(x).f'(x) - f(x).g'(x)}{(g(x))^2} = \frac{(1+\sqrt{x}).\frac{1}{2\sqrt{x}} - (\sqrt{x}-1).\frac{1}{2\sqrt{x}}}{(1+\sqrt{x})^2} = \frac{\frac{1+\sqrt{x} - \sqrt{x} + 1}{2\sqrt{x}}}{(1+\sqrt{x})^2} = \frac{2}{2\sqrt{x}.(1+\sqrt{x})^2} = \frac{1}{\sqrt{x}.(1+\sqrt{x})^2}$

$$\Rightarrow y = \tan^{-1} u, \quad \frac{dy}{du} = \frac{1}{1+u^2} = \frac{1}{1+\left(\frac{\sqrt{x}-1}{1+\sqrt{x}}\right)^2} = \frac{(1+\sqrt{x})^2}{(1+\sqrt{x})^2 + (\sqrt{x}-1)^2} = \frac{(1+\sqrt{x})^2}{1+2\sqrt{x}+x+x-2\sqrt{x}+1} = \frac{(1+\sqrt{x})^2}{2x+2} = \frac{(1+\sqrt{x})^2}{2(x+1)}$$

$$\therefore \frac{dy}{dx} = \frac{dy}{du} \times \frac{du}{dx} = \frac{(1+\sqrt{x})^2}{2(x+1)} \cdot \frac{1}{\sqrt{x}.(1+\sqrt{x})^2} = \frac{1}{2\sqrt{x}\,(x+1)} \quad \text{Ans.}$$

(b) $y = \tan^{-1}\left(\frac{\sin x + \cos x}{1 - \sin x \cos x}\right) = \tan^{-1}(\sin x) + \tan^{-1}(\cos x)$ $\quad \left[\text{formula,} \quad \tan^{-1}\left(\frac{A+B}{1-AB}\right) = \tan^{-1} A + \tan^{-1} B\right]$

or $\frac{dy}{dx} = \frac{1}{1+(\sin x)^2} \cdot \frac{d}{dx}(\sin x) + \frac{1}{1+(\cos x)^2} \cdot \frac{d}{dx}(\cos x)$ $\quad \left[\text{formula,} \quad y = \tan^{-1} u, \quad \frac{dy}{dx} = \frac{1}{1+u^2} \cdot \frac{du}{dx}\right]$

or $\frac{dy}{dx} = \frac{1}{1+\sin^2 x} \cdot \cos x + \frac{1}{1+\cos^2 x} \cdot (-\sin x) = \frac{\cos x\,(1+\cos^2 x) - \sin x\,(1+\sin^2 x)}{(1+\sin^2 x)(1+\cos^2 x)} = \frac{\cos x + \cos^3 x - \sin x - \sin^3 x}{1+\cos^2 x + \sin^2 x + \sin^2 x \cos^2 x}$

$$= \frac{(\cos x - \sin x) + (\cos^3 x - \sin^3 x)}{(1 + 1 + \sin^2 x \cos^2 x)}$$

or $\frac{dy}{dx} = \frac{(\cos x - \sin x) + [(\cos x - \sin x)^3 + 3\sin x \cos x\,(\cos x - \sin x)]}{(2 + \sin^2 x \cos^2 x)} = \frac{(\cos x - \sin x)[1 + (\cos x - \sin x)^2 + 3\sin x \cos x]}{(2 + \sin^2 x \cos^2 x)}$

or $\frac{dy}{dx} = \frac{(\cos x - \sin x)[1 + \cos^2 x + \sin^2 x - 2\sin x \cos x + 3\sin x \cos x]}{(2 + \sin^2 x \cos^2 x)}$

$$= \frac{(\cos x - \sin x)[2 + \sin x \cos x]}{(2 + \sin^2 x \cos^2 x)} \quad \text{Ans.} \quad [\,(a-b)^3 = a^3 - b^3 - 3ab(a-b)\,]$$

IInd Method: $-\ y = \tan^{-1}\left(\frac{\sin x + \cos x}{1 - \sin x \cos x}\right)$ $\quad$ Let $u = \frac{\sin x + \cos x}{1 - \sin x \cos x}$ $\quad \therefore \frac{du}{dx} = \frac{(\cos x - \sin x)[2 + \sin x \cos x]}{(1 - \sin x \cos x)^2}$

$$\Rightarrow y = \tan^{-1} u \quad \therefore \frac{dy}{du} = \frac{1}{1+u^2} = \frac{1}{1+\left(\frac{\sin x + \cos x}{1 - \sin x \cos x}\right)^2} = \frac{(1 - \sin x \cos x)^2}{(1 - \sin x \cos x)^2 + (\sin x + \cos x)^2}$$

$$\therefore \frac{dy}{dx} = \frac{dy}{du} \times \frac{du}{dx} = \frac{(1 - \sin x \cos x)^2}{(1 - \sin x \cos x)^2 + (\sin x + \cos x)^2} \cdot \frac{(\cos x - \sin x)[2 + \sin x \cos x]}{(1 - \sin x \cos x)^2} = \frac{(\cos x - \sin x)[2 + \sin x \cos x]}{(1 - \sin x \cos x)^2 + (\sin x + \cos x)^2} \text{(solve)}$$

$$= \frac{(\cos x - \sin x)[2 + \sin x \cos x]}{(2 + \sin^2 x \cos^2 x)} \quad \text{Ans.}$$

(c) $y = \tan^{-1}\left(\frac{6x}{1 - x^2}\right)$, $\quad$ Let $u = \frac{6x}{1 - x^2}$ $\quad \therefore \frac{du}{dx} = \frac{(1-x^2).\frac{d}{dx}(6x) - 6x.\frac{d}{dx}(1-x^2)}{(1-x^2)^2}$

$$\therefore \frac{dy}{dx} = \frac{6(1-x^2) - 6x.(-2x)}{(1-x^2)^2} = \frac{6 - 6x^2 + 12x^2}{(1-x^2)^2} = \frac{6(1-x^2)}{(1-x^2)^2} = \frac{6}{1-x^2}$$

Now, $y = \tan^{-1} u$ $\quad \therefore \frac{dy}{du} = \frac{1}{1+u^2} = \frac{1}{1+\left(\frac{6x}{1-x^2}\right)^2} = \frac{(1-x^2)^2}{(1-x^2)^2 + 36x^2} = \frac{(1-x^2)^2}{1+x^4 - 2x^2 + 36x^2} = \frac{(1-x^2)^2}{x^4 + 34x^2 + 1}$

$$\therefore \frac{dy}{dx} = \frac{dy}{du} \times \frac{du}{dx} = \frac{(1-x^2)^2}{(x^4 + 34x^2 + 1)} \cdot \frac{6}{1-x^2} = \frac{6(1-x^2)}{(x^4 + 34x^2 + 1)} \quad \text{Ans.}$$

(d) $y = \tan^{-1}\left(\frac{x+3}{1+x^2}\right) + \tan^{-1}\left(\frac{2x+1}{2-3x}\right) = \tan^{-1}\left(\frac{\frac{x+3}{1+x^2} + \frac{2x+1}{2-3x}}{1 - \frac{x+3}{1+x^2}\cdot\frac{2x+1}{2-3x}}\right) = \tan^{-1}\left(\frac{2x^2 - 2x^3 + 5x - 7}{3x^3 + 10x + 1}\right)$

$$\left[\text{use formula,} \quad \tan^{-1}\left(\frac{A+B}{1-AB}\right) = \tan^{-1} A + \tan^{-1} B \quad \text{and} \quad y = \tan^{-1} u, \quad \frac{dy}{dx} = \frac{1}{1+u^2} \cdot \frac{du}{dx}\right]$$

$$\therefore \frac{dy}{dx} = \frac{(3x^3 + 10x + 1)(5 + 4x - 6x^2) - (2x^2 - 2x^3 + 5x - 7)(9x^2 + 10)}{(3x^3 + 10x + 1)^2 + (2x^2 - 2x^3 + 5x - 7)^2} \quad \text{Ans.}$$

(e) $y = \cot^{-1}\left(\dfrac{\sqrt{1+x}-\sqrt{1-x}}{\sqrt{1-x}+\sqrt{1+x}}\right) = \cot^{-1}\left(\dfrac{\sqrt{1+x}-\sqrt{1-x}}{\sqrt{1-x}+\sqrt{1+x}} \times \dfrac{\sqrt{1+x}+\sqrt{1-x}}{\sqrt{1+x}+\sqrt{1-x}}\right)$

$y = \cot^{-1}\left(\dfrac{1+x-1+x}{1+x+1-x-2\sqrt{1-x^2}}\right) = \cot^{-1}\left(\dfrac{2x}{2-2\sqrt{1-x^2}}\right) = \cot^{-1}\left(\dfrac{x}{1-\sqrt{1-x^2}}\right)$

Put $x = \sin\theta \quad \therefore \dfrac{dx}{d\theta} = \cos\theta$ then $y = \cot^{-1}\left(\dfrac{\sin\theta}{1-\sqrt{1-\sin^2\theta}}\right) = \cot^{-1}\left(\dfrac{\sin\theta}{1-\cos\theta}\right)$

$y = \cot^{-1}\left(\dfrac{2\sin\frac{\theta}{2}\cos\frac{\theta}{2}}{1-\left(\cos^2\frac{\theta}{2}-\sin^2\frac{\theta}{2}\right)}\right) = \cot^{-1}\left(\dfrac{2\sin\frac{\theta}{2}\cos\frac{\theta}{2}}{1-\cos^2\frac{\theta}{2}+\sin^2\frac{\theta}{2}}\right) = \cot^{-1}\left(\dfrac{2\sin\frac{\theta}{2}\cos\frac{\theta}{2}}{\sin^2\frac{\theta}{2}+\sin^2\frac{\theta}{2}}\right) = \cot^{-1}\left(\dfrac{2\sin\frac{\theta}{2}\cos\frac{\theta}{2}}{2\sin^2\frac{\theta}{2}}\right)$

$y = \cot^{-1}\left(\dfrac{\cos\frac{\theta}{2}}{\sin\frac{\theta}{2}}\right) = \cot^{-1}\left(\cot\frac{\theta}{2}\right) = \dfrac{\theta}{2} \quad \therefore \dfrac{dy}{d\theta} = \dfrac{1}{2}$

$\therefore \dfrac{dy}{dx} = \dfrac{\frac{dy}{d\theta}}{\frac{dx}{d\theta}} = \dfrac{dy}{d\theta} \div \dfrac{dx}{d\theta} = \dfrac{\frac{1}{2}}{\cos\theta} = \dfrac{1}{2\cos\theta} = \dfrac{1}{2\sqrt{1-\sin^2\theta}} = \dfrac{1}{2\sqrt{1-x^2}}$ Ans. $\left[\text{Put } x = \sin\theta, \ \cos^2\theta = 1-\sin^2\theta \text{ or } \cos\theta = \sqrt{1-\sin^2\theta}\right]$

(f) $y = \sin^{-1}\dfrac{1}{\sqrt{1+x^2}} + \tan^{-1}\dfrac{x}{\sqrt{1-x^2}} = u + v$ (say)

$u = \sin^{-1}\dfrac{1}{\sqrt{1+x^2}}$, Put $x = \tan\theta$ then $\dfrac{dx}{d\theta} = \sec^2\theta$

$u = \sin^{-1}\left(\dfrac{1}{\sqrt{1+(\tan\theta)^2}}\right) = \sin^{-1}\left(\dfrac{1}{\sec\theta}\right) = \sin^{-1}(\cos\theta) = \sin^{-1}\left(\sin\left(\dfrac{\pi}{2}-\theta\right)\right) = \dfrac{\pi}{2}-\theta$

$\therefore \dfrac{du}{d\theta} = -1, \quad \dfrac{du}{dx} = \dfrac{du}{d\theta} \div \dfrac{dx}{d\theta} = -\dfrac{1}{\sec^2\theta} = -\dfrac{1}{1+\tan^2\theta} = -\dfrac{1}{1+x^2}$ [Put $\tan\theta = x$, $\sec^2\theta = 1+\tan^2\theta$]

$v = \tan^{-1}\dfrac{x}{\sqrt{1-x^2}}$, Put $x = \sin\theta$ then $\dfrac{dx}{d\theta} = \cos\theta$

$v = \tan^{-1}\dfrac{x}{\sqrt{1-x^2}} = \tan^{-1}\left(\dfrac{\sin\theta}{\sqrt{1-(\sin\theta)^2}}\right) = \tan^{-1}\left(\dfrac{\sin\theta}{\sqrt{1-\sin^2\theta}}\right) = \tan^{-1}\left(\dfrac{\sin\theta}{\cos\theta}\right) = \tan^{-1}(\tan\theta) = \theta$

$\therefore \dfrac{dv}{d\theta} = 1, \quad \dfrac{dv}{dx} = \dfrac{dv}{d\theta} \div \dfrac{dx}{d\theta} = \dfrac{1}{\cos\theta} = \dfrac{1}{\sqrt{1-\sin^2\theta}} = \dfrac{1}{\sqrt{1-x^2}}$ $\left[\text{Put } \sin\theta = x, \ \cos\theta = \sqrt{1-\sin^2\theta}\right]$

$y = u+v, \quad \dfrac{dy}{dx} = \dfrac{du}{dx} + \dfrac{dv}{dx} = -\dfrac{1}{1+x^2} + \dfrac{1}{\sqrt{1-x^2}} = \dfrac{1}{\sqrt{1-x^2}} - \dfrac{1}{1+x^2}$ Ans.

(6) (a) $y = \sin^{-1}\sqrt{x} + \sin^{-1}\sqrt{1-x}$ $\left\{\text{formula}, \ \sin^{-1}A \pm \sin^{-1}B = \sin^{-1}\left[A\sqrt{1-B^2} \pm B\sqrt{1-A^2}\right]\right\}$

$y = \sin^{-1}\left[\sqrt{x}.\sqrt{1-\left(\sqrt{1-x}\right)^2} + \sqrt{1-x}.\sqrt{1-\left(\sqrt{x}\right)^2}\right] = \sin^{-1}\left[\sqrt{x}.\sqrt{1-1+x} + \sqrt{1-x}.\sqrt{1-x}\right] = \sin^{-1}[x+1-x] = \sin^{-1}1 = \dfrac{\pi}{2}$

$$y = \dfrac{\pi}{2}, \ \dfrac{dy}{dx} = 0 \quad \text{Ans.}$$

(b) $y = \sin^{-1}(x^2-1) - \sin^{-1}(2x^2)$ $\left\{\text{formula}, \quad y = \sin^{-1}u \quad \therefore \dfrac{dy}{dx} = \dfrac{1}{\sqrt{1-u^2}}.\dfrac{du}{dx}\right\}$

$\dfrac{dy}{dx} = \dfrac{1}{\sqrt{1-(x^2-1)^2}}.2x - \dfrac{1}{\sqrt{1-(2x^2)^2}}.4x = \dfrac{2x}{\sqrt{1-(x^4-2x^2+1)}} - \dfrac{4x}{\sqrt{1-4x^4}} = \dfrac{2x}{\sqrt{1-x^4+2x^2-1}} - \dfrac{4x}{\sqrt{1-4x^4}} = \dfrac{2x}{\sqrt{2x^2-x^4}} - \dfrac{4x}{\sqrt{1-4x^4}}$

$$= \dfrac{2}{\sqrt{2-x^2}} - \dfrac{4x}{\sqrt{1-4x^4}} \quad \text{Ans.}$$

(c) $y = \cos^{-1}\dfrac{1}{1+x} + \cos^{-1}(1-x)$, $\dfrac{dy}{dx} = \dfrac{1}{-\sqrt{1-\left(\frac{1}{1+x}\right)^2}} \cdot \dfrac{d}{dx}\left(\dfrac{1}{1+x}\right) + \dfrac{1}{-\sqrt{1-(1-x)^2}} \cdot \dfrac{d}{dx}(1-x)$

$\therefore \dfrac{dy}{dx} = \dfrac{1+x}{-\sqrt{(1+x)^2-1}} \cdot \dfrac{-1}{(1+x)^2} + \dfrac{1}{-\sqrt{1-(1-2x+x^2)}} \cdot (-1) = \dfrac{1}{(1+x)\cdot\sqrt{1+2x+x^2-1}} + \dfrac{1}{\sqrt{1-1+2x-x^2}}$

$= \dfrac{1}{(1+x)\cdot\sqrt{2x+x^2}} + \dfrac{1}{\sqrt{2x-x^2}}$ Ans.

(d) $y = \cos^{-1}\dfrac{x}{\sqrt{1+x}} - \cos^{-1}\dfrac{1}{\sqrt{1-x}}$, $\left\{\text{formula,}\quad y = \cos^{-1}u \quad \therefore \dfrac{dy}{dx} = \dfrac{1}{-\sqrt{1-u^2}}\cdot\dfrac{du}{dx}\right\}$

$\dfrac{d}{dx}\left(\dfrac{x}{\sqrt{1+x}}\right) = \dfrac{\sqrt{1+x}\cdot\frac{d(x)}{dx} - x\cdot\frac{d(\sqrt{1+x})}{dx}}{\left(\sqrt{1+x}\right)^2} = \dfrac{\sqrt{1+x} - x\cdot\frac{1}{2\sqrt{1+x}}\cdot\frac{d(1+x)}{dx}}{\left(\sqrt{1+x}\right)^2} = \dfrac{2+2x-x}{2\sqrt{1+x}\cdot(1+x)} = \dfrac{2+x}{2(1+x)\sqrt{1+x}}$

$\dfrac{d}{dx}\left(\dfrac{1}{\sqrt{1-x}}\right) = \dfrac{\sqrt{1-x}\cdot\frac{d(1)}{dx} - 1\cdot\frac{d(\sqrt{1-x})}{dx}}{\left(\sqrt{1-x}\right)^2} = \dfrac{0 - 1\cdot\frac{1}{2\sqrt{1-x}}\cdot\frac{d(1-x)}{dx}}{(1-x)} = \dfrac{1}{2(1-x)\sqrt{1-x}}$

$\dfrac{dy}{dx} = \dfrac{1}{2}\left[\dfrac{1}{\sqrt{1-x}\cdot\sqrt{1-2x}} - \dfrac{2+x}{\sqrt{1+x}}\right]$ Ans.   (see above question and solve it.)

(e) $y = \sin^{-1}\left[\sqrt{x}\cdot\sqrt{2-x} + \sqrt{x-1}\cdot\sqrt{1-x}\right]$ $\left\{\text{formula,}\quad \sin^{-1}A + \sin^{-1}B = \sin^{-1}\left[A\sqrt{1-B^2} + B\sqrt{1-A^2}\right]\right\}$

$y = \sin^{-1}\left[\sqrt{x}\cdot\sqrt{1-\left(\sqrt{x-1}\right)^2} + \sqrt{x-1}\cdot\sqrt{1-\left(\sqrt{x}\right)^2}\right]$ here $A = \sqrt{x}$ and $B = \sqrt{x-1}$

$y = \sin^{-1}\sqrt{x} + \sin^{-1}\sqrt{x-1}$, $\dfrac{dy}{dx} = \dfrac{1}{\sqrt{1-\left(\sqrt{x}\right)^2}}\cdot\dfrac{d\left(\sqrt{x}\right)}{dx} + \dfrac{1}{\sqrt{1-\left(\sqrt{x-1}\right)^2}}\cdot\dfrac{d\left(\sqrt{x-1}\right)}{dx}$

$\therefore \dfrac{dy}{dx} = \dfrac{1}{\sqrt{1-x}}\cdot\dfrac{1}{2\sqrt{x}} + \dfrac{1}{\sqrt{2-x}}\cdot\dfrac{1}{2\sqrt{x-1}} = \dfrac{1}{2\cdot\sqrt{x}\cdot\sqrt{1-x}} + \dfrac{1}{2\cdot\sqrt{x-1}\cdot\sqrt{2-x}}$ Ans.

(f) $y = \cos^{-1}\left[\sqrt{x}\cdot\dfrac{1}{\sqrt{1+x^2}} - \sqrt{1-x}\cdot\sqrt{\dfrac{x^2}{1+x^2}}\right]$ $\left[\text{formula,}\quad \cos^{-1}\left(AB - \sqrt{1-A^2}\cdot\sqrt{1-B^2}\right) = \cos^{-1}A + \cos^{-1}B\right]$

$y = \cos^{-1}\left(\sqrt{x}\cdot\dfrac{1}{\sqrt{1+x^2}} - \sqrt{1-\left(\sqrt{x}\right)^2}\cdot\sqrt{1-\left(\dfrac{1}{\sqrt{1+x^2}}\right)^2}\right) = \cos^{-1}\sqrt{x} + \cos^{-1}\left(\dfrac{1}{\sqrt{1+x^2}}\right)$

Put $x = \tan\theta$, $\dfrac{dx}{d\theta} = \sec^2\theta$ then $y = \cos^{-1}\sqrt{\tan\theta} + \cos^{-1}\left(\dfrac{1}{\sqrt{1+\tan^2\theta}}\right) = \cos^{-1}\sqrt{\tan\theta} + \cos^{-1}\left(\dfrac{1}{\sqrt{\sec^2\theta}}\right)$

$y = \cos^{-1}\sqrt{\tan\theta} + \cos^{-1}\left(\dfrac{1}{\sec\theta}\right) = \cos^{-1}\sqrt{\tan\theta} + \cos^{-1}(\cos\theta) = \cos^{-1}\sqrt{\tan\theta} + \theta$

$\dfrac{dy}{d\theta} = \dfrac{1}{-\sqrt{1-\left(\sqrt{\tan\theta}\right)^2}}\cdot\dfrac{d\left(\sqrt{\tan\theta}\right)}{d\theta} + 1 = -\dfrac{1}{\sqrt{1-\tan\theta}}\cdot\dfrac{1}{2\sqrt{\tan\theta}}\cdot\dfrac{d(\tan\theta)}{d\theta} + 1 = -\dfrac{1}{\sqrt{1-\tan\theta}}\cdot\dfrac{1}{2\sqrt{\tan\theta}}\cdot\sec^2\theta + 1$

$\dfrac{dy}{d\theta} = -\dfrac{1+\tan^2\theta}{2\sqrt{\tan\theta}\cdot\sqrt{1-\tan\theta}} + 1 = -\dfrac{1+x^2}{2\sqrt{x}\cdot\sqrt{1-x}} + 1$ Put $\tan\theta = x$, $\dfrac{dx}{d\theta} = \sec^2\theta = 1+\tan^2\theta = 1+x^2$

$\dfrac{dy}{dx} = \dfrac{dy}{d\theta} \div \dfrac{dx}{d\theta} = \dfrac{-\dfrac{1+x^2}{2\sqrt{x}\cdot\sqrt{1-x}} + 1}{1+x^2} = \dfrac{2\sqrt{x}\cdot\sqrt{1-x} - (1+x^2)}{2\sqrt{x}\cdot\sqrt{1-x}\cdot(1+x^2)} = \dfrac{1}{1+x^2} - \dfrac{1}{2\sqrt{x}\cdot\sqrt{1-x}}$ Ans.

(7) (a) Differentiate $\tan^{-1}\dfrac{\sqrt{1-x^2}}{x}$ with respect to $\tan^{-1}x$

Let $y = \tan^{-1}\dfrac{\sqrt{1-x^2}}{x}$ and $z = \tan^{-1}x$ then find $\dfrac{dy}{dz} = \dfrac{dy}{dx} \div \dfrac{dz}{dx}$ Put $x = \cos\theta$ and $\dfrac{dz}{dx} = \dfrac{1}{1+x^2}$

$y = \tan^{-1}\dfrac{\sqrt{1-\cos^2\theta}}{\cos\theta} = \tan^{-1}\left(\dfrac{\sin\theta}{\cos\theta}\right) = \tan^{-1}(\tan\theta) = \theta = \cos^{-1}x$, $\qquad \dfrac{dy}{dx} = -\dfrac{1}{\sqrt{1-x^2}}$

$\dfrac{dy}{dz} = \dfrac{dy}{dx} \div \dfrac{dz}{dx} = \dfrac{-\dfrac{1}{\sqrt{1-x^2}}}{\dfrac{1}{1+x^2}} = -\dfrac{1+x^2}{\sqrt{1-x^2}}$ Ans.

(b) Differentiate $\sin^{-1}\sqrt{\dfrac{1+x}{1-x}}$ with respect to $\sqrt{x}$.

Let $y = \sin^{-1}\sqrt{\dfrac{1+x}{1-x} \times \dfrac{1+x}{1+x}} = \sin^{-1}\left(\dfrac{1+x}{\sqrt{1-x^2}}\right)$ and $z = \sqrt{x}$ then find $\dfrac{dy}{dz}$. $\quad \therefore \dfrac{dy}{dz} = \dfrac{dy}{dx} \div \dfrac{dz}{dx}$

$$\text{Put } x = \cos\theta\,, \quad \dfrac{dx}{d\theta} = -\sin\theta = -\sqrt{1-\cos^2\theta} = -\sqrt{1-x^2}$$

$y = \sin^{-1}\left(\dfrac{1+\cos\theta}{\sqrt{1-\cos^2\theta}}\right) = \sin^{-1}\left(\dfrac{1+\cos\theta}{\sin\theta}\right) = \sin^{-1}\left(\dfrac{1+\cos^2\frac{\theta}{2}-\sin^2\frac{\theta}{2}}{2\sin\frac{\theta}{2}\cos\frac{\theta}{2}}\right) = \sin^{-1}\left(\dfrac{2\cos^2\frac{\theta}{2}}{2\sin\frac{\theta}{2}\cos\frac{\theta}{2}}\right)$

$y = \sin^{-1}\left(\dfrac{\cos\frac{\theta}{2}}{\sin\frac{\theta}{2}}\right) = \sin^{-1}\left(\cot\frac{\theta}{2}\right)$ or $\sin y = \cot\frac{\theta}{2}$, $\cos y\dfrac{dy}{d\theta} = -\dfrac{1}{2}\csc^2\left(\dfrac{\theta}{2}\right)$ $\therefore \dfrac{dy}{d\theta} = -\dfrac{\csc^2\left(\frac{\theta}{2}\right)}{2\cos y}$

To be solve, $\dfrac{dy}{d\theta} = \dfrac{-(1+\cos\theta)}{\sin\theta.\sqrt{-2\cos^2\theta - 2\cos\theta}}$ then $\dfrac{dy}{dx} = \dfrac{dy}{d\theta} \times \dfrac{d\theta}{dx} = \dfrac{-(1+\cos\theta)}{\sin\theta.\sqrt{-2\cos^2\theta - 2\cos\theta}} \cdot -\dfrac{1}{\sin\theta}$

$\dfrac{dy}{dx} = \dfrac{(1+\cos\theta)}{(1-\cos^2\theta)\sqrt{-2\cos^2\theta - 2\cos\theta}} = \dfrac{(1+x)}{(1-x^2).\sqrt{-2x^2 - 2x}} = \dfrac{(1+x)}{(1-x^2).\sqrt{-2x(x+1)}} = \dfrac{(1+x)}{\sqrt{2}.\sqrt{x}(1-x^2)\sqrt{-x-1}}$

$z = \sqrt{x}\,, \quad \dfrac{dz}{dx} = \dfrac{1}{2\sqrt{x}} \quad \therefore \dfrac{dy}{dz} = \dfrac{dy}{dx} \div \dfrac{dz}{dx} = \dfrac{(1+x)}{\sqrt{2}.\sqrt{x}(1-x^2)\sqrt{-x-1}} . 2\sqrt{x} = \dfrac{\sqrt{2}.(1+x)}{(1+x)(1-x).\sqrt{-x-1}}$

$$\dfrac{dy}{dz} = \dfrac{\sqrt{2}}{(1-x).\sqrt{-x-1}} \quad \text{Ans.}$$

(c) Differentiate $x^{\cos^{-1}\sqrt{x}}$ with respect to $\cos^{-1}\sqrt{x}$.

Let $y = x^{\cos^{-1}\sqrt{x}}$ and $z = \cos^{-1}\sqrt{x}$ or $\cos z = \sqrt{x}$ or $x = \cos^2 z$

$y = x^z = (\cos^2 z)^z \quad \therefore \log y = \log(\cos^2 z)^z = z\log(\cos z)^2 = 2z\log(\cos z) = u.v$ (say)

then $\dfrac{1}{y}.\dfrac{dy}{dz} = 2z.\dfrac{d}{dz}(\log(\cos z)) + \log(\cos z).\dfrac{d}{dz}(2z) = 2z.\dfrac{1}{\cos z}.(-\sin z) + 2\log(\cos z) = 2\log(\cos z) - 2z.\tan z$

$\dfrac{dy}{dz} = y[2\log(\cos z) - 2z.\tan z] = 2(\cos^2 z)^z.[\log(\cos z) - z.\tan z]$ Ans.

(d) Differentiate $\cot^{-1}\dfrac{x}{\sqrt{1+x^2}}$ with respect to $\csc^{-1}\dfrac{1}{x^2+1}$.

Let $y = \cot^{-1}\dfrac{x}{\sqrt{1+x^2}}$ Put $x = \tan\theta\,,$ $\dfrac{dx}{d\theta} = \sec^2\theta = 1+\tan^2\theta = 1+x^2$

$y = \cot^{-1}\left(\dfrac{\tan\theta}{\sqrt{1+\tan^2\theta}}\right) = \cot^{-1}\left(\dfrac{\tan\theta}{\sec\theta}\right) = \cot^{-1}\left(\dfrac{\frac{\sin\theta}{\cos\theta}}{\frac{1}{\cos\theta}}\right) = \cot^{-1}(\sin\theta), \dfrac{dy}{d\theta} = -\dfrac{1}{1+\sin^2\theta}.\cos\theta = -\dfrac{\cos\theta}{1+\sin^2\theta}$

$$\therefore \frac{dy}{dx} = \frac{dy}{d\theta} \div \frac{dx}{d\theta} = -\frac{\cos\theta}{1+\sin^2\theta} \cdot \frac{1}{1+\tan^2\theta} \quad \text{if } x = \tan\theta \text{ then } \cos\theta = \frac{1}{\sqrt{1+x^2}} \text{ and } \sin\theta = \frac{x}{\sqrt{1+x^2}}$$

Put the value of $\cos\theta, \sin\theta$ and $\tan\theta$ in $\dfrac{dy}{dx} = -\dfrac{\cos\theta}{1+\sin^2\theta} \cdot \dfrac{1}{1+\tan^2\theta} = -\dfrac{1}{(1+2x^2) \cdot \sqrt{1+x^2}}$

Now, Let $z = \operatorname{cosec}^{-1}\dfrac{1}{x^2+1}$ $\quad \therefore \dfrac{dz}{dx} = -\dfrac{1}{\left|\dfrac{1}{x^2+1}\right| \cdot \sqrt{\left(\dfrac{1}{x^2+1}\right)^2 - 1}} \cdot \dfrac{d}{dx}\left(\dfrac{1}{x^2+1}\right) = -\dfrac{(x^2+1) \cdot |x^2+1|}{\sqrt{-x^4-2x^2}} \cdot -\dfrac{2x}{(x^2+1)^2}$

$$\frac{dz}{dx} = \frac{2(x^2+1) \cdot |x^2+1|}{\sqrt{-x^2-2}} \cdot \frac{1}{(x^2+1)^2} = \frac{2}{\sqrt{-x^2-2}} \quad \text{or} \quad \frac{dy}{dz} = \frac{dy}{dx} \div \frac{dz}{dx} = -\frac{1}{(1+2x^2) \cdot \sqrt{1+x^2}} \cdot \frac{\sqrt{-x^2-2}}{2} = -\frac{\sqrt{-x^2-2}}{2(1+2x^2) \cdot \sqrt{1+x^2}} \quad \text{Ans.}$$

(8) (a) Let $y = \log_{(1-x)} \cos^{-1}(1-x)$ and $z = 2^{2(1-x)}$ $\quad$ Put $1 - x = t$, $\dfrac{dx}{dt} = -1$

or $y = \log_t \cos^{-1} t$ and $z = 2^{2t} = 4^t$ $\quad \therefore \dfrac{dy}{dz} = \dfrac{dy}{dt} \times \dfrac{dt}{dz}$ or $\dfrac{dy}{dt} \div \dfrac{dz}{dt}$

$y = \log_t \cos^{-1} t = \dfrac{\log(\cos^{-1} t)}{\log t}$, $\dfrac{dy}{dt} = \dfrac{\log t \cdot \dfrac{d}{dt}[\log(\cos^{-1} t)] - \log(\cos^{-1} t) \cdot \dfrac{d}{dt}(\log t)}{(\log t)^2}$

$$\frac{dy}{dt} = \frac{\log t \cdot \dfrac{1}{\cos^{-1} t} \cdot \dfrac{1}{-\sqrt{1-t^2}} - \log(\cos^{-1} t) \cdot \dfrac{1}{t}}{(\log t)^2} = \frac{-t\log t - \sqrt{1-t^2} \cdot \cos^{-1} t \cdot \log(\cos^{-1} t)}{t \cdot \cos^{-1} t \cdot \sqrt{1-t^2} \cdot (\log t)^2} = \frac{-1}{\cos^{-1} t \cdot \sqrt{1-t^2} \cdot \log t} - \frac{\log(\cos^{-1} t)}{t \cdot (\log t)^2}$$

or $z = 4^t$, $\dfrac{dz}{dt} = 4^t \log 4$ or $\dfrac{dt}{dz} = \dfrac{1}{4^t \log 4}$

$$\frac{dy}{dz} = \frac{dy}{dt} \times \frac{dt}{dz} = \left[\frac{-1}{\cos^{-1} t \cdot \sqrt{1-t^2} \cdot \log t} - \frac{\log(\cos^{-1} t)}{t \cdot (\log t)^2}\right] \frac{1}{4^t \log 4} = \frac{1}{4^t \log 4}\left[\frac{-1}{\cos^{-1} t \cdot \sqrt{1-t^2} \cdot \log t} - \frac{\log(\cos^{-1} t)}{t \cdot (\log t)^2}\right] \quad \text{Ans.}$$

At $x = \dfrac{1}{2}$ then $t = 1 - x = 1 - \dfrac{1}{2} = \dfrac{1}{2}$ $\quad \therefore \log t = \log\left(\dfrac{1}{2}\right) = -\log 2$, $\cos^{-1} t = \cos^{-1}\left(\dfrac{1}{2}\right) = \dfrac{\pi}{3}$

$$\left(\frac{dy}{dz}\right)_{t=\frac{1}{2}} = \frac{1}{4^{\frac{1}{2}} \cdot \log 4}\left\{\frac{-1}{\cos^{-1}\left(\frac{1}{2}\right) \cdot \sqrt{1-\left(\frac{1}{2}\right)^2} \cdot \log\left(\frac{1}{2}\right)} - \frac{\log\left(\cos^{-1}\left(\frac{1}{2}\right)\right)}{\frac{1}{2} \cdot \left[\log\left(\frac{1}{2}\right)\right]^2}\right\} = \frac{1}{2\log 4}\left\{\frac{-1}{\frac{\pi}{3} \cdot \frac{\sqrt{3}}{2} \cdot (-\log 2)} - \frac{\log\left(\frac{\pi}{3}\right)}{\frac{1}{2} \cdot (\log 2)^2}\right\}$$

$$= \frac{1}{\log 16}\left\{\frac{6}{\sqrt{3} \cdot \pi \cdot \log 2} - \frac{2 \cdot \log\left(\frac{\pi}{3}\right)}{(\log 2)^2}\right\} \quad \text{Ans.}$$

(b) If $\sqrt{1+4x^2} - \sqrt{1-9y^2} = a(2x-3y)$ then find $\dfrac{dy}{dx}$.

or $\sqrt{1+4x^2} - \sqrt{1-9y^2} = a(2x-3y)$ $\quad \therefore \dfrac{1}{2\sqrt{1+4x^2}} \cdot 8x - \dfrac{1}{2\sqrt{1-9y^2}} \cdot 18y \cdot \dfrac{dy}{dx} = 2a - 3a \cdot \dfrac{dy}{dx}$

$$\Rightarrow \frac{dy}{dx}\left(3a - \frac{9y}{\sqrt{1-9y^2}}\right) = \left(2a - \frac{4x}{\sqrt{1+4x^2}}\right) \quad \text{or} \quad \frac{dy}{dx} = \frac{\left(2a - \dfrac{4x}{\sqrt{1+4x^2}}\right)}{\left(3a - \dfrac{9y}{\sqrt{1-9y^2}}\right)} = \frac{2a \cdot \sqrt{1+4x^2} - 4x}{\sqrt{1+4x^2}} \times \frac{\sqrt{1-9y^2}}{3a \cdot \sqrt{1-9y^2} - 9y}$$

$$\therefore \frac{dy}{dx} = \frac{2\sqrt{1-9y^2} \cdot \left(a\sqrt{1+4x^2} - 2x\right)}{3\sqrt{1+4x^2} \cdot \left(a\sqrt{1-9y^2} - 3y\right)} \quad \text{Ans.}$$

(c) $y = \tan^{-1}\dfrac{1}{1+x} + \cot^{-1}(1+x) = \tan^{-1}\dfrac{1}{1+x} + \tan^{-1}\dfrac{1}{1+x} = 2\tan^{-1}\dfrac{1}{1+x}$ $\quad \left[\text{Put } \tan^{-1} x = \cot^{-1}\left(\dfrac{1}{x}\right)\right]$

or $\dfrac{dy}{dx} = 2 \cdot \dfrac{1}{1+\left(\dfrac{1}{1+x}\right)^2} \cdot \dfrac{d}{dx}\left(\dfrac{1}{1+x}\right) = \dfrac{2(1+x)^2}{(1+x)^2+1} \cdot \left(-\dfrac{1}{(1+x)^2}\right) = \dfrac{-2}{(1+x)^2+1} = \dfrac{-2}{1+2x+x^2+1} = \dfrac{-2}{x^2+2x+2}$

Aganin, Differentiate with respect to x, we get

$$\frac{dy}{dx} = \frac{-2}{x^2 + 2x + 2}, \quad \frac{d^2y}{dx^2} = \frac{(x^2 + 2x + 2).\frac{d(-2)}{dx} - (-2).\frac{d}{dx}(x^2 + 2x + 2)}{(x^2 + 2x + 2)^2} = \frac{0 + 2(2x + 2)}{(x^2 + 2x + 2)^2} = \frac{4(x + 1)}{(x^2 + 2x + 2)^2} \quad \text{Ans.}$$

(d) $x = a(1 - \cos t), \ y = a(1 + \sin t)$

$$\frac{dx}{dt} = -a(-\sin t) = a \sin t \ \text{ and } \ \frac{dy}{dt} = a \cos t \quad \therefore \ \frac{dy}{dx} = \frac{dy}{dt} \div \frac{dx}{dt} = \frac{a \cos t}{a \sin t} = \frac{\cos t}{\sin t} = \cot t \quad \text{Ans.}$$

Independent of t, $x = a(1 - \cos t)$ or $\frac{x}{a} = 1 - \cos t$ or $\cos t = 1 - \frac{x}{a} \quad \therefore \ \cos t = \frac{a - x}{a}$

and $y = a(1 + \sin t)$ or $\frac{y}{a} = 1 + \sin t$ or $\sin t = \frac{y}{a} - 1 \quad \therefore \ \sin t = \frac{y - a}{a}$

Put value of $\sin t, \cos t$ in $\frac{dy}{dx}$ then $\frac{dy}{dx} = \cot t = \frac{\cos t}{\sin t} = \frac{\frac{a - x}{a}}{\frac{y - a}{a}} = \frac{a - x}{y - a}$ (independent of t) Ans.

(e) $x = a\left[\sin t - \log\left(\tan\frac{t}{2}\right)\right], \ y = a \cos t$

$$\frac{dx}{dt} = a\left[\cos t - \frac{1}{\tan(t/2)}.\sec^2(t/2).\frac{1}{2}\right] = a\left[\cos t - \frac{\frac{1}{\cos^2(t/2)}}{2\frac{\sin(t/2)}{\cos(t/2)}}\right] = a\left[\cos t - \frac{1}{2\sin(t/2).\cos(t/2)}\right] = a\left[\cos t - \frac{1}{\sin t}\right] = a[\cos t - \csc t]$$

and $y = a \cos t \quad \therefore \ \frac{dy}{dt} = -a \sin t$

$$\frac{dy}{dx} = \frac{dy}{dt} \div \frac{dx}{dt} = \frac{dy}{dt} \times \frac{dt}{dx} = -a \sin t.\frac{1}{a(\cos t - \csc t)} = -\frac{\sin t}{(\cos t - \csc t)} \ \text{ or } \ \frac{\sin t}{\csc t - \cos t} \quad \text{Ans.}$$

At $t = \frac{\pi}{4}$ then $\frac{dy}{dx} = \frac{\sin t}{\csc t - \cos t} = \frac{\sin\frac{\pi}{4}}{\csc\frac{\pi}{4} - \cos\frac{\pi}{4}} = \frac{\frac{1}{\sqrt{2}}}{\sqrt{2} - \frac{1}{\sqrt{2}}} = \frac{\frac{1}{\sqrt{2}}}{\frac{2 - 1}{\sqrt{2}}} = 1$ Ans.

(9) (a) $x = t^2 + \frac{1}{t^2} \ldots\ldots\ldots$ (i) and $y = t^2 - \frac{1}{t^2} \ldots\ldots\ldots\ldots$ (ii)

Adding equation (i) & (ii), we have $\Rightarrow x + y = t^2 + \frac{1}{t^2} + t^2 - \frac{1}{t^2} = 2t^2 \quad \therefore \ t^2 = \frac{x + y}{2}$ or $t = \sqrt{\frac{x + y}{2}}$

Now, $x = t^2 + \frac{1}{t^2} \quad \therefore \ \frac{dx}{dt} = 2t - \frac{2}{t^3} = \frac{2t^4 - 2}{t^3}$ and $y = t^2 + \frac{1}{t^2} \quad \therefore \ \frac{dy}{dt} = 2t + \frac{2}{t^3} = \frac{2t^4 + 2}{t^3}$

Then, $\frac{dy}{dx} = \frac{dy}{dt} \div \frac{dx}{dt} = \frac{dy}{dt} \times \frac{dt}{dx} = \frac{2t^4 + 2}{t^3} \times \frac{t^3}{2t^4 - 2} = \frac{2(t^4 + 1)}{2(t^4 - 1)} = \frac{t^4 + 1}{t^4 - 1} \quad \therefore \ \frac{d^2y}{dx^2} = 0$ Ans.

$\therefore \ \frac{dy}{dx}$ is independent of t, $\frac{dy}{dx} = \frac{t^4 + 1}{t^4 - 1}$ Put $t^2 = \frac{x + y}{2}$ or $t = \sqrt{\frac{x + y}{2}}$

Then, $\frac{dy}{dx} = \frac{(t^2)^2 + 1}{(t^2)^2 - 1} = \frac{\left(\frac{x + y}{2}\right)^2 + 1}{\left(\frac{x + y}{2}\right)^2 - 1} = \frac{\frac{(x + y)^2 + 4}{4}}{\frac{(x + y)^2 - 4}{4}} = \frac{(x + y)^2 + 4}{(x + y)^2 - 4}$ (independent of t) Ans.

(b) Do yourself. $x = \frac{1 - t^2}{t^3}$ and $y = \frac{3}{2t^3} - \frac{2}{t^2} \quad \therefore \ \frac{dx}{dt} = \frac{t^2 - 3}{t^4}, \ \frac{dy}{dt} = \frac{8t - 9}{2t^4}$

$\therefore \ \frac{dy}{dx} = \frac{dy}{dt} \div \frac{dx}{dt} = \frac{dy}{dt} \times \frac{dt}{dx} = \frac{8t - 9}{2t^4} \times \frac{t^4}{t^2 - 3} = \frac{8t - 9}{2(t^2 - 3)}$ at $t = 2 \quad \therefore \ \frac{dy}{dx} = \frac{16 - 9}{2(4 - 3)} = \frac{7}{2}$ Ans.

(c) $x = 2\sin t + \sin 2t$, $y = 2\cos t + \cos 2t$

$\therefore \dfrac{dx}{dt} = 2\cos t + 2\cos 2t = 2(\cos t + \cos 2t)$ and $\dfrac{dy}{dt} = -2\sin t - 2\sin 2t = -2(\sin t + \sin 2t)$

$\therefore \dfrac{dy}{dx} = \dfrac{dy}{dt} \div \dfrac{dx}{dt} = \dfrac{dy}{dt} \times \dfrac{dt}{dx} = \dfrac{-2(\sin t + \sin 2t)}{2(\cos t + \cos 2t)} = \dfrac{-(\sin t + \sin 2t)}{(\cos t + \cos 2t)}$

At $t = \dfrac{\pi}{2}$ then $\dfrac{dy}{dx} = \dfrac{-\left(\sin\left(\frac{\pi}{2}\right) + \sin\left(2.\frac{\pi}{2}\right)\right)}{\left(\cos\left(\frac{\pi}{2}\right) + \cos\left(2.\frac{\pi}{2}\right)\right)}$ $\therefore \left(\dfrac{dy}{dx}\right)_{t=\frac{\pi}{2}} = \dfrac{-(1 + \sin\pi)}{0 + \cos\pi} = \dfrac{-(1 + 0)}{0 + (-1)} = \dfrac{-1}{-1} = 1$ Ans.

(d) $x = \sin^{-1}\dfrac{1}{\sqrt{1 + t^2}}$, $y = \cos^{-1}\dfrac{t}{\sqrt{1 + t^2}}$ Put $t = \cot\theta$ $\therefore \dfrac{dt}{d\theta} = -\operatorname{cosec}^2\theta$

$x = \sin^{-1}\dfrac{1}{\sqrt{1 + (\cot\theta)^2}} = \sin^{-1}(\sin\theta) = \theta$ and $y = \cos^{-1}\dfrac{\cot\theta}{\sqrt{1 + (\cot\theta)^2}} = \cos^{-1}(\cos\theta) = \theta$

$\dfrac{dx}{d\theta} = 1$ and $\dfrac{dy}{d\theta} = 1$ $\therefore \dfrac{dy}{dx} = \dfrac{dy}{d\theta} \div \dfrac{dx}{d\theta} = \dfrac{dy}{d\theta} \times \dfrac{d\theta}{dx} = \dfrac{1}{1} = 1$ (independent of t) Proved.

(10) (a) $x = \dfrac{\sin^2 t}{\sqrt{\cos 2t}}$ and $y = \dfrac{\cos^2 t}{\sqrt{\sin 2t}}$ $\qquad \left[\text{formula,} \quad y = \dfrac{u}{v} \quad \therefore \dfrac{dy}{dx} = \dfrac{v.\frac{du}{dx} - u.\frac{dv}{dx}}{v^2} \right]$

$\dfrac{dx}{dt} = \dfrac{\sqrt{\cos 2t}.\frac{d}{dt}(\sin^2 t) - \sin^2 t.\frac{d}{dt}\left(\sqrt{\cos 2t}\right)}{\left(\sqrt{\cos 2t}\right)^2} = \dfrac{\sqrt{\cos 2t}.2\sin t.\cos t - \sin^2 t.\frac{1}{2\sqrt{\cos 2t}}.(-\sin 2t).2}{\cos 2t} = \dfrac{\dfrac{\sin 2t.\cos 2t + \sin 2t.\sin^2 t}{\sqrt{\cos 2t}}}{\cos 2t}$

$= \dfrac{\sin 2t.(\cos 2t + \sin^2 t)}{\cos 2t.\sqrt{\cos 2t}}$

$\dfrac{dy}{dt} = \dfrac{\sqrt{\sin 2t}.\frac{d}{dt}(\cos^2 t) - \cos^2 t.\frac{d}{dt}\left(\sqrt{\sin 2t}\right)}{\left(\sqrt{\sin 2t}\right)^2} = \dfrac{\sqrt{\sin 2t}.2\cos t.(-\sin t) - \cos^2 t.\frac{1}{2.\sqrt{\sin 2t}}.\cos 2t.2}{\sin 2t}$

$\dfrac{dy}{dt} = \dfrac{\dfrac{-\sin 2t.\sin 2t - \cos 2t.\cos^2 t}{\sqrt{\sin 2t}}}{\sin 2t} = \dfrac{-(\sin^2 2t + \cos 2t.\cos^2 t)}{\sin 2t.\sqrt{\sin 2t}}$

$\dfrac{dy}{dx} = \dfrac{dy}{dt} \div \dfrac{dx}{dt} = \dfrac{dy}{dt} \times \dfrac{dt}{dx} = \dfrac{-(\sin^2 2t + \cos 2t.\cos^2 t)}{\sin 2t.\sqrt{\sin 2t}} \times \dfrac{\cos 2t.\sqrt{\cos 2t}}{\sin 2t.(\cos 2t + \sin^2 t)}$

At $t = \dfrac{\pi}{4}$ then $\dfrac{dy}{dx} = \dfrac{-\left(\sin^2\left(2.\frac{\pi}{4}\right) + \cos\left(2.\frac{\pi}{4}\right).\cos^2\left(\frac{\pi}{4}\right)\right)}{\sin\left(2.\frac{\pi}{4}\right).\sqrt{\sin\left(2.\frac{\pi}{4}\right)}} \times \dfrac{\cos\left(2.\frac{\pi}{4}\right).\sqrt{\cos\left(2.\frac{\pi}{4}\right)}}{\sin\left(2.\frac{\pi}{4}\right).\left(\cos\left(2.\frac{\pi}{4}\right) + \sin^2\left(\frac{\pi}{4}\right)\right)} = \dfrac{-(1 + 0).0}{1\left(0 + \frac{1}{2}\right)}$

$\dfrac{dy}{dx} = \dfrac{2.0}{1} = 0$ Ans.

(b) Do yourself. $x = \tan\theta.\sqrt{\cos 2\theta}$ and $y = \cot\theta.\sqrt{\sin 2\theta}$ $\qquad \left[\text{formula, } y = u.v \quad \therefore \dfrac{dy}{dx} = u.\dfrac{dv}{dx} + v.\dfrac{du}{dx} \right]$

see above question and solve it , $\dfrac{dx}{d\theta} = \dfrac{\cos 2\theta.\sec^2\theta - \sin 2\theta.\tan\theta}{\sqrt{\cos 2\theta}}$ and $\dfrac{dy}{d\theta} = \dfrac{\cos 2\theta.\cot\theta - \sin 2\theta.\operatorname{cosec}^2\theta}{\sqrt{\sin 2\theta}}$

$\dfrac{dy}{dx} = \dfrac{dy}{d\theta} \div \dfrac{dx}{d\theta} = \dfrac{dy}{d\theta} \times \dfrac{d\theta}{dx} = \dfrac{\cos 2\theta.\cot\theta - \sin 2\theta.\operatorname{cosec}^2\theta}{\sqrt{\sin 2\theta}} \times \dfrac{\sqrt{\cos 2\theta}}{\cos 2\theta.\sec^2\theta - \sin 2\theta.\tan\theta}$

At $\theta = \dfrac{\pi}{6}$ then $\dfrac{dy}{dx} = \dfrac{\sqrt{\cos\left(\frac{\pi}{3}\right)}}{\sqrt{\sin\left(\frac{\pi}{3}\right)}}\left[\dfrac{\cos\left(\frac{\pi}{3}\right).\cot\left(\frac{\pi}{6}\right) - \sin\left(\frac{\pi}{3}\right).\operatorname{cosec}^2\left(\frac{\pi}{6}\right)}{\cos\left(\frac{\pi}{3}\right).\sec^2\left(\frac{\pi}{6}\right) - \sin\left(\frac{\pi}{3}\right).\tan\left(\frac{\pi}{6}\right)}\right] = -\dfrac{9.\sqrt{3}}{3^{\frac{1}{4}}} = -\left(3^2.3^{\frac{1}{2}}.3^{-\frac{1}{4}}\right) = -(3)^{\frac{9}{4}}$ Ans.

(c) $x^3 - 2xy + y^3 = a^3$ $\Rightarrow$ $3x^2 - 2\left(x.\frac{dy}{dx} + y.1\right) + 3y^2.\frac{dy}{dx} = 0$

$\Rightarrow$ $3y^2.\frac{dy}{dx} - 2x.\frac{dy}{dx} = 2y - 3x^2$ or $\frac{dy}{dx}(3y^2 - 2x) = (2y - 3x^2)$ $\therefore$ $\frac{dy}{dx} = \frac{2y - 3x^2}{3y^2 - 2x}$ Ans.

$$\frac{d^2y}{dx^2} = \frac{(3y^2 - 2x).\frac{d}{dx}(2y - 3x^2) - (2y - 3x^2).\frac{d}{dx}(3y^2 - 2x)}{(3y^2 - 2x)^2} = \frac{(3y^2 - 2x).\left(2.\frac{dy}{dx} - 6x\right) - (2y - 3x^2).\left(6y.\frac{dy}{dx} - 2\right)}{(3y^2 - 2x)^2}$$

$\therefore$ Put $\frac{dy}{dx} = \frac{2y - 3x^2}{3y^2 - 2x}$ in $\frac{d^2y}{dx^2} = \frac{(3y^2 - 2x).\left(2.\frac{2y - 3x^2}{3y^2 - 2x} - 6x\right) - (2y - 3x^2).\left(6y.\frac{2y - 3x^2}{3y^2 - 2x} - 2\right)}{(3y^2 - 2x)^2}$

$$\therefore \frac{d^2y}{dx^2} = \frac{(3y^2 - 2x)(6x^2 - 18xy^2 + 4y) - (2y - 3x^2)(6y^2 - 18x^2y + 4x)}{(3y^2 - 2x)^3}$$ Ans.

(d) $x = \cos^{-1}\left(2t\sqrt{1 - t^2}\right)$ and $y = \frac{\pi}{2} + \sin^{-1}t$ Put $t = \sin\theta$

Now, $x = \cos^{-1}\left(2\sin\theta.\sqrt{1 - \sin^2\theta}\right)$ and $y = \frac{\pi}{2} + \sin^{-1}(\sin\theta) = \frac{\pi}{2} + \theta$ .... (i)

$x = \cos^{-1}(2\sin\theta.\cos\theta) = \cos^{-1}(\sin 2\theta) = \cos^{-1}\left(\cos\left(\frac{\pi}{2} - 2\theta\right)\right) = \frac{\pi}{2} - 2\theta$ .... (ii)

solve the equation (i) & (ii) , we have

$x = \frac{\pi}{2} - 2\theta$ and $y = \frac{\pi}{2} + \theta$ or $\theta = y - \frac{\pi}{2}$ $\therefore$ $x = \frac{\pi}{2} - 2\left(y - \frac{\pi}{2}\right) = \frac{\pi}{2} - 2y + \pi$ or $x + 2y = \frac{\pi}{2} + \pi$

or $x + 2y = \frac{3\pi}{2}$ $\therefore$ $1 + 2\frac{dy}{dx} = 0$ or $2\frac{dy}{dx} = -1$ $\therefore$ $\frac{dy}{dx} = -\frac{1}{2}$ Ans. $\therefore$ $\frac{d^2y}{dx^2} = 0$ Ans.

(11) (a) $y = (\cos^{-1}x)^x$ $\therefore$ $\log y = x\log(\cos^{-1}x)$ $\therefore$ $\frac{1}{y}.\frac{dy}{dx} = x.\frac{1}{\cos^{-1}x}.\frac{1}{-\sqrt{1 - x^2}} + \log(\cos^{-1}x).1$

$\therefore$ $\frac{dy}{dx} = y\left[\log(\cos^{-1}x) - \frac{x}{\cos^{-1}x.\sqrt{1 - x^2}}\right] = (\cos^{-1}x)^x.\log(\cos^{-1}x) - (\cos^{-1}x)^x.\frac{x}{\cos^{-1}x.\sqrt{1 - x^2}}$ Ans.

(b) $y = x^{x^2}$ $\Rightarrow$ $\log y = \log x^{x^2} = x^2\log x$ $\therefore$ $\frac{1}{y}.\frac{dy}{dx} = x^2.\frac{1}{x} + \log x.2x = x + 2x.\log x = x(1 + 2\log x)$

$\therefore$ $\frac{dy}{dx} = y.x(1 + 2\log x) = x.x^{x^2}(1 + 2\log x) = x^{x^2+1}(1 + 2\log x)$ or $x^{x^2+1}(1 + \log x^2)$ Ans.

(c) $y = (\sin x)^{\log x}$ $\Rightarrow$ $\log y = \log x.\log(\sin x)$ $\therefore$ $\frac{1}{y}.\frac{dy}{dx} = \log x.\frac{1}{\sin x}.\cos x + \log(\sin x).\frac{1}{x}$

$\therefore$ $\frac{dy}{dx} = y\left[\log x.\cot x + \frac{\log(\sin x)}{x}\right] = (\sin x)^{\log x}\left[\log x.\cot x + \frac{\log(\sin x)}{x}\right]$ Ans.

(d) Do yourself (see above question), $y = (\cos x)^{\log x}$ $\therefore$ $\frac{dy}{dx} = (\cos x)^{\log x}\left[\frac{\log(\cos x)}{x} - \log x.\tan x\right]$ Ans.

(e) Do yourself (see above question), $y = (\tan x)^{\log x}$ $\therefore$ $\frac{dy}{dx} = (\tan x)^{\log x}\left[\frac{2x.\log x + \sin 2x.\log(\tan x)}{x.\sin 2x}\right]$ Ans.

(f) $y = (\log x)^{\log(\tan x)}$ $\Rightarrow$ $\log y = \log(\tan x).\log(\log x)$ $\therefore$ $\frac{1}{y}.\frac{dy}{dx} = \log(\tan x).\frac{1}{\log x}.\frac{1}{x} + \log(\log x).\frac{1}{\tan x}.\sec^2 x$

or $\frac{dy}{dx} = y\left[\log(\tan x).\frac{1}{x.\log x} + \log(\log x).\frac{1}{\sin x.\cos x}\right] = (\log x)^{\log(\tan x)}\left[\frac{\log(\tan x)}{x.\log x} + \frac{2\log(\log x)}{2\sin x.\cos x}\right]$

or $\frac{dy}{dx} = (\log x)^{\log(\tan x)}\left[\frac{\log(\tan x)}{x.\log x} + \frac{2\log(\log x)}{\sin 2x}\right] = (\log x)^{\log(\tan x)}\left[\frac{\sin 2x.\log(\tan x) + 2x.\log x.\log(\log x)}{x.\sin 2x.\log x}\right]$ Ans.

(g) $y = (\sin x)^{\cos x} + (\tan x)^{\cot x} = u + v$ (say)   where $u = (\sin x)^{\cos x}$ and $v = (\tan x)^{\cot x}$

$\Rightarrow u = (\sin x)^{\cos x}$ $\therefore \log u = \cos x \cdot \log(\sin x)$ $\therefore \frac{1}{u} \cdot \frac{du}{dx} = \cos x \cdot \frac{1}{\sin x} \cdot \cos x + \log(\sin x) \cdot (-\sin x)$

$\therefore \frac{du}{dx} = u \left[\frac{\cos^2 x}{\sin x} - \sin x \cdot \log(\sin x) \right] = (\sin x)^{\cos x} \left[\frac{\cos^2 x - \sin^2 x \cdot \log(\sin x)}{\sin x} \right]$

$\Rightarrow v = (\tan x)^{\cot x}$ $\therefore \log v = \cot x \cdot \log(\tan x)$ $\therefore \frac{1}{v} \cdot \frac{dv}{dx} = \cot x \cdot \frac{1}{\tan x} \cdot \sec^2 x + \log(\tan x) \cdot (-\cosec^2 x)$

$\therefore \frac{dv}{dx} = v \left[\frac{\cos^2 x}{\sin^2 x} \cdot \frac{1}{\cos^2 x} - \frac{1}{\sin^2 x} \cdot \log(\tan x) \right] = (\tan x)^{\cot x} \left[\frac{1 - \log(\tan x)}{\sin^2 x} \right]$

$\therefore \frac{dy}{dx} = \frac{du}{dx} + \frac{dv}{dx} = (\sin x)^{\cos x} \left[\frac{\cos^2 x - \sin^2 x \cdot \log(\sin x)}{\sin x} \right] + (\tan x)^{\cot x} \left[\frac{1 - \log(\tan x)}{\sin^2 x} \right]$ Ans.

(h) $y = (\cot x)^{\cos x}$ $\therefore \log y = \cos x \cdot \log(\cot x)$ $\therefore \frac{1}{y} \cdot \frac{dy}{dx} = \cos x \cdot \frac{1}{\cot x} \cdot (-\cosec^2 x) + \log(\cot x) \cdot (-\sin x)$

$\therefore \frac{dy}{dx} = -y \left[\frac{1}{\sin x} + \sin x \cdot \log(\cot x) \right] = -(\cot x)^{\cos x} \left[\frac{1 + \sin^2 x \cdot \log(\cot x)}{\sin x} \right]$ Ans. or $\frac{dy}{dx} = -(\cot x)^{\cos x} [\cosec x + \sin x \cdot \log(\cot x)]$ Ans.

(i) $y = (\tan x)^{\sin x}$ $\therefore \log y = \sin x \cdot \log(\tan x)$ $\therefore \frac{1}{y} \cdot \frac{dy}{dx} = \sin x \cdot \frac{1}{\tan x} \cdot \sec^2 x + \log(\tan x) \cdot \cos x$

$\therefore \frac{dy}{dx} = y \left[\frac{1}{\cos x} + \cos x \cdot \log(\tan x) \right] = (\tan x)^{\sin x} \left[\frac{1 + \cos^2 x \cdot \log(\tan x)}{\cos x} \right]$ or $(\tan x)^{\sin x} [\sec x + \cos x \cdot \log(\tan x)]$ Ans.

(j) $y = e^{x \cos x^2} + (\cot x)^x = u + v$ (say)   where $u = e^{x \cos x^2}$ and $v = (\cot x)^x$

$\Rightarrow u = e^{x \cos x^2}$ $\therefore \frac{du}{dx} = e^{x \cos x^2} [\cos x^2 \cdot 1 - x \cdot \sin x^2 \cdot 2x] = e^{x \cos x^2} [\cos x^2 - 2x^2 \cdot \sin x^2]$

$\Rightarrow v = (\cot x)^x$ $\therefore \log v = x \log(\cot x)$ $\therefore \frac{1}{v} \cdot \frac{dv}{dx} = x \cdot \frac{1}{\cot x} \cdot (-\cosec^2 x) + \log(\cot x) \cdot 1$

$\therefore \frac{dv}{dx} = v \left[\log(\cot x) - \frac{x}{\sin x \cos x} \right] = (\cot x)^x \left[\frac{\sin 2x \cdot \log(\cot x) - 2x}{\sin 2x} \right]$

$\therefore \frac{dy}{dx} = \frac{du}{dx} + \frac{dv}{dx} = e^{x \cos x^2} [\cos x^2 - 2x^2 \cdot \sin x^2] + (\cot x)^x \left[\frac{\sin 2x \cdot \log(\cot x) - 2x}{\sin 2x} \right]$ Ans.

(12) (a) $y = e^{x + x^2 + x^3 + \cdots \ldots \ldots \infty} = e^{\left(\frac{x}{1-x}\right)}$ $\left[\begin{array}{l} \text{formula, } x + x^2 + x^3 + \cdots \ldots \ldots \infty \text{ are in G. P, common ratio } (r) = \frac{x^2}{x} = \frac{x^3}{x^2} = \cdots \ldots = x \\ \text{and first term } (a) = x \text{ then } s_n = \frac{x}{1-x} \end{array} \right]$

$\therefore y = e^{\left(\frac{x}{1-x}\right)}$ or $\log y = \log e^{\left(\frac{x}{1-x}\right)} = \frac{x}{1-x}$ $\therefore \frac{1}{y} \cdot \frac{dy}{dx} = \frac{(1-x) - x(-1)}{(1-x)^2} = \frac{1 - x + x}{(1-x)^2} = \frac{1}{(1-x)^2}$

$\therefore \frac{dy}{dx} = y \left[\frac{1}{(1-x)^2} \right] = e^{\left(\frac{x}{1-x}\right)} \left[\frac{1}{(1-x)^2} \right] = \frac{e^{\left(\frac{x}{1-x}\right)}}{(1-x)^2}$ Ans.

(b) $y = e^{(\sin x + \cos x)}$ $\therefore \log y = \log e^{(\sin x + \cos x)} = \sin x + \cos x$ $\therefore \frac{1}{y} \cdot \frac{dy}{dx} = \cos x - \sin x$

$\therefore \frac{dy}{dx} = y(\cos x - \sin x) = e^{(\sin x + \cos x)}(\cos x - \sin x)$ Ans.

(c) $y = \sqrt{\sin x \sqrt{\sin x \sqrt{\sin \ldots \ldots \ldots \infty}}}$,   Let $y = \sqrt{\sin x \sqrt{\sin \ldots \ldots \ldots \infty}}$ then $y = \sqrt{\sin xy}$

$\Rightarrow y^2 = \sin xy$ $\quad \therefore 2y.\dfrac{dy}{dx} = \cos xy.\left(x.\dfrac{dy}{dx} + y.1\right)$ $\quad \therefore \dfrac{dy}{dx}(2y - x\cos xy) = y\cos xy$ $\quad \therefore \dfrac{dy}{dx} = \dfrac{y.\cos xy}{(2y - x\cos xy)}$ Ans.

(d) $y = \sqrt{\sin x + \sqrt{\sin x + \sqrt{\sin x + \sqrt{\ldots\ldots\ldots\infty}}}}$, $\quad$ Let $y = \sqrt{\sin x + \sqrt{\sin x + \sqrt{\ldots\ldots\ldots\infty}}}$

then $y = \sqrt{\sin x + y}$ $\quad$ squaring both of side , we have $\quad \therefore y^2 = \sin x + y$

$\therefore 2y.\dfrac{dy}{dx} = \cos x + \dfrac{dy}{dx}$ $\quad \therefore \dfrac{dy}{dx}(2y - 1) = \cos x$ $\quad \therefore \dfrac{dy}{dx} = \dfrac{\cos x}{2y - 1}$ Ans.

(e) $y = e^{\sin x + \sin^2 x + \sin^3 x + \cdots\ldots\ldots\infty}$, Let $s_n = \sin x + \sin^2 x + \sin^3 x + \cdots\ldots\ldots\ldots\infty$ are in G.P, $a = \sin x$, $r = \sin x$

then $s_n = \dfrac{a}{1 - r} = \dfrac{\sin x}{1 - \sin x}$ $\quad \therefore y = e^{\left(\frac{\sin x}{1 - \sin x}\right)}$ $\quad \therefore \log y = \dfrac{\sin x}{1 - \sin x}.\log e = \dfrac{\sin x}{1 - \sin x}$

$\therefore \dfrac{1}{y}.\dfrac{dy}{dx} = \dfrac{(1 - \sin x).\cos x - \sin x.(-\cos x)}{(1 - \sin x)^2} = \dfrac{\cos x - \sin x\cos x + \sin x\cos x}{(1 - \sin x)^2} = \dfrac{\cos x}{(1 - \sin x)^2}$

$\therefore \dfrac{dy}{dx} = y\left[\dfrac{\cos x}{(1 - \sin x)^2}\right] = e^{\left(\frac{\sin x}{1 - \sin x}\right)}\left[\dfrac{\cos x}{(1 - \sin x)^2}\right]$ Ans.

(f) $y = (\sin x)^{(\sin x)^{\sin x \cdots\cdots\cdots\infty}}$, Let $y = (\sin x)^{\sin x \cdots\cdots\cdots\infty}$ then $y = (\sin x)^y$ $\quad \therefore \log y = y.\log(\sin x)$

$\therefore \dfrac{1}{y}.\dfrac{dy}{dx} = y.\dfrac{1}{\sin x}.\cos x + \log(\sin x).\dfrac{dy}{dx}$ $\quad \therefore \dfrac{dy}{dx}\left(\dfrac{1}{y} - \log(\sin x)\right) = y\cot x$ $\quad \therefore \dfrac{dy}{dx} = \dfrac{y\cot x}{\dfrac{1}{y} - \log(\sin x)}$

$\therefore \dfrac{dy}{dx} = \dfrac{y^2\cot x}{1 - y\log(\sin x)} = \dfrac{\left[(\sin x)^{(\sin x)^{\sin x \cdots\cdots\cdots\infty}}\right]^2.\cot x}{1 - \left[(\sin x)^{(\sin x)^{\sin x \cdots\cdots\cdots\infty}}\right].\log(\sin x)}$ Ans.

(g) $y = (\log x)^{(\log x)^{\log x \cdots\cdots\cdots\infty}}$, Let $y = (\log x)^{\log x \cdots\cdots\cdots\infty}$ then $y = (\log x)^y$ $\quad \therefore \log y = y\log(\log x)$

$\therefore \dfrac{1}{y}.\dfrac{dy}{dx} = y.\dfrac{1}{\log x}.\dfrac{1}{x} + \log(\log x).\dfrac{dy}{dx}$ $\quad \therefore \dfrac{dy}{dx}\left(\dfrac{1}{y} - \log(\log x)\right) = \dfrac{y}{x\log x}$ $\quad \therefore \dfrac{dy}{dx} = \dfrac{\dfrac{y}{x\log x}}{\dfrac{1}{y} - \log(\log x)}$

$\therefore \dfrac{dy}{dx} = \dfrac{y^2}{x\log x.[1 - y\log(\log x)]}$ Ans.

(h) $y = e^{(\sqrt{x})^{e^{(\sqrt{x})^{\cdots\cdots\cdots\infty}}}}$, Let $y = e^{(\sqrt{x})^{\cdots\cdots\cdots\infty}}$ then $y = e^{(\sqrt{x})^y}$ $\quad \therefore \log y = (\sqrt{x})^y.\log_e e = (\sqrt{x})^y$

$\therefore \log(\log y) = y\log(\sqrt{x})$ $\quad \therefore \dfrac{1}{\log y}.\dfrac{1}{y}.\dfrac{dy}{dx} = y.\dfrac{1}{\sqrt{x}}.\dfrac{1}{2\sqrt{x}} + \log(\sqrt{x}).\dfrac{dy}{dx}$

$\therefore \dfrac{dy}{dx}\left(\dfrac{1}{y\log y} - \log(\sqrt{x})\right) = \dfrac{y}{2x}$ $\quad \therefore \dfrac{dy}{dx} = \dfrac{\dfrac{y}{2x}}{\dfrac{1 - y\log y.\log(\sqrt{x})}{y\log y}} = \dfrac{y^2\log y}{2x[1 - y\log y.\log(\sqrt{x})]}$ Ans.

(i) $y = \log(\log(\log(\cos x)))$, Let $u = \log(\cos x)$ $\quad \therefore \dfrac{du}{dx} = \dfrac{1}{\cos x}.(-\sin x) = -\tan x$

$\Rightarrow y = \log(\log u)$, Let $v = \log u$ $\quad \therefore \dfrac{dv}{du} = \dfrac{1}{u} = \dfrac{1}{\log(\cos x)}$ $\quad$ [Put $u = \log(\cos x)$]

$\Rightarrow y = \log v$ $\quad \therefore \dfrac{dy}{dv} = \dfrac{1}{v} = \dfrac{1}{\log u} = \dfrac{1}{\log(\log(\cos x))}$ $\quad$ [Put $v = \log u$ and $u = \log(\cos x)$]

$$\therefore \frac{dy}{dx} = \frac{dy}{dv} \times \frac{dv}{du} \times \frac{du}{dx} = \frac{1}{\log(\log(\cos x))} \times \frac{1}{\log(\cos x)} \times (-\tan x) = -\frac{\tan x}{\log(\cos x) . \log(\log(\cos x))} \quad \text{Ans.}$$

(j) Do yourself $\left(\text{see question no. } 11. (g)\right)$.

$y = x^x + x^{\sin x} = u + v$ (say) where $u = x^x$ and $v = x^{\sin x}$

$$\therefore \frac{du}{dx} = x^x(1 + \log x) \quad \text{and} \quad \frac{dv}{dx} = x^{\sin x}\left(\log x . \cos x + \frac{\sin x}{x}\right) \quad \therefore \frac{dy}{dx} = \frac{du}{dx} + \frac{dv}{dx} = x^x(1 + \log x) + x^{\sin x}\left(\log x . \cos x + \frac{\sin x}{x}\right) \quad \text{Ans.}$$

(k) Do yourself (see above question). $y = a^x + a^{\sin x} = u + v$ (say) where $u = a^x$ and $v = a^{\sin x}$

$$\therefore \frac{du}{dx} = a^x \log a \quad \text{and} \quad \frac{dv}{dx} = a^{\sin x} . \log a . \cos x \quad \therefore \frac{dy}{dx} = \frac{du}{dx} + \frac{dv}{dx} = a^x \log a + a^{\sin x} . \log a . \cos x \quad \therefore \frac{dy}{dx} = \log a \left(a^x + a^{\sin x} . \cos x\right) \quad \text{Ans.}$$

(l) $e^y = e^{x-y}$ or $e^y = \frac{e^x}{e^y}$ or $e^{2y} = e^x$ $\therefore e^{2y} . 2.\frac{dy}{dx} = e^x$ $\therefore \frac{dy}{dx} = \frac{e^x}{2e^{2y}} = \frac{e^x}{2e^x}$ [Put $e^{2y} = e^x$] $\therefore \frac{dy}{dx} = \frac{1}{2}$ Ans.

(m) $x^m y^n = (x - y)^{m-n}$ $\therefore x^m . \frac{d(y^n)}{dx} + y^n . \frac{d(x^m)}{dx} = (m - n) . (x - y)^{m-n-1}\left(1 - \frac{dy}{dx}\right)$

$$\therefore x^m . ny^{n-1} . \frac{dy}{dx} + y^n . mx^{m-1} = (m - n)(x - y)^{m-n-1} - (m - n) . (x - y)^{m-n-1} . \frac{dy}{dx}$$

$$\therefore \frac{dy}{dx}(x^m . ny^{n-1} + (m - n) . (x - y)^{m-n-1}) = (m - n)(x - y)^{m-n-1} - y^n . mx^{m-1}$$

$$\therefore \frac{dy}{dx} = \frac{(m - n)(x - y)^{m-n-1} - y^n . mx^{m-1}}{x^m . ny^{n-1} + (m - n) . (x - y)^{m-n-1}} \quad \text{or} \quad \frac{y(my - nx)}{x(nx + my - 2ny)} \quad \text{Ans.}$$

IInd Method: $-$ $x^m y^n = (x - y)^{m-n}$ $\therefore \log(x^m y^n) = \log(x - y)^{m-n}$

$\therefore \log x^m + \log y^n = (m - n) \log(x - y)$ Differentiate with respect to x, we have

$$\therefore \frac{m}{x} + \frac{n}{y} . \frac{dy}{dx} = \frac{m - n}{x - y} . \left(1 - \frac{dy}{dx}\right) \quad \therefore \frac{dy}{dx}\left(\frac{n}{y} + \frac{m - n}{x - y}\right) = \frac{m - n}{x - y} - \frac{m}{x} \quad \therefore \frac{dy}{dx} = \frac{\frac{m - n}{x - y} - \frac{m}{x}}{\frac{n}{y} + \frac{m - n}{x - y}}$$

$$\therefore \frac{dy}{dx} = \frac{\frac{mx - nx - mx + my}{(x - y)x}}{\frac{nx - ny + my - ny}{y(x - y)}} = \frac{my - nx}{nx + my - 2ny} \times \frac{y}{x} = \frac{y}{x}\left[\frac{my - nx}{nx + my - 2ny}\right] \quad \text{Ans.}$$

(n) $x^y + y^x = (x - y)^{x-y}$ $\therefore yx^{y-1} + xy^{x-1} . \frac{dy}{dx} = (x - y)(x - y)^{x-y-1}\left(1 - \frac{dy}{dx}\right)$

$$\therefore xy^{x-1} . \frac{dy}{dx} + (x - y)(x - y)^{x-y-1} . \frac{dy}{dx} = (x - y)(x - y)^{x-y-1} - yx^{y-1}$$

$$\therefore \frac{dy}{dx}(xy^{x-1} + (x - y)(x - y)^{x-y-1}) = (x - y)(x - y)^{x-y-1} - yx^{y-1} \quad \therefore \frac{dy}{dx} = \frac{(x - y)(x - y)^{x-y-1} - yx^{y-1}}{[xy^{x-1} + (x - y)(x - y)^{x-y-1}]} \quad \text{Ans.}$$

(13) (a) $y = x^2 \log\left(\frac{a + x}{a - x}\right) = u . v$ (say), where $u = x^2$ and $v = \log\left(\frac{a + x}{a - x}\right)$

or $\frac{du}{dx} = 2x$ and $\frac{dv}{dx} = \frac{1}{\frac{a + x}{a - x}} . \frac{(a - x) . 1 - (a + x)(-1)}{(a - x)^2} = \frac{(a - x)2a}{(a + x)(a - x)^2} = \frac{2a}{a^2 - x^2}$

$$\therefore \frac{dy}{dx} = u . \frac{dv}{dx} + v . \frac{du}{dx} = x^2 . \frac{2a}{a^2 - x^2} + \log\left(\frac{a + x}{a - x}\right) . 2x = 2x\left[\frac{ax}{a^2 - x^2} + \log\left(\frac{a + x}{a - x}\right)\right] \quad \text{Proved.}$$

(b) $y = \left(x - \sqrt{x^2 - a^2}\right)^n$, Put $x = a \sec \theta$ $\therefore \frac{dx}{d\theta} = a \tan \theta \sec \theta$

$$\Rightarrow \quad y = \left(a\sec\theta - \sqrt{(a\sec\theta)^2 - a^2}\right)^n = \left(a\sec\theta - \sqrt{a^2(\sec^2\theta - 1)}\right)^n = \left(a\sec\theta - \sqrt{a^2.\tan^2\theta}\right)^n = (a\sec\theta - a\tan\theta)^n = [a(\sec\theta - \tan\theta)]^n$$
$$= a^n.(\sec\theta - \tan\theta)^n$$

$$\therefore \quad \frac{dy}{d\theta} = n.[a(\sec\theta - \tan\theta)]^{n-1}.a(\tan\theta\sec\theta - \sec^2\theta) = na.a^{n-1}(\sec\theta - \tan\theta)^{n-1}.\sec\theta(\tan\theta - \sec\theta)$$

$$\therefore \quad \frac{dy}{dx} = \frac{dy}{d\theta} \times \frac{d\theta}{dx} = na^n.\sec\theta.(\sec\theta - \tan\theta)^{n-1}.(\tan\theta - \sec\theta) \times \frac{1}{a\tan\theta\sec\theta} = \frac{na^{n-1}.(\sec\theta - \tan\theta)^{n-1}.(\tan\theta - \sec\theta)}{\tan\theta}$$
$$= \frac{-na^{n-1}.(\sec\theta - \tan\theta)^{n-1}.(\sec\theta - \tan\theta)}{\tan\theta}$$

$$\therefore \quad \frac{dy}{dx} = -\frac{na^{n-1}.(\sec\theta - \tan\theta)^n.(\sec\theta - \tan\theta)}{\tan\theta.(\sec\theta - \tan\theta)} = -\frac{na^{n-1}.(\sec\theta - \tan\theta)^n}{\tan\theta} \quad \text{Ans.}$$

(c) $y = \dfrac{1 - \sqrt{x-1}}{1 + \sqrt{x-1}} + \dfrac{\sqrt{x-1}}{\sqrt{x}}$, Put $x = \sec^2\theta$ $\quad \therefore \quad \dfrac{dx}{d\theta} = 2\sec\theta.\tan\theta.\sec\theta = 2\sec^2\theta.\tan\theta$

$$\therefore \quad y = \frac{1 - \sqrt{\sec^2\theta - 1}}{1 + \sqrt{\sec^2\theta - 1}} + \frac{\sqrt{\sec^2\theta - 1}}{\sqrt{\sec^2\theta}} = \frac{1 - \tan\theta}{1 + \tan\theta} + \frac{\tan\theta}{\sec\theta} = \frac{1 - \dfrac{\sin\theta}{\cos\theta}}{1 + \dfrac{\sin\theta}{\cos\theta}} + \frac{\dfrac{\sin\theta}{\cos\theta}}{\dfrac{1}{\cos\theta}} = \frac{\cos\theta - \sin\theta}{\cos\theta + \sin\theta} + \sin\theta$$

$$\therefore \quad \frac{dy}{d\theta} = \frac{(\cos\theta + \sin\theta)(-\sin\theta - \cos\theta) - (\cos\theta - \sin\theta)(-\sin\theta + \cos\theta)}{(\cos\theta + \sin\theta)^2} + \cos\theta = \frac{-(\cos\theta + \sin\theta)^2 - (\cos\theta - \sin\theta)^2}{(\cos\theta + \sin\theta)^2} + \cos\theta$$

$$\therefore \quad \frac{dy}{d\theta} = \frac{-(\cos^2\theta + \sin^2\theta + 2\sin\theta\cos\theta) - (\cos^2\theta + \sin^2\theta + 2\sin\theta\cos\theta)}{(\cos\theta + \sin\theta)^2} + \cos\theta$$
$$= \frac{-\cos^2\theta - \sin^2\theta - 2\sin\theta\cos\theta - \cos^2\theta - \sin^2\theta + 2\sin\theta\cos\theta}{(\cos\theta + \sin\theta)^2} + \cos\theta$$

$$\therefore \quad \frac{dy}{d\theta} = \frac{-2(\sin^2\theta + \cos^2\theta)}{(\cos\theta + \sin\theta)^2} + \cos\theta = \frac{-2}{(\cos\theta + \sin\theta)^2} + \cos\theta = \frac{-2}{(\sin^2\theta + \cos^2\theta + 2\cos\theta\sin\theta)} + \cos\theta = \frac{-2}{1 + \sin 2\theta} + \cos\theta$$
$$= \frac{-2 + \cos\theta + \sin 2\theta\cos\theta}{1 + \sin 2\theta}$$

$$\therefore \quad \frac{dy}{dx} = \frac{dy}{d\theta} \times \frac{d\theta}{dx} = \frac{-2 + \cos\theta + \sin 2\theta\cos\theta}{1 + \sin 2\theta} \times \frac{1}{2\sec^2\theta.\tan\theta} = \frac{\cos\theta + \sin 2\theta\cos\theta - 2}{1 + \sin 2\theta} \times \frac{1}{2\dfrac{1}{\cos^2\theta}.\dfrac{\sin\theta}{\cos\theta}}$$
$$= \frac{(\cos\theta + \sin 2\theta\cos\theta - 2)}{1 + \sin 2\theta} \times \frac{\cos^3\theta}{2\sin\theta} = \frac{\cos^4\theta(1 + \sin 2\theta) - 2\cos^3\theta}{2\sin\theta(1 + \sin 2\theta)} = \frac{\cos^3\theta[\cos\theta + \sin 2\theta.\cos\theta - 2]}{2\sin\theta(1 + \sin 2\theta)}$$
$$= \frac{\cos^3\theta}{\sin\theta}\left[\frac{\cos\theta}{2} - \frac{1}{1 + \sin 2\theta}\right] \quad \text{Ans.}$$

(d) $\cos y = a\cos(x+y)$ $\quad \therefore \quad \cos y = a(\cos x\cos y - \sin x\sin y)$ Divide both of sides by $\cos y$, we have

$$\therefore \quad 1 = a\cos x - a\sin x.\tan y \quad \therefore \quad -a\sin x - a\left[\sin x.\sec^2 y.\frac{dy}{dx} + \tan y.\cos x\right] = 0$$

or $a\sin x.\sec^2 y.\dfrac{dy}{dx} = -a\tan y.\cos x - a\sin x$ $\quad \therefore \quad \dfrac{dy}{dx} = \dfrac{-a(\tan y.\cos x + \sin x)}{a\sin x.\sec^2 y} = \dfrac{-\left(\dfrac{\sin y}{\cos y}.\cos x + \sin x\right)}{\sin x.\dfrac{1}{\cos^2 y}}$

$$\therefore \quad \frac{dy}{dx} = \frac{-(\sin y\cos x + \sin x\cos y)}{\sin x} \times \cos y = \frac{-\cos y(\sin x\cos y + \cos x\sin y)}{\sin x} = -\frac{\cos y.\sin(x+y)}{\sin x} \quad \text{Ans.}$$

or $\dfrac{dy}{dx} = -\cos^2 y - \cot x.\sin y\cos y$ or $\dfrac{dy}{dx} = -\dfrac{a\cos(x+y).\sin(x+y)}{\sin x} = -\dfrac{a\sin 2(x+y)}{2\sin x}$ Ans.

(e) $y = 1 + \dfrac{Ax^2}{x-A} + \dfrac{Bx}{x-B} + \dfrac{C}{x-C}$, $\dfrac{dy}{dx} = 0 + \dfrac{(x-A).2Ax - Ax^2}{(x-A)^2} + \dfrac{(x-B).B - Bx}{(x-B)^2} + \dfrac{(x-C).0 - C}{(x-C)^2}$

$$\therefore \quad \frac{dy}{dx} = \frac{2Ax^2 - 2xA^2 - Ax^2}{(x-A)^2} + \frac{Bx - B^2 - Bx}{(x-B)^2} - \frac{C}{(x-C)^2} = \frac{Ax(x-2A)}{(x-A)^2} - \frac{B^2}{(x-B)^2} - \frac{C}{(x-C)^2} \quad \text{Ans.}$$

(f) Do yourself. $y = \tan^{-1}\left(\dfrac{\sqrt{a^2 + b^2}\,\cos x}{b - a\cos x}\right)$ $\therefore \tan y = \dfrac{\sqrt{a^2 + b^2}\,\cos x}{b - a\cos x}$ $\therefore \dfrac{dy}{dx} = \dfrac{-b.\sqrt{a^2 + b^2}\,.\sin x}{(b - a\cos x)^2 + (a^2 + b^2).\cos^2 x}$ Ans.

(g) $y\sqrt{x^2 + 1} = \log\left(\sqrt{x^2 + 1} - x\right)$ or $y = \dfrac{\log\left(\sqrt{x^2 + 1} - x\right)}{\sqrt{x^2 + 1}}$ Put $x = \tan\theta$ $\therefore \dfrac{dx}{d\theta} = \sec^2\theta = 1 + \tan^2\theta = 1 + x^2$

$\Rightarrow y = \dfrac{\log\left(\sqrt{(\tan\theta)^2 + 1} - \tan\theta\right)}{\sqrt{(\tan\theta)^2 + 1}} = \dfrac{\log(\sqrt{\tan^2\theta + 1} - \tan\theta)}{\sqrt{\tan^2\theta + 1}} = \dfrac{\log(\sec\theta - \tan\theta)}{\sec\theta}$

$\therefore \dfrac{dy}{d\theta} = \dfrac{\sec\theta.\dfrac{1}{(\sec\theta - \tan\theta)}.(\sec\theta\tan\theta - \sec^2\theta) - \log(\sec\theta - \tan\theta).\sec\theta\tan\theta}{\sec^2\theta}$

$= \dfrac{\dfrac{\sec^2\theta\,(\tan\theta - \sec\theta)}{(\sec\theta - \tan\theta)} - \log(\sec\theta - \tan\theta).\sec\theta\tan\theta}{\sec^2\theta} = \dfrac{-\sec^2\theta - \log(\sec\theta - \tan\theta).\sec\theta\tan\theta}{\sec^2\theta}$

$= \dfrac{-\sec\theta\,[\sec\theta + \log(\sec\theta - \tan\theta).\tan\theta]}{\sec^2\theta}$

$\therefore \dfrac{dy}{d\theta} = -\dfrac{[\sec\theta + \log(\sec\theta - \tan\theta).\tan\theta]}{\sec\theta}$ Put $x = \tan\theta$, $1 + \tan^2\theta = \sec^2\theta$ $\therefore \sec\theta = \sqrt{1 + x^2}$

$\therefore \dfrac{dy}{dx} = \dfrac{dy}{d\theta} \times \dfrac{d\theta}{dx} = -\dfrac{[\sec\theta + \log(\sec\theta - \tan\theta).\tan\theta]}{\sec\theta} \times \dfrac{1}{\sec^2\theta} = -\dfrac{[\sec\theta + \log(\sec\theta - \tan\theta).\tan\theta]}{\sec\theta.\sec^2\theta}$

$$\therefore \dfrac{dy}{dx} = -\dfrac{[\sqrt{1 + x^2} + x.\log(\sqrt{1 + x^2} - x)]}{(1 + x^2).\sqrt{1 + x^2}} \quad \text{Ans.}$$

(h) $y = 2\tan^{-1}\left(\dfrac{x}{\sqrt{1 - x^2}}\right) + \log\left(\dfrac{1 + 2\sqrt{x} + x}{1 - 2\sqrt{x} + x}\right) = u + v$ (say)

where $u = 2\tan^{-1}\left(\dfrac{x}{\sqrt{1 - x^2}}\right)$, Put $x = \sin\theta$ $\therefore \dfrac{dx}{d\theta} = \cos\theta$ and $v = \log\left(\dfrac{1 + 2\sqrt{x} + x}{1 - 2\sqrt{x} + x}\right)$

$u = 2\tan^{-1}\left(\dfrac{\sin\theta}{\sqrt{1 - (\sin\theta)^2}}\right) = 2\tan^{-1}\left(\dfrac{\sin\theta}{\sqrt{1 - \sin^2\theta}}\right) = 2\tan^{-1}\left(\dfrac{\sin\theta}{\sqrt{\cos^2\theta}}\right) = 2\tan^{-1}\left(\dfrac{\sin\theta}{\cos\theta}\right) = 2\tan^{-1}(\tan\theta) = 2\theta$

$\therefore \dfrac{du}{d\theta} = 2$ $\therefore \dfrac{du}{dx} = \dfrac{du}{d\theta} \times \dfrac{d\theta}{dx} = 2.\dfrac{1}{\cos\theta} = \dfrac{2}{\sqrt{1 - \sin^2\theta}} = \dfrac{2}{\sqrt{1 - x^2}}$ $\left[\text{Put } x = \sin\theta,\ \cos\theta = \sqrt{1 - \sin^2\theta}\right]$

Now, $v = \log\left(\dfrac{1 + 2\sqrt{x} + x}{1 - 2\sqrt{x} + x}\right) = \log\left[\dfrac{(1 + \sqrt{x})^2}{(1 - \sqrt{x})}\right]^2 = 2\log\left(\dfrac{1 + \sqrt{x}}{1 - \sqrt{x}}\right)$

$\therefore \dfrac{dv}{dx} = 2.\dfrac{1}{\left(\dfrac{1 + \sqrt{x}}{1 - \sqrt{x}}\right)}.\dfrac{(1 - \sqrt{x}).\dfrac{1}{2\sqrt{x}} - (1 + \sqrt{x}).\left(-\dfrac{1}{2\sqrt{x}}\right)}{(1 - \sqrt{x})^2} = \dfrac{2\left[\dfrac{1 - \sqrt{x} + 1 + \sqrt{x}}{2\sqrt{x}}\right]}{(1 + \sqrt{x})(1 - \sqrt{x})} = \dfrac{2}{\sqrt{x}.(1 - x)}$

$\therefore \dfrac{dy}{dx} = \dfrac{du}{dx} + \dfrac{dv}{dx} = \dfrac{2}{\sqrt{1 - x^2}} + \dfrac{2}{\sqrt{x}.(1 - x)} = 2\left(\dfrac{1}{\sqrt{1 - x^2}} + \dfrac{1}{\sqrt{x}.(1 - x)}\right)$ Ans.

(14) (a) $y = x + \dfrac{1}{x + \dfrac{1}{x + \dfrac{1}{x + \cdots \ldots \infty}}}$, Let $y = x + \dfrac{1}{x + \dfrac{1}{x + \cdots \ldots \infty}}$ then $y = x + \dfrac{1}{y}$ or $y^2 - xy = 1$

$\therefore 2y.\dfrac{dy}{dx} - \left[x.\dfrac{dy}{dx} + y\right] = 0$ or $2y.\dfrac{dy}{dx} - x.\dfrac{dy}{dx} = y$ or $\dfrac{dy}{dx}(2y - x) = y$ or $\dfrac{dy}{dx} = \dfrac{y}{2y - x}$ Ans.

$\therefore \dfrac{dy}{dx} = \dfrac{y}{2y - x} \times \dfrac{y}{y} = \dfrac{y^2}{2y^2 - xy} = \dfrac{y^2}{2y^2 - y^2 + 1} = \dfrac{y^2}{y^2 + 1}$ Ans. $[\text{Put } y^2 - xy = 1,\ xy = y^2 - 1]$

(b) $y = \sin x + \cfrac{1}{\sin x + \cfrac{1}{\sin x + \cdots \ldots \infty}}$, Let $y = \sin x + \cfrac{1}{\sin x + \cdots \ldots \infty}$ then $y = \sin x + \dfrac{1}{y}$ or $y^2 - y\sin x = 1$

$\therefore 2y\dfrac{dy}{dx} - \left[y\cos x + \sin x\dfrac{dy}{dx}\right] = 0$ or $2y\dfrac{dy}{dx} - \sin x\dfrac{dy}{dx} = y\cos x$ or $\dfrac{dy}{dx}(2y - \sin x) = y\cos x$

$\therefore \dfrac{dy}{dx} = \dfrac{y\cos x}{2y - \sin x} \times \dfrac{y}{y} = \dfrac{y^2\cos x}{2y^2 - y\sin x} = \dfrac{y^2\cos x}{2y^2 - (y^2 - 1)} = \dfrac{y^2\cos x}{2y^2 - y^2 + 1} = \dfrac{y^2\cos x}{y^2 + 1}$ Ans.

(c) $x^2 + y^2 + 2xy + x + y + 5 = 0$ or $2x + 2y\dfrac{dy}{dx} + 2\left(x\dfrac{dy}{dx} + y\right) + 1 + \dfrac{dy}{dx} = 0$

or $\dfrac{dy}{dx}(2y + 2x + 1) = -(2y + 2x + 1)$ $\therefore \dfrac{dy}{dx} = -\dfrac{(2y + 2x + 1)}{(2y + 2x + 1)} = -1$ $\therefore \dfrac{d^2y}{dx^2} = 0$ Ans.

(d) $y = (\log_{\sin x}\cos x)(\log_{\tan x}\cot x) + \cos^{-1}\dfrac{x}{1 + x} = \dfrac{\log(\cos x)}{\log(\sin x)} \cdot \dfrac{\log(\cot x)}{\log(\tan x)} + \cos^{-1}\dfrac{x}{1 + x}$

$\Rightarrow\ y = \dfrac{\log(\cos x) \cdot \log(\cot x)}{\log(\sin x) \cdot \log(\tan x)} + \cos^{-1}\dfrac{x}{1 + x} = u + v$ (say)   where $u = \dfrac{\log(\cos x) \cdot \log(\cot x)}{\log(\sin x) \cdot \log(\tan x)}$ and $v = \cos^{-1}\dfrac{x}{1 + x}$

$\therefore \dfrac{d}{dx}[\log(\cos x) \cdot \log(\cot x)] = \log(\cos x) \cdot \dfrac{1}{\cot x} \cdot (-\operatorname{cosec}^2 x) + \log(\cot x) \cdot \dfrac{1}{\cos x} \cdot (-\sin x) = -\left[\dfrac{\log(\cos x)}{\sin x \cos x} + \tan x\log(\cot x)\right]$

$\therefore \dfrac{d}{dx}[\log(\sin x) \cdot \log(\tan x)] = \log(\sin x) \cdot \dfrac{1}{\tan x} \cdot \sec^2 x + \log(\tan x) \cdot \dfrac{1}{\sin x} \cdot \cos x = \left\{\dfrac{\log(\cos x)}{\sin x \cos x} + \cot x\log(\tan x)\right\}$

$\dfrac{du}{dx} = \dfrac{-[\log(\sin x) \cdot \log(\tan x)] \cdot \left[\dfrac{\log(\cos x)}{\sin x \cos x} + \tan x\log(\cot x)\right] - [\log(\cos x) \cdot \log(\cot x)]\left[\dfrac{\log(\cos x)}{\sin x \cos x} + \cot x\log(\tan x)\right]}{[\log(\sin x) \cdot \log(\tan x)]^2}$

and $v = \cos^{-1}\dfrac{x}{1 + x}$ $\therefore \dfrac{dv}{dx} = \dfrac{1}{-\sqrt{1 - \left(\dfrac{x}{1 + x}\right)^2}} \cdot \dfrac{(1 + x) \cdot 1 - x \cdot 1}{(1 + x)^2} = -\dfrac{(1 + x)}{\sqrt{2x + 1} \cdot (1 + x)^2} = -\dfrac{1}{(1 + x) \cdot \sqrt{2x + 1}}$

$\therefore \dfrac{dy}{dx} = \dfrac{du}{dx} + \dfrac{dv}{dx}$

$\dfrac{dy}{dx} = \dfrac{-[\log(\sin x) \cdot \log(\tan x)] \cdot \left[\dfrac{\log(\cos x)}{\sin x \cos x} + \tan x\log(\cot x)\right] - [\log(\cos x) \cdot \log(\cot x)]\left[\dfrac{\log(\cos x)}{\sin x \cos x} + \cot x\log(\tan x)\right]}{[\log(\sin x) \cdot \log(\tan x)]^2} - \dfrac{1}{(1 + x) \cdot \sqrt{2x + 1}}$

At $x = \dfrac{\pi}{4}$, $\dfrac{dy}{dx} = 0 - \dfrac{1}{\left(1 + \dfrac{\pi}{4}\right) \cdot \sqrt{2 \cdot \dfrac{\pi}{4} + 1}} = -\dfrac{1}{\left(\dfrac{4 + \pi}{4}\right) \cdot \sqrt{\dfrac{\pi + 2}{2}}} = -\dfrac{4\sqrt{2}}{(4 + \pi) \cdot \sqrt{\pi + 2}}$ Ans.

(e) Differentiate $f(x) = \log_x \cos x^2 + \cos x^2$ with respect to $\sqrt{x + 1}$

Let $y = f(x) = \log_x \cos x^2 + \cos x^2 = \dfrac{\log(\cos x^2)}{\log x} + \cos x^2$ and $z = g(x) = \sqrt{x + 1}$

$\dfrac{dy}{dx} = \dfrac{\log x \cdot \dfrac{1}{\cos x^2} \cdot (-\sin x^2) \cdot 2x - \log(\cos x^2) \cdot \dfrac{1}{x}}{(\log x)^2} + (-\sin x^2) \cdot 2x = \dfrac{-[2x^2 \cdot \tan x^2 \cdot \log x + \log(\cos x^2)]}{x(\log x)^2} - 2x\sin x^2$

$\qquad = \dfrac{-[2x^2 \cdot \tan x^2 \cdot \log x + \log(\cos x^2) + 2x^2\sin x^2 \cdot (\log x)^2]}{x(\log x)^2}$

now $\dfrac{dz}{dx} = \dfrac{1}{2\sqrt{x + 1}}$ $\therefore \dfrac{dy}{dz} = \dfrac{dy}{dx} \times \dfrac{dx}{dz} = \dfrac{-[2x^2 \cdot \tan x^2 \cdot \log x + \log(\cos x^2) + 2x^2\sin x^2 \cdot (\log x)^2]}{x(\log x)^2} \times 2\sqrt{x + 1}$ Ans.

(f) $y = (\cos^{-1}x + \sin^{-1}x)x = \dfrac{\pi}{2}x$ $\left[\text{formula, } \sin^{-1}x + \cos^{-1}x = \dfrac{\pi}{2}\right]$ $\therefore \dfrac{dy}{dx} = \dfrac{\pi}{2}$ Proved. $\therefore \dfrac{d^2y}{dx^2} = 0$ Proved.

(g) $Ax^2 + By^2 = 1$ or $Ax^2 + By^2 - 1 = 0$ $\therefore$ $2Ax + 2By\dfrac{dy}{dx} = 0$ or $2By\dfrac{dy}{dx} = -2Ax$

$\therefore \dfrac{dy}{dx} = -\dfrac{2Ax}{2By} = -\dfrac{Ax}{By}$ Again differentiate with respect to x, we have

$\therefore \dfrac{d^2y}{dx^2} = -\dfrac{By \cdot A - Ax \cdot B\dfrac{dy}{dx}}{(By)^2}$ or $\dfrac{d^2y}{dx^2} = \dfrac{AB\left(x\dfrac{dy}{dx} - y\right)}{B^2 y^2} = \dfrac{Ax \cdot \dfrac{dy}{dx} - Ay}{By^2}$ or $By^2 \dfrac{d^2y}{dx^2} = Ax \cdot \dfrac{dy}{dx} - Ay$

$\therefore$ $By^2 \dfrac{d^2y}{dx^2} - Ax \cdot \dfrac{dy}{dx} + Ay = 0$ or $By^2 y'' - Axy' + Ay = 0$ Proved.

or $(1 - Ax^2)y'' - Axy' + Ay = 0$ Proved.   [Put $Ax^2 + By^2 = 1$ or $By^2 = 1 - Ax^2$]

(h), (i) and (j) Do yourself.

(15) (a) $y = \cos x \cdot \cos 3x \cdot \cos 5x = \dfrac{2\cos x \cdot \cos 3x \cdot \cos 5x}{2}$ or $2y = 2\cos x \cdot \cos 3x \cdot \cos 5x$

$\qquad\qquad$ [use formula, $2\cos A \cos B = \cos(A + B) + \cos(A - B)$ and $\cos(-\theta) = \cos\theta$]

or $2y = (2\cos x \cdot \cos 3x) \cdot \cos 5x = [\cos(x + 3x) + \cos(x - 3x)] \cdot \cos 5x = \cos 4x \cos 5x + \cos 2x \cos 5x$

$\therefore$ $2\dfrac{dy}{dx} = \cos 4x \cdot (-\sin 5x) \cdot 5 + \cos 5x \cdot (-\sin 4x) \cdot 4 + \cos 2x \cdot (-\sin 5x) \cdot 5 + \cos 5x \cdot (-\sin 2x) \cdot 2$

$\qquad\qquad = -5\cos 4x \sin 5x - 4\cos 5x \sin 4x - 5\cos 2x \sin 5x - 2\cos 5x \sin 2x$

$\therefore$ $\dfrac{dy}{dx} = -\dfrac{1}{2}[5\cos 4x \sin 5x + 5\cos 2x \sin 5x + 4\cos 5x \sin 4x + 2\cos 5x \sin 2x]$

$\qquad\qquad = -\dfrac{1}{2}[5\sin 5x\,(\cos 4x + \cos 2x) + 2\cos 5x\,(2\sin 4x + \sin 2x)]$ Ans.

or $\dfrac{dy}{dx} = -\dfrac{5}{2}\sin 5x \cdot 2\cos\left(\dfrac{4x + 2x}{2}\right) \cdot \cos\left(\dfrac{4x - 2x}{2}\right) - 2\cos 5x \sin 4x - \cos 5x \sin 2x$

$\qquad\qquad = -5\sin 5x \cdot \cos 3x \cdot \cos x - 2\cos 5x \cdot \sin 4x - \cos 5x \cdot \sin 2x$ Ans.

Useful formula, $2\sin A \sin B = \cos(A - B) - \cos(A + B)$, $2\sin A \cos B = \sin(A + B) + \sin(A - B)$

$\qquad\qquad 2\cos A \cos B = \cos(A + B) + \cos(A - B)$, $2\cos A \sin B = \sin(A + B) - \sin(A - B)$

$\qquad\qquad \sin C + \sin D = 2\sin\dfrac{C + D}{2} \cdot \cos\dfrac{C - D}{2}$, $\sin C + \sin D = 2\cos\dfrac{C + D}{2} \cdot \sin\dfrac{C - D}{2}$

$\qquad\qquad \cos C + \cos D = 2\cos\dfrac{C + D}{2} \cdot \cos\dfrac{C - D}{2}$, $\cos C + \cos D = 2\sin\dfrac{C + D}{2} \cdot \sin\dfrac{D - C}{2}$

$\qquad\qquad \sin(-\theta) = -\sin\theta$ and $\cos(-\theta) = \cos\theta$

(b) $y = 8\sin x \cdot \sin 2x \cdot \sin 4x = 4(2\sin x \cdot \sin 2x) \cdot \sin 4x = 4\sin 4x\,[\cos(x - 2x) - \cos(x + 2x)]$ (above formula)

or $y = 4\sin 4x\,[\cos x - \cos 3x] = 4\sin 4x \cos x - 4\sin 4x \cos 3x = 2(2\sin 4x \cos x) - 2(2\sin 4x \cos 3x)$

or $y = 2[\sin(4x + x) + \sin(4x - x)] - 2[\sin(4x + 3x) + \sin(4x - 3x)] = 2(\sin 5x + \sin 3x) - 2(\sin 7x + \sin x)$

or $y = 2\sin 5x + 2\sin 3x - 2\sin 7x - 2\sin x$

$\therefore$ $\dfrac{dy}{dx} = 2\cos 5x \cdot 5 + 2\cos 3x \cdot 3 - 2\cos 7x \cdot 7 - 2\cos x = 10\cos 5x + 6\cos 3x - 14\cos 7x - 2\cos x$ Ans.

(c) $y = \cos 5x \cdot \sin 3x = \dfrac{2\cos 5x \cdot \sin 3x}{2} = \dfrac{\sin(5x + 3x) - \sin(5x - 3x)}{2} = \dfrac{\sin 8x - \sin 2x}{2} = \dfrac{1}{2}(\sin 8x - \sin 2x)$

$\therefore$ $\dfrac{dy}{dx} = \dfrac{1}{2}(\cos 8x \cdot 8 - \cos 2x \cdot 2) = \dfrac{1}{2}(8\cos 8x - 2\cos 2x) = \dfrac{1}{2} \cdot 2(4\cos 8x - \cos 2x) = 4\cos 8x - \cos 2x$ Ans.

(d) $y = \sin 2x + \cos 3x$, $\dfrac{dy}{dx} = \cos 2x \cdot 2 + (-\sin 3x \cdot 3) = 2\cos 2x - 3\cos 3x$ Ans.

(e) $y = 2\tan 2\theta + 3\cot 2\theta$ $\therefore \dfrac{dy}{d\theta} = 2\sec^2 2\theta \cdot 2 + 3(-\text{cosec}^2 2\theta) \cdot 2 = 4\sec^2 2\theta - 6\,\text{cosec}^2 2\theta$ Ans.

or $\dfrac{dy}{d\theta} = \dfrac{4}{\cos^2 2\theta} - \dfrac{6}{\sin^2 2\theta} = \dfrac{2(2\sin^2 2\theta - 3\cos^2 2\theta)}{\sin^2 2\theta \cdot \cos^2 2\theta}$ Ans.

(f) Do yourself. $y = \sec 2\theta \cdot \text{cosec}\, 3\theta$ $\therefore \dfrac{dy}{d\theta} = \sec 2\theta \cdot \text{cosec}\, 3\theta\,(2\tan 2\theta - 3\cot 3\theta) = y(2\tan 2\theta - 3\cot 3\theta)$ Ans.

(g) $y = \cos 3\theta + \sec 2\theta$ $\therefore \dfrac{dy}{d\theta} = -\sin 3\theta \cdot 3 + 2\tan 2\theta \sec 2\theta = 2\tan 2\theta \sec 2\theta - 3\sin 3\theta$ Ans.

(h) $y = \sin 2\theta + \text{cosec}\, 3\theta$ $\therefore \dfrac{dy}{d\theta} = \cos 2\theta \cdot 2 + 3(-\cot 3\theta\,\text{cosec}\,3\theta) = 2\cos 2\theta - 3\cot 3\theta\,\text{cosec}\,3\theta$ Ans.

(i) $y = \log_2 \log_2 \log_2 x^2$, Let $u = \log_2 x^2$ $\therefore \dfrac{du}{dx} = \dfrac{1}{x^2} \cdot 2x = \dfrac{2}{x}$

Now, $y = \log_2 \log_2 u$ Let $v = \log_2 u$ $\therefore \dfrac{dv}{du} = \dfrac{1}{u} = \dfrac{1}{\log_2 x^2}$ [Put $u = \log_2 x^2$]

Now, $y = \log_2 v$ $\therefore \dfrac{dy}{dv} = \dfrac{1}{v} = \dfrac{1}{\log_2 u} = \dfrac{1}{\log_2 \log_2 x^2}$ [Put $v = \log_2 u$ and $u = \log_2 x^2$]

Then, $\dfrac{dy}{dx} = \dfrac{dy}{dv} \times \dfrac{dv}{du} \times \dfrac{du}{dx} = \dfrac{1}{\log_2 \log_2 x^2} \times \dfrac{1}{\log_2 x^2} \times \dfrac{2}{x} = \dfrac{2}{x.\log_2 x^2 . \log_2 \log_2 x^2}$ Ans.

(j) $y = 4\log_e \log_e \log_e \log_e x$, Let $u = \log_e x$ $\therefore \dfrac{du}{dx} = \dfrac{1}{x}$

Now, $y = 4\log_e \log_e \log_e u$ Let $v = \log_e u$ $\therefore \dfrac{dv}{du} = \dfrac{1}{u} = \dfrac{1}{\log_e x}$ [Put $u = \log_e x$]

Now, $y = 4\log_e \log_e v$ Let $w = \log_e v$ $\therefore \dfrac{dw}{dv} = \dfrac{1}{v} = \dfrac{1}{\log_e u} = \dfrac{1}{\log_e \log_e x}$ [Put $v = \log_e u$ and $u = \log_e x$]

Now, $y = 4\log_e w$ $\therefore \dfrac{dy}{dw} = \dfrac{4}{w} = \dfrac{4}{\log_e v} = \dfrac{4}{\log_e \log_e \log_e x}$ [Put $w = \log_e v$, $v = \log_e u$ and $u = \log_e x$]

Then, $\dfrac{dy}{dx} = \dfrac{dy}{dw} \times \dfrac{dw}{dv} \times \dfrac{dv}{du} \times \dfrac{du}{dx} = \dfrac{4}{\log_e \log_e \log_e x} \times \dfrac{1}{\log_e \log_e x} \times \dfrac{1}{\log_e x} \times \dfrac{1}{x} = \dfrac{4}{x.\log_e x.\log_e \log_e x.\log_e \log_e \log_e x}$

$\therefore \dfrac{dy}{dx} = \dfrac{4}{x.\log_e x.\log x.\log_e \log x}$ Ans. $\left[e^{\log_e x} = x \right]$

(k) $f(x) = \log_{x^3} \log x^2 = \dfrac{1}{3}\log_x \log x^2$, Let $f(u) = \log x^2$ $\therefore f'(u) = \dfrac{1}{x^2} \cdot 2x = \dfrac{2}{x}$

Now, $f(x) = \dfrac{1}{3}\log_x f(u) = \dfrac{\log f(u)}{3\log x}$ $\therefore f'(x) = \dfrac{1}{3}\left[\dfrac{\log x \cdot \dfrac{1}{f(u)} \cdot f'(u) - \log f(u) \cdot \dfrac{1}{x}}{(\log x)^2}\right]$ Put value of $f(u)$ and $f'(u)$

$\therefore f'(x) = \dfrac{1}{3}\left[\dfrac{\log x \cdot \dfrac{1}{\log x^2} \cdot \dfrac{2}{x} - \log \log x^2 \cdot \dfrac{1}{x}}{(\log x)^2}\right] = \dfrac{1}{3}\left[\dfrac{\log x \cdot \dfrac{1}{2\log x} \cdot \dfrac{2}{x} - \log \log x^2 \cdot \dfrac{1}{x}}{(\log x)^2}\right] = \dfrac{1}{3}\left[\dfrac{1 - \log \log x^2}{x(\log x)^2}\right]$

At $x = e$ then $f'(e) = \dfrac{1}{3}\left[\dfrac{1 - \log \log e^2}{e(\log e)^2}\right] = \left(\dfrac{1 - \log 2}{3e}\right)$ Ans.

(16) (a) $y = \sin^m x \cdot \cos^n x$ $\therefore \dfrac{dy}{dx} = \sin^m x \cdot n(\cos x)^{n-1}(-\sin x) + \cos^n x \cdot m(\sin x)^{m-1} \cdot \cos x$

$$\therefore \frac{dy}{dx} = m(\sin x)^{m-1}(\cos x)^{n+1} - n(\sin x)^{m+1}(\cos x)^{n-1} \qquad \text{Ans.}$$

(b) $y = \sin^n x \cdot \cos mx \qquad \therefore \frac{dy}{dx} = \sin^n x \cdot (-\sin mx) \cdot m + \cos mx \cdot n(\sin x)^{n-1} \cdot \cos x$

$$\therefore \frac{dy}{dx} = n \cdot \cos x \cdot \cos mx \cdot (\sin x)^{n-1} - m \cdot \sin mx \cdot \sin^n x \qquad \text{Ans.}$$

(c) $y = f\big(f(\log x)\big)$ where $f(x) = \log x, \ f(\log x) = \log\log x, \ f\big(f(\log x)\big) = \log[\log(\log x)]$

Now, $y = \log[\log(\log x)] \qquad$ Let $u = \log x \qquad \therefore \frac{du}{dx} = \frac{1}{x}$

Now, $y = \log(\log u) \qquad$ Let $v = \log u \qquad \therefore \frac{dv}{du} = \frac{1}{u} = \frac{1}{\log x} \qquad [\text{ Put } u = \log x]$

Now, $y = \log v \qquad \therefore \frac{dy}{dv} = \frac{1}{v} = \frac{1}{\log u} = \frac{1}{\log\log x} \qquad [\text{ Put } v = \log u \text{ and } u = \log x]$

Then, $\dfrac{dy}{dx} = \dfrac{dy}{dv} \times \dfrac{dv}{du} \times \dfrac{du}{dx} = \dfrac{1}{\log\log x} \times \dfrac{1}{\log x} \times \dfrac{1}{x} = \dfrac{1}{x \cdot \log x \cdot \log(\log x)} \qquad \text{Ans.}$

(d) $y = f\big(f(\sin x)\big)$, where $f(x) = \sin x, \ f(\sin x) = \sin(\sin x), \ f\big(f(\sin x)\big) = \sin[\sin(\sin x)]$

Now, $y = f\big(f(\sin x)\big) = \sin[\sin(\sin x)] \qquad$ Let $u = \sin x \qquad \therefore \frac{du}{dx} = \cos x$

Now, $y = \sin(\sin u) \qquad$ Let $v = \sin u \qquad \therefore \frac{dv}{du} = \cos u = \cos(\sin x) \qquad [\text{ Put } u = \sin x]$

Now, $y = \sin v \qquad \therefore \frac{dy}{dv} = \cos v = \cos(\sin u) = \cos[\sin(\sin x)] \qquad [\text{ Put } v = \sin u \text{ and } u = \sin x]$

Then, $\dfrac{dy}{dx} = \dfrac{dy}{dv} \times \dfrac{dv}{du} \times \dfrac{du}{dx} = \cos[\sin(\sin x)] \times \cos(\sin x) \times \cos x = \cos x \cdot \cos(\sin x) \cdot \cos[\sin(\sin x)] \quad \text{Proved.}$

(16) Do yourself. Question No. $-$ (e), (f) and (g).